Worlds Without End

Worlds Without End

THE MANY LIVES OF THE MULTIVERSE

. . . in which are discussed pre-, early-, and postmodern multiple-worlds cosmologies; the sundry arguments for and against them; the striking peculiarities of their adherents and detractors; the shifting boundaries of science, philosophy, and religion; and the stubbornly persistent question of whether creation has been "designed"

Mary-Jane Rubenstein

Columbia University Press *New York*

Columbia University Press
Publishers Since 1893
New York   Chichester, West Sussex
cup.columbia.edu

Copyright © 2014 Columbia University Press

Library of Congress Cataloging-in-Publication Data

Rubenstein, Mary-Jane.
Worlds without end : the many lives of the multiverse . . . in which are discussed
pre-, early-, and postmodern multiple-worlds cosmologies : the sundry arguments for
and against them : the striking peculiarities of their adherents and detractors : the
shifting boundaries of science, philosophy, and religion : and the stubbornly
persistent question of whether creation has been "designed" / Mary-Jane Rubenstein.
pages cm
Includes bibliographical references and index.
ISBN 978-0-231-15662-2 (cloth : alk. paper)
ISBN 978-0-231-52742-2 (e-book)
1. Cosmology—Popular works.   I. Title.

QB982.R83 2014
523.1—dc23
2013027560

♾

Columbia University Press books are printed on permanent and durable acid-free paper.
This book is printed on paper with recycled content.
Printed in the United States of America

c 10 9 8 7 6 5 4 3 2 1

COVER ART AND DESIGN: Kenan Rubenstein

References to Web sites (URLs) were accurate at the time of writing. Neither the author
nor Columbia University Press is responsible for URLs that may have expired or changed
since the manuscript was prepared.

If the grandeur of a planetary world . . . fills the understanding with wonder, with what astonishment are we transported when we behold the infinite multitude of worlds and systems which fill the extension of the Milky Way! But how is this astonishment increased, when we become aware that all these immense orders of star-worlds again form one of a number whose termination we do not know! . . . There is here no end but an abyss of a real immensity, in presence of which all the capability of human conception sinks exhausted.

IMMANUEL KANT,
UNIVERSAL NATURAL HISTORY AND THEORY OF THE HEAVENS

"All philosophy," I told her, "is based on two things only: curiosity and poor eyesight. . . . The trouble is, we want to know more than we can see."

BERNARD LE BOVIER DE FONTENELLE,
CONVERSATIONS ON THE PLURALITY OF WORLDS

CONTENTS

ILLUSTRATIONS

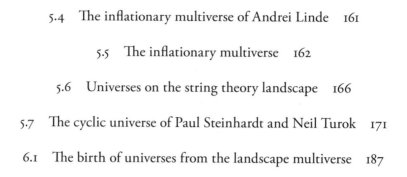

ACKNOWLEDGMENTS

This book is the product—and in a certain sense, the record—of a multitude of fortuitous conversations and collaborations. Thanks are due, first of all, to Clayton Crockett, whose stubborn insistence that I find *something* to say at a conference on energy threw me into the apophatic expanse of cosmic acceleration, which opened in turn onto the strange worlds of the multiverse. Having had very little advanced training in the natural sciences, I am grateful for the guidance and gentle corrections of Marcelo Gleiser, Laura Mersini-Houghton, Jamie Hinderks, Joe Rouse, Bill Herbst, Kirk Wegter-McNelly, Whitney Bauman, and especially Brian Wecht, who read the introduction and physics chapters with more care than I could have dared to request.

Other friends and colleagues provided invaluable orientation as I found myself wandering through their fields of expertise; my thanks to Elizabeth Castelli, Eirene Visvardi, Tushar Irani, and Andy Szegedy-Maszak for their help with the classical material; Constance Leidy for her mathematical breakdown of infinity; Michael Granada for clarifying late Renaissance cosmology; Jenna Supp-Montgomerie for her work on "worlds" in the American technoreligious imagination; Cate Williamson for introducing me to multiverse cosmologies in narrative theory; Liz Lerman for getting me to dance out the stages of cosmic expansion; Charles Jencks for talking me through his glorious Garden of Cosmic Speculation at

Portrack, Scotland; and Alistair Clark, head gardener, for walking me through it.

This admittedly strange and syncretic project would have remained an inchoate set of potential connections were it not for the patience, energy, and brilliance of my students at Wesleyan—especially the members of my "Worlding" seminars in the fall semesters of 2010 and 2011. My research has been supported by a grant from the Andrew W. Mellon Foundation and a fellowship at Wesleyan University's Center for the Humanities, then under the directorship of Jill Morawski. It has benefitted immensely from the generosity and solidarity of my colleagues in the Religion Department and the Feminist, Gender, and Sexuality Studies Program; from the assistance of Rhonda Kissinger; and from the scrutiny of a number of working groups on campus—many thanks in this regard to Margot Weiss, Matthew Garrett, Joseph Fitzpatrick, Nima Bassiri, Isaac Kamola, Liza McAlister, Annalise Glauz-Todrank, Laura Harrington, Justine Quijada, Attiya Ahmad, Jennifer Tucker, Christina Crosby, and Lori Gruen. I am also grateful to Kendall Hobbs, Susanne Javorski, Diane Klare, Suzy Taraba, and Katherine Wolfe in the libraries at Wesleyan; to John Wareham and Kevin Wiliarty in Information Technology Services; to Annie Barva for her meticulous copyediting; and to Wendy Lochner, Christine Dunbar, and Irene Pavitt at Columbia University Press for their careful attention to the manuscript throughout the publication process.

I managed to clarify the prose of particularly tricky sections thanks to scrupulous readings by Lisa Cohen, Elizabeth Salzer, Andrew Aghapour, Wayne Proudfoot, and Helen Ashley. Veronica Warren provided enthusiastic and invaluable assistance with subtitles, scansion, and epigraphs. Isaac Rubenstein did his best to keep the prose from being too boring. Kenan Rubenstein drew and designed the cover; illustrated the "quilted," "inflationary," and "connected" multiverses; and talked me through so many conceptual and stylistic difficulties that I no longer know which ideas were his to begin with. And I relied throughout the course of this project on the patience and encouragement of Keera Bhandari, who found articles, read and improved chapters, and augmented my own excitement about the cosmological entanglement of physics and metaphysics.

Finally, I am indebted beyond measure to Catherine Keller, whose work has shaped nearly every part of this project, who responded with characteristic clarity to recalcitrant paragraphs and numerous drafts, and whose tireless encouragement made this whole thing seem, impossibly, possible.

Worlds Without End

INTRODUCTION

How to Avoid the G-word

If you don't want God, you'd better have a multiverse.

BERNARD CARR, QUOTED IN TIM FOLGER, "SCIENCE'S ALTERNATIVE TO AN
INTELLIGENT CREATOR: THE MULTIVERSE THEORY"

Back to the Multiverse

Although the idea of multiple worlds is hardly a new one, it has been bubbling up with increasing regularity over the past fifteen years. No longer merely the stuff of science fiction, the notion is now under consideration as a scientific hypothesis, with some of the earth's most highly respected physicists and cosmologists suggesting that our whole universe—from our perspective, all that *is*—might be just a negligible part of a vast, perhaps infinite, "multiverse." At first, this idea was confined mainly to specialized journals and edited volumes, with the exception of a few crucial issues of *Scientific American*.[1] Then, more or less all at once, the multiverse hit the magazines, the blogosphere, radio shows, televised documentaries, and the popular-science bookshelves—most notably in the form of Brian Greene's latest best-selling monograph, *The Hidden Reality: Parallel Universes and the Deep Laws of the Cosmos*.[2]

This multiversal explosion has in turn led to a renewed proliferation of pop-cultural explorations of hidden dimensions, parallel universes, and copycat cosmoi. At the time of my writing this introduction, for example, a Google news alert set for the term *multiverse* is turning up more links to a forthcoming *Family Guy* video game for PlayStation® and Xbox® called *Back to the Multiverse* than to scholarly books or articles—by a factor of five. The game takes its lead from a season 8 episode that aired in September 2009, "Road to the Multiverse."[3] Early in the episode, the preternaturally intelligent, British-accented toddler Stewie Griffin asks the family's preternaturally intelligent, upright dog, "You ever heard of the multiverse theory, Brian?"

"Well, of course I have," Brian responds haltingly, "but—I'm wondering if *you* have."

"Oh my God," sighs Stewie, "so transparent. Well, the theory states that there are an infinite number of universes coexistent with ours on parallel dimensional planes."

". . . dimensional planes, right."

"Oh, don't do that," Stewie retorts. "Don't, don't—don't *repeat* the last two words like you already *kind* of knew what I was talking about; you have *no idea* what I'm talking about. Now in each of these alternate universes, the reality is different than our own. Sometimes only slightly, sometimes quite radically. The point is, every possible eventuality *exists*."

A quick riff on the Many-Worlds Interpretation (MWI) of quantum mechanics, with a few elements from inflationary cosmology, superstring theory, modal realism, and *The Twilight Zone* thrown in, Stewie's explanation is both mystifying and intriguing enough to get Brian hooked. So with the aid of a genre-appropriate remote-control device, the two characters go traveling through this mash-up multiverse. The first world they visit is one in which "Christianity never existed, which means the dark ages of scientific repression never occurred, and thus humanity is a thousand years more advanced." The second is a world populated by *Flintstones* characters, speaking with *Family Guy* voices about "rock sex." The third is a world in which "the United States never dropped the atomic bomb on Hiroshima. So the Japanese—just never quit," conquering, among other things, the town of Quahog, Rhode Island, where the Griffin family lives. The worlds go on, and the racist and scatological jokes multiply, until finally Stewie and Brian find their way back to their own universe.

Again, this sort of madcap fascination with multiple worlds is neither surprising nor new. What *is* remarkable is that the impetus for this recent bout of popular fascination is coming from the heights (and depths) of contemporary cosmology and particle physics. What is remarkable, in other words, is that "the multiverse" has suddenly and dramatically become scientifically *thinkable*. And although most physicists, astronomers, and cosmologists still remain highly skeptical of the idea, its proponents include so many scientific luminaries (including Brian Greene, Martin Rees, Alan Guth, and, in his own way, Stephen Hawking) that no one at this point can simply ignore it. Even the physicist and historian of science Helge Kragh, a notorious multiverse skeptic, concedes that among a "not insignificant" number of theoretical physicists, "what used to be a philosophical speculation is now claimed to be a new [cosmological] paradigm."[4]

In order to account for the emergence of the multiverse as what Lorraine Daston would call a "scientific object,"[5] this book looks back to the earliest documented sources on multiple worlds, beginning with the Greek "Atomist" philosophers in the fifth century B.C.E. and then gradually working its way through to the present.[6] The volume's task is fourfold: first, to give a historical account of the ebbs and flows of multiple-world cosmologies; second, to map contemporary models of the multiverse in relation to their philosophical, mythological, and even theological precedents; third, to ask how, why, and to whom the multiverse has become a particularly attractive hypothesis at this historical juncture; and fourth, to mark multiverse cosmologies as the site of a constructive reconfiguration of the boundaries between "science" and "religion." Each of these endeavors contributes to the book's central philosophical project, which is to find a way to come to terms conceptually with the multiverse. This project is a challenging one because the first question a philosopher tends to ask (What is it?) is in this case remarkably difficult to answer. If, traditionally speaking, "the universe" has meant "all that is," then what on earth does it mean to posit more than one of them? What is the multiverse?

The One and the Many

The term *multiverse* seems to have been coined by the psychologist-philosopher William James in 1895.[7] In "Is Life Worth Living?" an address to the Young Men's Christian Association at Harvard University, James expressed his sense, which he imagined his audience shared, that "visible nature is all plasticity and indifference, a *multiverse*, as one might call it, and not a universe." What he was trying to evoke with this term was the post-Darwinian feeling that the natural world around us, suddenly figured as female, is changeable and cruel; James goes so far as to call "her" a "harlot" to whom "we owe no moral allegiance."[8] As Wayne Proudfoot has explained, the term *moral* for James "is not restricted to ethics" but extends out to the scope of human existence itself. In this sense, "to ask whether the universe is moral or unmoral is to ask whether it is shaped to human thought and action"—that is, whether there is any inherent order in the universe that might make human life meaningful.[9] And as far as we can tell by examining nature alone, James tells his audience, there is no such order; nature on "her" own appears to be a disorganized aggregate of incoherent and often treacherous events and processes that lend no meaning at all to human existence.[10] It is only in relation to a "supplementary unseen or *other* world," a "truer, more eternal world" (the world, we will learn, of a male God),[11] that the natural world gains coherence and the thoughtful human being finds a reason to live. James therefore recommends that the young men do everything in their powers to *make* themselves believe—to muster up faith in a "truer," unseen world that might endow this visible one with an ounce of sense. "Believe that life *is* worth living," he tells them, "and your belief will help create the fact."[12] Belief will help unify the disparate and give order to the chaos; belief, in short, will make a universe out of the multiverse.

More than a decade later, in a lecture series James delivered at Oxford University in 1908 and 1909, it became clear that he had reevaluated his multiverse substantially—thanks to a philosophical and theological shift in his thinking.[13] Rather than locating the order of the universe in some external, divine realm, he now finds it *within* the world; in fact, he affirms pantheism as "the only opinio[n] quite worthy of arresting our attention."[14] As far as James can see, pantheism has two "subspecies": *monism*, which thinks of the world as a perfectly unified whole (or uni-verse), and *pluralism*, which affirms that "things are 'with' one another in many ways, but [that] nothing includes everything, or dominates over everything. . . . Something always escapes."[15] For the pluralist, then, the world is irreducibly a "multiverse"—a set of different phenomena, relations, and connections that cannot be assembled under a single principle. But unlike the younger James, this pantheistic James does not regard his multiverse as a cause for alarm or dismay. Nor does the world's irreducible manyness amount to cosmic fragmentation or moral senselessness. For although the pluralist may not see "all things" as immediately connected or co-implicated, she also does not see them as unrelated or independent. Rather, she sees "each thing" as connected to other things, which are themselves connected to other things, forming a chain of associations and interdependencies that eventually connects back to each thing. So there is coherence in the multiverse—just not an "all-is-oneness." Its unity, James says, is of the "strung-along type," never "absolutely complete."[16] Ultimately, then, he can affirm that "our 'multiverse' still makes a 'universe'; for every part, tho it

4

may not be in actual or immediate connexion, is nevertheless in some possible or mediate connexion, with every other part however remote, through the fact that each part hangs together with its very next neighbors in inextricable interfusion."[17] And so these "strung-along" interfusions keep the things of the pluralist's world both separate and unified, both many and one.

These days the term *multiverse* has taken on what may or may not be a vastly different meaning, depending on whom you ask. For James, the many things of our one, visible world constitute a "multiverse," whereas the coherence among those things is the "universe." For contemporary physicists, by contrast, our one, visible world constitutes the universe (a sphere 40 billion light years in radius also called our "Hubble volume" or "observable universe"), whereas the greater ensemble of unseen worlds constitutes the multiverse (sometimes called "metaverse" or "megaverse.") In terms of number alone, then, these two meanings of *multiverse* might seem exactly opposite: for James, this one world constitutes a multiverse, whereas for the physicists, many worlds constitute a multiverse. When it comes to the question of coherence, however, the relationship is less clear. In fact, it is a point of ongoing contention whether the physicists' multiverse constitutes what the early James would call a multiverse—that is, a chaos of unconnected phenomena—or whether it ultimately constitutes what the *later* James would call a multiverse: a causally interrelated, complexly connected system that is coherent yet never "absolutely complete."

Moreover, there is no scientific consensus yet as to whether there is a multiverse at all. Astronomer Royal Martin Rees suggests that if there were such a consensus, then insofar as "universe" typically designates "all that is," we should call the multiverse "the universe" and the universe something such as "the metagalaxy." But "so long as this whole [multiverse] idea remains speculative," he admits, "it is probably best to continue to denote what cosmologists observe as 'the universe' and to introduce a new term, 'multiverse,' for the whole hypothetical ensemble."[18]

To complicate matters further, there is more than one multiverse—more than one "whole hypothetical ensemble" of universes. As we shall see in chapters 5 and 6, some of these ensembles are totally incompatible with one another.[19] Others are said to be nested inside one another in an ascending cosmic hierarchy of infinities within infinities.[20] And other models might ultimately prove to be different ways of expressing the same ensemble.[21] In any event, contemporary models of the multiverse tend to come in four major "types." The first type configures universes *spatially*, with an infinite number of different worlds separated either by gargantuan expanses of ordinary space-time or by a rapidly expanding sea of energy.[22] Although some of the details are quite different, these spatial multiverses resonate with ancient Greek "Atomist" philosophy, which posited an infinite number of worlds strewn throughout an infinite spatial void. The second type configures universes *temporally*, so that a universe or part of a universe "collapses" in order to form a new universe, a process repeated throughout infinite time.[23] These "cyclical" multiverses resemble the cosmology of the Stoics (rivals to the Atomists), who argued that the universe is periodically consumed in fire and then reborn out of the ashes of the old world. A third type of multiverse is based on the Many-Worlds Interpretation (MWI) of quantum mechanics, which suggests that the universe separates into different branches every time a subatomic particle "decides" on a position.[24] Such quantum models lack direct philosophical precedents in the ancient or early modern worlds, but they have produced numerous popular outcroppings in the five decades since they first emerged: one might think here of the Choose Your Own Adventure children's series, which produced numerous different endings depending on the course of action the readers chose to pursue, or, more recently, an episode of the television series *Community*, in which one character rolls a die and produces six different paths for the episode to follow.[25] Neither quite spatial nor exactly temporal, these quantum universes exist as different branches of the same quantum "wave function" in infinite-dimensional space.

Finally, there is the "modal" type of multiverse, which stipulates that all possible worlds must actually exist and, moreover, that an infinite number of each possible world must actually exist.[26] This is what Brian Greene calls the "Ultimate Multiverse" because it contains all other multiverses within it (Paul Davies calls it "a multiverse with a vengeance").[27] Most vigorously defended by the theoretical physicist Max Tegmark, this idea finds its early modern roots in the philosophy of Gottfried Leibniz, who argued that insofar as God chose among an infinite number of "possible worlds" at the moment of creation, and insofar as God is supremely good, our world must be "the best of all possible worlds."[28] The major difference between these theories—apart from God—is that Leibniz's other worlds were only possible, whereas in the modal multiverse all the possible worlds are *actual*. Nevertheless, they are utterly inaccessible to one another—at least physically. Neither spatially nor temporally arrayed *nor* branching off the same wave function, the ensemble of these universes exists, as Tegmark puts it, "outside space and time."[29]

This, then, is where the outer reaches of theoretical cosmology lie these days: infinite copies of every mathematically possible world, physically existing, outside of space and time. To be sure, most multiverse theorists stop somewhere short of Tegmark's proposal, affirming one or more of these models, but not *all* of them. After all, some of the central tenets of modern scientific practice include observability, testability, and simplicity, all of which seem to be rather dramatically violated by an infinite number of all possible worlds actually existing outside space and time. That having been said, the very possibility that there might be more matter and space-time beyond our visible universe does set one on what Martin Rees admits is a "slippery slope,"[30] from other *stuff* to other *worlds* to other worlds within metaworlds and eventually to the ultimate multiverse. But what, one might wonder, put modern cosmology on this particular slope in the first place?

Whence the Modern Multiverse?

There are two ways to respond to this question. The first is that the multiverse has emerged fairly inexorably from developments in both subatomic physics and cosmology, and the second is that the multiverse is a matter of philosophical expedience.

The first scientifically significant multiverse hypothesis emerged in the late 1950s, when Hugh Everett first posited the MWI as an alternative to the then-standard "Copenhagen Interpretation" by Niels Bohr and Werner Heisenberg.[31] The second early multiverse hypothesis emerged in the mid-1980s, when theoretical physicists such as Alexander Vilenkin, Andrei Linde, Alan Guth, Andreas Albrecht, and Paul Steinhardt proposed that the "inflationary" process that generated our universe might eternally generate more universes.[32] The details of these scenarios will become clearer in chapters 5 and 6, but what is important to recognize at this stage is simply that these theoretical models were proposed decades before the current multiverse craze and that they emerged from both the subatomic and the cosmic ends of the physical scale. As Vilenkin has explained, however, very few people paid attention to these many-worlds scenarios until the turn of the millennium,[33] which brings us to the second, more complicated reason for the scientific turn to the multiverse.

The more physicists learn about the fundamental laws of the universe, the more remarkable it seems that we exist at all. The constants of nature—which, among others, include the strength of gravity, the mass of the electron, and the strength of the nuclear forces—have values that seem precisely calculated to allow life to emerge somewhere in the universe. For example, if the nuclear force were just a little bit stronger, "then all the hydrogen atoms in the infant universe would have fused with other hydrogen atoms to make helium, and there would be no hydrogen left. No hydrogen means no water. . . . On the other hand, if the nuclear force were substantially weaker . . . then the complex atoms needed for biology could not hold together."[34] If gravity were any stronger, it would "crush anything larger than an insect," whereas if it were weaker, planets and stars could never have formed.[35] "And if we double the mass of the electron," Linde intones what has become a common litany, "life as we know it will disappear. If we change the strength of the interaction between protons and electrons, life will disappear. . . . If we had four space dimensions and one time dimension, then planetary systems would be unstable and our version of life would be impossible. If we had two space dimensions and one time dimension, we would not exist."[36] All these scenarios are instances of what physicists call the "fine-tuning problem": each of the constants of nature seems as if it has been set just right to condition our existence.[37]

One quality that theoretical physicists share with philosophers (and that both groups share with children) is a tireless capacity to ask "why." Theoretical physicists cannot or, at the very least, *do* not simply accept these parameters as such.[38] Why, they ask, are the parameters so delicately calibrated? Why does each of these constants have the value it has, rather than any other value? Why is the universe configured in such a way that it allows the existence of theoretical physicists who ask why it's configured this way? One possible answer, of course, is that an all-powerful deity set each of the controls just right, so that not only life but *conscious* life might emerge. To be sure, this argument is not properly "scientific," yet it has encroached on physicists' terrain with increasing insistence over the past fifty years or so, provoking frustration in most of them, acceptance in a few, and abject rage in others.

Most commonly called "intelligent design theory," the attribution of fine-tuning to divine activity has its roots in Thomas Aquinas's fifth proof of the existence of God: the "teleological proof" or the "argument from ends."[39] It finds its more colloquial, Enlightenment-age articulation through Cleanthes, a character in David Hume's *Dialogues Concerning Natural Religion* (1776). A spokesperson for "natural religion" (as distinct from "revealed religion," whose spokesperson is Demea, and from skepticism, whose spokesperson is Philo), Cleanthes argues that any careful observation of nature will lead one to affirm the existence, omnipotence, and benevolence of God. "Look round the world," he entreats his interlocutors,

> You will find it to be nothing but one great machine, subdivided into an infinite number of lesser machines, which again admit of subdivisions to a degree beyond what human senses and faculties can trace and explain. All these various machines, and even their most minute parts, are adjusted to each other with an accuracy which ravages into admiration all men who have ever contemplated them. The curious adapting of means to ends, throughout all nature, resembles exactly, though it much exceeds, the productions of human contrivance; of human design, thought, wisdom, and intelligence. Since therefore the effects resemble each other, we are led to infer, by all the rules of analogy, that the causes [also] resemble, and that the Author of Nature is somewhat similar to the mind of man, though possessed of much larger faculties, proportioned to the grandeur of the work which he has executed.[40]

The major premise of this argument is that similar causes produce similar effects. The minor premise is that the universe (the effect) looks like a set of carefully interrelated machines. The conclusion, then, is that we can suppose the cause of the universe to resemble the cause of machines. The characters in the *Dialogues* go on to consider the examples of a ship and a house. If one were to stumble upon a beautifully built structure, one would immediately believe it to be the work of a highly intelligent and well-intentioned craftsperson (16). Similarly, when we contemplate the beauty and economy of the universe, we should believe it to be the work of a highly intelligent and well-intentioned universe builder.[41]

As the *Dialogues* progress, Philo the skeptic unleashes a relentless flood of objections to Cleanthes's argument from design. First, Philo points out that the analogy does not hold: the universe looks almost nothing like a ship, a house, *or* a machine. Because the effects are so dissimilar, we have no reason to liken the cause of the universe to an architect or a manufacturer. Second, Philo wonders why Cleanthes assumes that the cause of the universe must be *intelligent*. There are far more powerful forces in the universe than intelligence—heat, for example, or attraction, repulsion, refraction—any of which could presumably have brought the world into being. "Why select so minute, so weak, so bounded a principle as the reason and design of animals is [*sic*] found to be upon this planet?" Philo asks. "*What peculiar privilege has this little agitation of the brain which we call 'thought,'* that we must thus make it the model of the whole universe?" (19, emphasis added). Third, Philo channels Hume himself to argue that we learn about causes and effects only through repeated experience.[42] I know that dropping a book causes it to fall downward because I have done it countless times. Similarly, we know that houses have human builders not because of any inherent property of houses, but because we have seen houses being built again and again. But when, Philo asks, has any of us seen even a single world come into being, much less the thousands we would need in order to understand the causation of one? "Have worlds ever been formed under your eye," he asks Cleanthes, "and have you had leisure to observe the whole progress of the phenomenon, from the first appearance of order to its final consummation? If you have, then cite your appearance and deliver your theory" (22).[43]

As if this weren't enough, Philo piles on even more objections to the argument from design in the chapters that follow. If the world has been caused by something outside it, he suggests, then why not ask what caused the Cause? And then why not ask what caused the Supercause that caused that Cause? One might go on heaping unknown causes onto unknown causes from now until eternity, *or* one might just stop with this world and refrain from adding an unknown cause to it in the first place (31–32). And by the way, Philo continues, "what shadow of an argument can you produce from your hypothesis to prove the *unity* of the Deity? A great number of men join in building a house or a ship . . . [so] why may not several deities combine in contriving and framing a world?" (36, emphasis added). For all we know of the world, it might have been created by committee or generated through divine copulation or woven by a great cosmic spider. For all we know, this purportedly harmonious creation might be "only the first rude essay of some infant deity who afterwards abandoned it, ashamed of his lame performance" (37). Or it might be the most recent effort in a long *series* of creative attempts; as Philo imagines this scenario, "many worlds might have been botched and bungled, throughout an eternity, ere this system was struck out; much labor lost; many fruitless trials made; and a slow but continued improvement carried on during infinite ages in the art of world-making" (36). Maybe this world is just the first one that finally worked at all.

Most important, there is the problem of evil. To Cleanthes's harmonious cosmos of interlocking machines, Philo opposes a world full of violence, greed, treachery, and anger, an earth that convulses with quakes, hurricanes, tornadoes, and floods. What kind of a God, Philo wonders, would make a universe with *mosquitoes* in it? How could there be a whole class of insects designed merely to "vex and molest" other animals (animals, by the way, that exist by destroying other animals and plants)? Far from this being the best of all possible worlds, "the course of nature tends not to human or animal felicity," setting all against all in an indifferent universe (65). It seems, then, either that God lacks the power to make a better world or that he lacks the will to do so: "Is he willing to prevent evil, but not able? Then he is impotent. Is he able, but not willing? Then he is malevolent. Is he both able and willing? Whence then is evil?" (63).

"Here I triumph," declares Philo. There is no way, he insists, to look at the evils plaguing this world and know that a benevolent, omnipotent God created it. Now, of course, one can always *believe* such a God exists in the face of evil—one might say that God permits evil to exist because God gives humans free will or that God permits evil so that humans can choose the good—but such positions, he insists, would ultimately be rooted in "faith alone" (66).[44] In other words, Philo has not *disproved* the existence or attributes of God; he has simply shown that the argument from design does not work to establish them. There is no way, simply from observing the world, to prove that it is the work of an omnipotent, benevolent, extra-cosmic, anthropomorphic, single male God.

As exhaustive as Philo's counterarguments might seem, however, very few of them seem to have convinced William Paley, whose *Natural Theology* (1802) resurrects the argument from design, defends it in unprecedented detail, and serves as the most direct template for contemporary intelligent design theory. "In crossing a heath," the book begins—directed most likely to any heath-en who may have opened it—"suppose I pitched my foot against a *stone*, and were asked how the stone came to be there, I might possibly answer that, for any thing I knew to the contrary, it had lain there for ever." So far, so good. "But suppose I had found a *watch* upon the ground. . . . I should hardly think of the answer which I had given before, that for any thing I knew, the watch might always have been there." The difference between a watch and a stone, Paley says, is that a watch is clearly an object of "contrivance"; any decent examination of the mechanisms of wheels, cogs, and teeth will indicate that it had a maker—and an intelligent one at that—who intended that his object fulfill its purpose of keeping time. Paley then claims that "every manifestation of design, which existed in the watch, exists in the works of nature; with the difference, on the side of nature, of being greater and more, and that in a degree which exceeds all competition."[45] Nature, in other words, is as thoroughly designed in each of its components and relations as the aforementioned watch. It must therefore have an anthropomorphic Maker, whose grandeur is apportioned to the baffling scope of his creative work.

In order to convey this grandeur, Paley spends the majority of the *Natural Theology* enumerating various features of the cosmos that could not possibly have arisen by accident. From a lengthy meditation on the complexity and delicate function of the eye ("What does chance ever do for us?" he asks. Chance produces "a wen, a wart, a mole, a pimple, but never an eye"),[46] he proceeds over nearly six hundred pages to analyze the musculoskeletal and circulatory systems of animals, the structures and functions of vegetables and fruits, the harmonious composition of the elements, and the elliptical orbits of planets and comets—stopping along the way to scrutinize such "peculiar organizations" as "the air bladder . . . of a *fish*," "the *fang of a viper*," "the *bag* [pouch] *of the opossum*," "the *stomach of the camel*," "the tongue of the wood-pecker," "the proboscis of the bee," "the abdomen of the silkworm," and "the hinges in the wings of an *earwig*."[47] His point throughout this encyclopedic tour is to demonstrate the extent to which all these specific "contrivances" are suited toward particular *purposes*; the camel has to retain water, but the fish does not; the waterfowl needs webbed feet, but the land fowl does not. Half a century before the publication of Darwin's *On the Origins of Species* (1859), Paley insists through each of these examples that there is no other way to account for the delicate adjustment of a creature's means to its ends than to affirm that "there is a God; a perceiving, intelligent, designing Being, at the head of creation, and from whose will it proceeded."[48]

This is the sort of argument that has returned in the face of the finely tuned parameters of modern physics. How else, the theists ask, is one to account for the precarious perfection of these constants? As Francis Collins, director of the National Institutes of Health, said in 2011, "To get our kind of universe, with all of its potential for complexities or any kind of potential for any kind of life-form, everything has to be precisely defined on this knife edge of probability. . . . *You have to see the hands of a creator* who set the parameters to be just so because the creator was interested in something a little more complicated than random particles."[49] Most theoretical physicists are hardly satisfied by this argument. Nor, however, can they shrug off the order of the universe as the product of chance; in the words of Paul Davies, "[M]ost scientists concede that there are features of the observed Universe which appear . . . ingeniously and felicitously arranged in their relationship to the existence of biological organisms in general and intelligent observers in particular."[50] The trick, then, is to account for this "ingenious and felicitous arrangement" without reference to a creator; as Davies writes elsewhere, "[M]any scientists who are struggling to construct a fully comprehensive theory of the physical universe openly admit that part of the motivation is to finally get rid of God, whom they view as a dangerous and infantile delusion."[51] And the most promising, most controversial strategy for finding such a theory involves "the anthropic principle."

Coined by theoretical physicist Brandon Carter in 1974, this principle states that "what we can expect to observe must be restricted by the conditions necessary for our presence as observers"; that is, unless the universe were compatible with life, we would not be here to ask why it is the way it is.[52] This principle comes in a variety of strains—chief among them, the weak anthropic principle (WAP) and the strong anthropic principle (SAP).[53] WAP arguments simply assert that "since there are observers in our universe, its characteristics . . . must be consistent with the presence of such observers."[54] In other words, if the universe did not have the parameters necessary to produce life, then there would be no one around to ask the question. SAP arguments, by contrast, assert that the universe *must* possess those parameters necessary to bring about intelligent life—that "the world is fine-tuned so that we can be here to observe it."[55]

For decades, most physicists were reluctant to go anywhere near the anthropic principle (Andrei Linde testifies that they avoided saying "the 'A'-word" as fervently as Harry Potter's friends avoid saying "Voldemort").[56] After all, the SAP is just a repackaged argument from design: rather than God's tuning the universe so that life can emerge, the universe tunes the universe so that life can emerge. It is the same claim, just with a strictly pantheist deity (that is, a God who *is* the universe). And, of course, one can always reject this pantheist deity and tag an external creator back onto a strongly anthropic universe, saying that it is God who configures the universe so that the universe can configure itself so that life can emerge. For this reason, whereas some theologians reject the anthropic principle as unnecessary,[57] others have adopted it as what physicist and Anglican priest John Polkinghorne calls "a cumulative case for theism."[58] Evangelical theologian William Lane Craig has even gone so far as to say that John Barrow and Frank Tipler's *The Anthropic Cosmological Principle* (a 1986 compendium of works on cosmic fine-tuning) became "for the design argument in the twentieth century what Paley's *Natural Theology* was in the nineteenth," causing the design argument to come "roaring back into prominence during the latter half of the last century."[59] Carter himself has expressed frustration that the anthropic principle, particularly in its "stronger" forms, has been harnessed to support the argument from design,[60] and this widespread theistic co-optation has obviously done little to warm secular physicists to the idea.

As far as the WAP goes, its least controversial formulations verge on tautologies along the lines of "we are here because we can be here." To say that the conditions of any universe in which an observer finds herself must be consistent with her existence is not really to answer the question of why the conditions are as they are in the first place. Poking fun at the WAP, philosopher of religion Richard Swinburne offers a colorful analogy:

> On a certain occasion the firing squad aim their rifles at the prisoner to be executed. There are twelve expert marksmen in the firing squad, and they fire twelve rounds each. However, on this occasion all 144 shots miss. The prisoner laughs and comments that the event is not something requiring any explanation because if the marksmen had not missed, he would not be here to observe them having done so. But of course the prisoner's comment is absurd; the marksmen all having missed is indeed something requiring explanation; and so too is what goes with it—the prisoner being alive to observe it. And the explanation will be either that it was an accident (a most unusual chance event) or that it was planned (e.g. all the marksmen had been bribed to miss).[61]

In short, the WAP still leaves us baffled by the extent to which the forces of nature have not obliterated us—more precisely, by the extent to which they have not prevented our existence altogether. It tells us we are here because if we were not, no one would ask, "Why are we here?" So, one is left asking . . . "Why are we here?" As Swinburne suggests, God and dumb luck seem to be the only options—and dumb luck is not really an answer.[62]

The Rise of the Dark Lord

Then fifteen years ago, the fine-tuning problem got worse. Ever since the widespread acceptance of the big bang hypothesis in the mid-1960s (see chap. 5, sec. "Let There Be Light"), cosmologists had assumed that although the universe was still expanding from its initial outward burst, this expansion had been slowing down ever since then. The question became: *How much* was the expansion slowing down? Would the universe continue to increase in size forever—just at a slower and slower rate? Would it one day stop expanding, reverse its direction, and race back inward to an apocalyptic "big crunch"? By using Type Ia supernovae to measure the brightness, and therefore the distance, of faraway galaxies, two independent research teams set out in 1998 to measure the "deceleration parameter"—that is, the rate at which cosmic expansion is slowing down. But to everyone's surprise, even horror, both teams found that the universe's expansion is not slowing down; it is speeding up.[63] Faraway galaxies are moving away from us faster and faster as time goes on.

When it comes to the expansion of the universe, there seem to be two major forces at work, engaged in a cosmic tug-of-war. Gravity exerts an attractive "pull" on the fabric of space-time, while something else provides a repulsive "push." This "something else," whose tortured history we glimpse in chapters 4 and 5, is called the "cosmological constant" or, more popularly, "dark energy." On the broadest possible level, our 13.8-billion-year-old cosmos has undergone three major stages since its initial eruption: first, a rapid expansion dominated by searing radiation; second, a gradual slowing down and cooling off as gravity took hold; and third, a reacceleration as dark energy's push overcame gravity's pull, about 7.5 billion years after the bang.

Represented by the Greek letter lambda (Λ), the cosmological constant seems to be the energy of empty space itself. As this brief description might already make clear, however, *empty* is a misleading adjective; one of the remarkable discoveries of quantum mechanics is that empty space is not really empty at all. Rather, what physicists call the "vacuum" is "alive with virtual particles . . . flashing into and out of existence."[64] When quantum field theorists add up the effects of these virtual particles, they can determine the vacuum's overall energy. But when they do so, they infamously produce a number that is 10^{120} times larger than the value of dark energy observed (that is, 10 with 120 zeroes after it; for comparison, there are "only" about 10^{80} atoms in the entire visible universe).[65] Michael Turner, who coined the term *dark energy*, calls this miscalculation "the most embarrassing number in physics"; Sean Carroll judges it "a complete fiasco"; and Lee Smolin muses that "it must just qualify as the worst prediction ever made by a scientific theory . . . something is badly wrong here."[66] The baffling smallness of the value of dark energy constitutes the so-called cosmological constant problem.[67]

But the problem, to return to our central concern, is not just a calculative one; rather, the strength of the cosmological constant is "the most striking" example yet of a finely tuned parameter.[68] When it comes to the mass of the electron, the nuclear forces, and most of the other constants of nature, any value other than those observed would have prevented the emergence of "life as we know it" in the universe. When it comes to the cosmological constant, any value other than that observed would have prevented the emergence of the universe itself.[69] If the cosmological constant were greater than it is, it would have pushed space-time apart before planets and stars could form. If it were smaller, gravity would have drawn the early world into a fiery collapse.[70] As Leonard Susskind has put it, the discovery of dark energy therefore forced physicists to confront "the elephant in the room," which is to say the reviled anthropic principle.[71] How is it possible that the universe happens to have a vacuum energy small enough to let us live?

The answer, as Steven Weinberg began to insist in 2000, requires a multiverse.[72] If there is only one universe, with only one value for each fundamental parameter, then it is impossible to explain how the cosmological constant (Λ) is so improbably small without appealing to either God or dumb luck. But if there are *a whole slew of universes*, each with a different value for lambda, then every possible value is out there somewhere. Some universes collapse under a weak lambda, some blow apart under a strong lambda, and some have a lambda that is just the right value to allow stars and planets to form. We live in one of those "Goldilocks" universes, where the constants are just right.

The discovery and miscalculation of dark energy therefore prompted a widespread reexamination of some of the multiverse theories that had existed for decades—most notably, the "new" and "chaotic" eternal inflationary scenarios that had emerged in the 1980s. According to these models, our universe is one of an infinite number of "bubbles" that form in a rapidly expanding sea of energy, and each of these bubbles might have different physical laws from the others.[73] So at the turn of the millennium, as strange as it had seemed a decade earlier, eternal inflation began to look like a promising way to produce the scores of *other* parameters needed to make our strangely improbable parameters inevitable. The anthropic multiverse hypothesis was further assisted by developments in string theory—in particular, by the realization that there might be upward of 10^{500} or even $10^{1,000}$

different string vacuum states, each corresponding to a different type of universe.[74] By the turn of the millennium, then, quantum mechanics, modern cosmology, and string theory had independently collided with one or another version of the multiverse. It is here that we begin to see the merging of scientific developments and philosophical expedience: although the multiverse can be said to have emerged "naturally" from developments in both astro- and particle physics, it has become a more broadly viable hypothesis in the scientific community because it provides a way out of the fine-tuning problem.[75]

The multiverse, in short, redeems the anthropic argument—saving it from tautology *and* from God. We are reduced neither to saying, "We're here because we're here" nor to postulating a benevolent, omnipotent, transcendent creator who must have set everything just right so that life might emerge in the universe. After all, if there are an infinite number of worlds that take on all possible parameters throughout infinite time, then strange as our specific parameters may seem, they were bound to emerge at some point. Lawrence Krauss compares the process of universe selection to throwing "zillions and zillions of darts" at a dartboard; a few of them will eventually hit the bull's-eye.[76] For this reason, Stephen Hawking and Leonard Mlodinow compare the twenty-first-century emergence of the multiverse hypothesis with the nineteenth-century emergence of natural selection: "Just as Darwin and Wallace explained how the apparently miraculous design of living forms could appear without intervention by a supreme being, the multiverse concept can explain the fine-tuning of physical law without the need for a benevolent creator who made the universe for our benefit."[77]

There is, then, a profoundly nontheistic (sometimes even antitheistic) motivation behind the scientific turn to many-worlds scenarios.[78] As Bernard Carr has summarized the matter, "If you don't want God, you'd better have a multiverse."[79] Of course, the two hypotheses are not totally incompatible; one can always argue that a transcendent God created the multiverse that created the universe.[80] But the *necessity* for a God figure is gone; as the Atomists realized 2,500 years ago, the multiverse hypothesis does not disprove God's existence—it just takes his most significant job away through the twin powers of infinity and accident. That having been said, the multiverse replaces God with what is perhaps an equally baffling article of faith: the actual existence of an infinite number of worlds, eternally generated yet forever inaccessible to us. The multiverse, then, becomes its own kind of theological postulate even as aims to unsettle all theological postulates. This may be the function of the question to which the multiverse is responding in the first place; one might argue that the moment one asks *why* the universe is the way it is or why there is something rather than nothing, one is already on metaphysical ground (or in the metaphysical clouds). In any event and for better or worse, multiple-world cosmologies consistently rearrange the boundaries between and among philosophy, theology, astronomy, and physics—ultimately, I will argue, opening the possibility of new relationships across these boundaries.

Mapping the Multiverse

The word *cosmos* (*kosmos* in Greek, pl. *kosmoi*) can be traced back to Homer, who used it to refer to the order of soldiers on a battlefield or of rowers in their boats.[81] Herodotus later called on the term in reference to the highly organized state of Sparta and to designate order in general.[82] At some point, the term *kosmos* came to refer to the whole world, first designating "the order exhibited by the universe, and then, by extension . . . the universe itself as a well-ordered system."[83] Although it is not certain who initially proposed this "cosmic" meaning of *kosmos* (it may have been either Pythagoras in the sixth century B.C.E. or Protagoras in the fifth),[84] the association was firmly in place by Plato's time (429–347 B.C.E.). In fact, Plato called on a number of terms to signify "the universe": not only *kosmos*, but also *to pan* (the all), *onta* (all things), *ouranos* (the heavens), and *to olon* (the whole).[85]

This book is concerned with the manyness at the heart of this "whole"— with the multiple ways the world(s) can be said to be multiple. It begins in chapter 1 with a reading of Plato's *Timaeus*, which insists that this world must be both singular and everlasting and which establishes a blueprint for 2,400 years of subsequent Western cosmology. But even in the *Timaeus*, we shall see, the singular world is not simply singular. Whether by design or in spite of himself, Plato offers a unique and undivided cosmos that is nevertheless composed of difference, mixtures, and pluralities. The Platonic cosmos can thus be regarded as an interplay of order and disorder, of the singular and the plural, and of unity and difference. I propose that this interplay might best be called *multiplicity*. Neither the all-is-oneness of singularity (what William James referred to as "monism") nor the disconnected indifference of plurality, multiplicity would name that mixing of order and chaos through which worlds emerge—both in the midst of multiplicity and as themselves multiple.[86]

The chapters that follow trace the theme of cosmic multiplicity through the tradition that runs from Plato to string theory. Chapter 1 concludes with Aristotle's consolidation of a singular, geocentric cosmos—a world whose oneness and eternity he believed Plato had failed to secure. Chapter 2 then explores the two cosmologies that posed the greatest threat to Platono-Aristotelian oneness and eternity—the spatial multiplicity of the Atomists and the temporal multiplicity of the Stoics. Of course, neither of these cosmologies "won"; from Aristotle through Einstein, the mainline philosophical and scientific tradition maintained that the world was singular and unchanging. But even the most loathsome ideas resurge periodically, thwarting every effort to repress them, and it is these resurgences that the remaining chapters consider. Chapter 3 begins with Thomas Aquinas's final attempt to defend cosmic singularity along Aristotelian lines—a position condemned a few years later when Bishop Etienne Tempier of Paris declared anathema the position "that God cannot create more worlds than one."[87] The decades and centuries that followed therefore witnessed a surge of treatises on the possibility of multiple worlds—a trajectory that produced Nicholas of Cusa's centerless multiverse in the fifteenth century and culminated in the infinite worlds of Giordano Bruno, whom the Roman Catholic Church executed in 1600.

Chapter 4 begins after the death of Bruno, when a combination of traditional theological restraint and modern scientific rigor prompted thinkers such as Galileo Galilei, Johannes Kepler, and Isaac Newton to confine their vision to the one world they could see—a restraint that held among natural scientists until the first decade of the third millennium. Among those whom Kepler called "mad philosophers," however, the question of multiple worlds persisted, flourishing in the wake of René Descartes's "vortex" cosmology and reaching its apex in the extravagant, confused, almost entirely ignored, and eventually disavowed multiversal vision of a young Immanuel Kant. Chapter 5 opens in the early twentieth century, by which time the singularity and eternity of the cosmos went without saying. Although the latter principle was discredited with the midcentury rise of the hot big bang hypothesis, the former would not be seriously challenged until the discovery of dark energy in 1998. Since then, physicists have produced a flood of new cosmologies of the multiverse, which chapters 5 and 6 map in relation to one another and to the models that surfaced before them.

Finally, the last chapter, "Unendings," focuses on the debate over the scientific status of multiverse cosmologies: How far can physics speculate about other universes without colliding with "mad philosophy"—or, worse, with theology? The question seems ultimately to hinge on relation: if different universes are totally separate from one another, then scientists are unjustified in relying on them as an explanatory principle. If, however, they somehow bear the imprints of one another, then cosmologists have something to measure and observe, so the multiverse might be a "proper" scientific object after all. "Mad philosophy," in the meantime, will have a great deal to come to terms with: a multi-universe that is neither one nor many but a many-one, a singular plurality that renders our one world multiple, entangled, and vulnerable.

But first things first . . .

A SINGLE, COMPLETE WHOLE

> Don't therefore be surprised, Socrates, if on many matters concerning the gods and the whole world of change we are unable in every respect and on every occasion to render a consistent and accurate account. You must be satisfied if our account is as likely as any, remembering that both I and you who are sitting in judgment on it are merely human, and should not look for anything more than a likely story in such matters.
>
> PLATO, *TIMAEUS*

"So Let Us Begin Again . . .": Plato's *Timaeus*

A Likely Story

In the beginning was chaos. This is as likely a story as any: cosmos from chaos, order from disorder, the whole world assembled from some formless flux. But what kind of a starting point is chaos, when chaos confounds all points? How do boundaries emerge from the boundless, things from the no-thing, information from the noise?

The most ancient of Greeks addressed this problem by positing an organizing force internal to the precosmic flux. In Hesiod's *Theogony*, for example, a primordial Chaos gives birth spontaneously to the earth, who in turn gives birth to heaven, through whom she conceives the ocean, time, and countless violent gods.[1] As the early Ionian philosophers sought to replace such mythology with a purely "rational" cosmogony, they transformed Chaos from a primordial goddess into one of the natural elements, which produced all things and would reclaim them in time.[2] This primary substance was water for Thales,[3] air for Anaximenes,[4] fire for Heraclitus,[5] and a pre-elemental "boundless" for Anaximander.[6] For Empedocles, the primary substance was all four elements (earth, air, fire, and water), which are separate from one another during a periodic stage of cosmic "strife" and indistinguishable from one another during the opposing periodic stage of "love."[7] The cosmos we inhabit must exist somewhere between these two stages; for Empedocles, *both* the total separation and the total union of the elements amount to chaos, which means that order itself is a fragile mixture of sameness and difference.

Then, with the rise of Atomism in the late fifth century B.C.E., Chaos was shattered into countless invisible bodies moving aimlessly in a void. Each atom (from the Greek *atomos* [uncuttable]) was said to collide haphazardly with other atoms, their collisions forming an infinite number of worlds in random motion. As one world crashed into another, both would unravel again into a primal swarm of atoms, and the whole process would repeat.[8] A century later, the Stoic school would posit a "Phoenix" universe that was born, destroyed, and reborn in fire.[9] Whether mythos or logos, however, these are all similar tales: out of the primordial mess comes the whole world, and back into it the world will one day be absorbed. Water to water, atoms to atoms, chaos to chaos.

Hesiod, the Ionians, and the Atomists provide most of the cosmogonies that would have been familiar to Timaeus, the astronomer and primary speaker in the Platonic dialogue that bears his name.[10] Like his philosophic predecessors, Timaeus describes the birth of the cosmos as an ordering of disorder, telling us that in the beginning the universe was a mess of forces in unruly motion.[11] The difference is that Timaeus does not attribute the ordering of this mess to an internal principle; rather, he explains it as the work of a god (*demiourgos*). For the first time in Greek philosophy, cosmogony is narrated as creation:[12] far from emerging out of the primordial element (or *being* it, as the Stoics will argue), the god of the *Timaeus* is external to the unformed world, "finding the visible universe in a state not of rest but of inharmonious and disorderly motion" (30a).[13] Immediately upon "finding" it thus—as if stumbling on it one day, shocked to come across anything so material, so unintelligible, so out of control— the god "reduces" the world "to order [*taxis*] from disorder," using as his model the idea of a "perfect living creature"—perfect because it "comprises all intelligible beings" (30a, 31b). The perfect living creature, in other words, is something like the Form of Forms—the Idea that contains all other ideas within it. In this light, the demiurge looks remarkably like the *Republic*'s philosopher-king, who puts the city in order by fixing his eyes on the Good.[14] With his hands in the chaos and his gaze on the Forms, the god of the *Timaeus* shapes the material world into a "visible, tangible, corporeal" image of the ideal and eternal (28c).

Against the (unnamed) Atomists and their randomly colliding worlds, Timaeus argues that the intelligently designed cosmos cannot be one of many *kosmoi*, let alone an infinite number of them. Rather, it must be singular—as unique as the Form of the "perfect living creature" it mirrors:

> Are we right then to speak of one universe [*ouranos*], or would it be more correct to speak of a plurality or infinity? ONE is right, if it was manufactured according to its pattern; for that which comprises all intelligible beings cannot have a double. . . . In order therefore that our universe should resemble the perfect living creature in being unique, the maker did not make two universes or an infinite number, but our universe was and is and will continue to be his only creation. (31a–b)

The argument here is that the universe must be singular in order to resemble its model most perfectly. The model itself must be singular because it must include "all intelligible beings" (if there were more than one such model, then they would simply be copies of the *genuine* Form of the perfect living creature). In short, because singularity is more perfect than plurality, there can only be one model, and because there is only one model, there is only one world.

Just as the Platonic cosmos must be singular, it must also be indestructible. Timaeus's reasoning is as follows: a body can be destroyed only by something inside it or something outside it. If something inside the world were to destroy it, it would do so either by violence or by gradual decay. But violence will never destroy this cosmos because, as we experience all the time, the elements exist in perfect harmony. And decay will never undo it because the death of any part of the cosmos—whether animal or vegetable—nourishes the whole (32b–33b). Even the periodic destruction of whole cities by flood and fire does not damage the world as a whole; rather, new civilizations will arise in their place (39d). Internally, then, the world is "ageless and free from disease" (33a). And externally, Timaeus assures us, the world is also free from danger because the god in his wisdom "used up the whole of each of [the] four elements" in creating it (29e, 32c). In other words (and, again, implicitly against the Atomists), *there is nothing outside the world*—neither other worlds nor extraneous particles that might vex it from without. The only power who *might* undo the cosmos would be the god himself, but his untrammeled goodness prevents his committing any such crime. As the demiurge tells the lesser gods he goes on to create, "Anything bonded together can of course be dissolved, though only an evil will would consent to dissolve anything whose composition and state were good. Therefore, since you have been created, you are not entirely immortal and indissoluble, *but you will never be dissolved nor taste death*, as you will find my will a stronger and more *sovereign bond* than those with which you were bound at your birth" (41a–b, emphasis added).[15]

Thanks to its benevolent sovereign, the world once delivered from chaos will remain perfectly ordered forever. Made in the image of the eternal and secured by divine will, the cosmos is unique and indestructible: a *"single, complete whole"* (33a, emphasis added).

Mixing the Multiple

Considered in this manner, the *Timaeus* seems to install cosmologically what eventually becomes the usual Western philosophical privileges: order over chaos, stasis over change, the intelligible over the sensible, and the singular over the plural. The model is one, so the world is one, and its order endures forever. As the story goes on, however, we find that this reading both does and does not hold. A strange dance between the one and the many begins the moment Timaeus, having established the singularity and indestructibility of the cosmos, goes on to explain *how* it was so made.

First, Timaeus details the god's meticulous formation of the world's "body" as a proportional balance of water and air, fire and earth (32b). Then he moves on to the world's "soul," telling his interlocutors that "god *created the soul before the body* . . . gave it precedence both in time and value, and made it the dominating and controlling partner" (43c, emphasis added). With this description, the reader stumbles: If the god created the soul before the body, then why did Timaeus tell the story the other way around? Why would he begin with the body if the god began with the soul? In his authoritative commentary on the *Timaeus*, classicist Francis M. Cornford skips quickly over this conundrum, saying that Timaeus starts his own story with the body "for convenience."[16] But why would it be more convenient to begin with the body, only to have to explain that it does not come first? What does it mean for the story to grant primacy to the body when the cosmogony grants primacy to the soul—especially considering Timaeus's own recognition that cosmogony can only be a "likely story" (*eikota mythos*, 29c–d) to begin with?[17] Is Timaeus telling the story badly?[18] Is he reversing his own privileging of the soul over the body—

whether playfully or unknowingly? Or is he simply demonstrating the difficulty of getting a solid beginning out of chaos?

Whatever the reason, Timaeus continues the story unself-consciously. The demiurge made the soul first so that, just as reason rules the passions and the guardians rule the merchants, the world's soul might rule the body it precedes. Of what, then, is this regal soul made? Timaeus has just asserted its superiority over the body, so we might expect him to sustain this privilege and proclaim the soul to be immaterial. But, again, Timaeus throws us off course, saying that "he [the demiurge] composed [the world soul] in the following way. . . . From the indivisible, eternally unchanging Existence and the divisible, changing Existence of the physical world he mixed [*synekerasato*] a third kind of Existence intermediate between them" (35a). This very mixing, however, undermines Timaeus's privilege of the soul over the body: How can the world's soul be superior to the physical world if it is *composed of the physical world*? How can the soul have preceded the body "in time" if the soul is partially made of the body? And how can the world be singular, eternal, and indivisible if its soul is irreducibly mixed? Yet this seems to be the soul of the "one" world: a commingling of the indivisible and unchanging (the Forms), the divisible and changing (the physical world), and a third substance that is neither divisible nor indivisible—or perhaps a little bit of both. And the god has not finished yet. After mixing the (in)divisible and the (un)changing, "again with the Same and the Different he made, in the same way, compounds intermediate between their indivisible element and their physical and divisible element: and taking these three components he mixed them into a single unity" (35a). A very bizarre "single unity" indeed, this world soul is now a conglomeration of indivisible existence, divisible existence, indivisible sameness, divisible sameness, indivisible difference, divisible difference, neither divisible nor indivisible existence, neither divisible nor indivisible sameness, and neither divisible nor indivisible difference.

By means of this many-layered mixing, Timaeus explains, the demiurge was "forcing the Different, which was by nature allergic to mixture, into union with the Same, and mixing both with Existence" (35a). This, too, seems a strange thing to say: Surely sameness is more "allergic" to mixing than difference is? But if we grant Timaeus this odd proposition— that difference is in some way inimical to mixing—then we must conclude that the resulting mixture, by definition more plural than "the Same," must also be different from "the Different." The mixture is something like difference *and* nondifference or something *different* from difference and nondifference, and this mixture alone constitutes the "unity" of the world—this assemblage of parts its "indivisibility." Perhaps predictably, this indivisibility is no sooner established than it is divided: after forcing the Same and the Different into "a single whole," the god finishes the soul by making "appropriate subdivisions, each containing a mixture of Same and Different and Existence" (35a). And so the soul is a "union," but it is a mixture. It is a "single whole," but it has subdivisions, each of which is itself a mixture. This cosmic intermingling of the singular and the plural, different as it is from difference and irreducibly mixed in its sameness, is an instance of what I am proposing to call the *multiple*. Every time the *Timaeus* tries to assert the Oneness of the world, the text collides with something like multiplicity.

This peculiar tango of the one and the many finds inimitable elaboration in Michel Serres's philosophical meditation on chaos. "For the *Timaeus*," he muses, "the world is a harmony, the world is a mix, the world is unitary, formed, comparable, thinkable, but it is only mixed with mixtures, it is even a mix of mixes of mixes, but . . ." Serres imagines Plato as a jester-sailor, tacking in a different direction each time we think that we have cornered him: "The ruse goes on, and Plato pulls to the right to the one-ward, then he pulls to the left to the multiple-ward, then pulls to the right again . . . and so forth." What this means for Serres is that Plato himself, unlike the reliably dualistic "Platonists," does not exclude the different, the changing, and the material; rather, "he negotiates it," sailing between the cosmic and the chaotic, the same and the other, the thinkable and the unthinkable.[19] Although it is not clear whether such a "negotiation" is performed willingly or begrudgingly, one cannot help but agree that the will toward oneness in the *Timaeus* finds itself multiply interrupted. At every turn, we find the same shot through with the different, spirit with matter, existence with nonexistence, *and all of them with a strange set of terms in between*. Even in the hands of the most intelligent and benevolent god, cosmos emerges only by means of chaos, as a mixture of itself and what is not itself, of different and same, of "both/and" and "neither/nor": what Serres calls "a pure multiplicity of ordered multiplicities and pure multiplicities."[20]

From this mixture of mixtures, this multiplicity of multiplicities that composes the body and soul, Timaeus goes on to derive a number of the world's components. Mininarratives explain the emergence of time, the sun and moon, the planets and their orbits, the gods, the human soul and body, light and vision, hearing and sound. And then, out of nowhere, Timaeus tells us that he is going to have to start again. Right after elaborating the gifts of harmony, speech, and rhythm, for no discernible reason at all, he interrupts his catalog to clarify that all the actions he has been describing are the work of divine intelligence (*nous*). But intelligence, he explains, is only half of the story. For although this god is perfectly rational, he is not free to dream up anything he likes. Rather, his intelligence must work together with "necessity" (*ananke*, 47e–48a). In other words, all the mixings we have encountered at the hands of intelligence must themselves be mixed with the possibilities and constraints of their materials—materials that come from the "inharmonious and disorderly" flux the demiurge initially "found." Halfway into the cosmos, then, Timaeus heads back to chaos.

A Likelier Story

"So let us begin again," Timaeus suggests (48e). Once again he throws the text's will toward oneness into stunning disarray, for this story about The Beginning begins twice. Perhaps concerned that his audience might get up and leave, Timaeus assures us that this second beginning will be a better one than the first. This beginning will be "more rather than less likely" because it will call on a more original origin, allowing us to glimpse the precosmic chaos itself (48e, 52d–53b). But if this is the case, wonders the somewhat agitated reader, then why was this "more likely," *more primordial* beginning not Timaeus's first beginning? Why must the first beginning come second—the more original origin springing up in the middle of things? Why do we get a clear view of chaos only halfway through the formation of the cosmos?

We encountered a similar problem in Timaeus's attempt to insist that the soul precedes the body that constitutes it. The difficulty in both cases is presumably that neither of these elements exists independently of the other, so, strictly speaking, there is no good place to start. One way through this morass would be to abandon the notion that the cosmos has a temporal beginning at all. This was the preferred interpretation of Xenocrates and the Academicians, who saw in the *Timaeus* an account of the *eternal* creation of the cosmos rendered in linear format.[21] In fact, the only people who seem to have taken Timaeus's "beginning" to be temporal (with a few exceptions)[22] were those who sought to oppose it—the Peripatetics, Epicureans, and Stoics.[23] In remarkably different ways that this book will address in time, each of these schools maintained against "Plato" that the cosmos had to be eternal in the direction of the past as well as in the direction of the future and so could not have had the temporal origin that "Plato" ascribed to it. But with most of these opponents long gone, Francis Cornford could state in the 1930s that "it is now generally agreed" that the *Timaeus* does not describe a beginning in time. Rather, he argued, the dialogue dramatizes the *eternal* creation of order out of disorder by means of Reason. As such, the very figure of chaos is "an abstraction—a picture of some part of the cosmos . . . with the works of Reason left out."[24]

Along this interpretation, chaos would be something like the political philosopher's "state of nature"—that is, a picture of the universe with the order subtracted from it. If this is the case, then the reason Timaeus cannot simply begin with chaos is that chaos already presupposes cosmos. In other words, "pure" chaos has never existed. At the same time, insofar as the term *cosmos* names the ordering *of disorder*, "pure" cosmos has never existed, either. In sum, the most we can say of cosmos and chaos is that each relies on the other. Chaos and cosmos are each other's conditions of possibility as well as conditions of impossibility—that is to say, each is itself as never quite itself, or each is itself only in and through the other, which nevertheless undoes it.[25] Although it is doubtful that Cornford would have extended his analysis quite this far, he did offer a fleeting suggestion "that chaos, if it never existed before cosmos, must stand for some element that is now and always present in the working of the universe."[26] As it turns out, this eternal "element" is none other than that "necessity" that prompts Timaeus to start his tale over in the first place (second place). So let us begin again.

Timaeus's second beginning opens by introducing a new character: *khôra*, the eternal "receptacle" (*hypodoche*) in which the cosmos becomes itself.[27] Admitting that all of this is "difficult and obscure," Timaeus tries to explain that whereas his first story distinguished "two forms of reality" (the visible world and its invisible archetype), this second story will "add a third" (49a, 48e). Strictly speaking, however, he is not really adding a third; he is calling attention to that on which the first two relied. Unsurprisingly, this heretofore excluded condition of possibility, this container that is awfully hard to see from the visible world it bears, is figured in consistently feminine terms: *khôra* is named "the nurse of all becoming," the spatial "mother" to the Formal "father" and the cosmic "son" (49a, 50d).[28] But insofar as being belongs primarily to the Forms and only secondarily to the material realm, *khôra* herself "is" nothing. She receives but is not; she provides a place but remains "devoid of all character" (50c). In short, *khôra* functions as "a kind of neutral plastic material" onto which copies of the Forms are inscribed, taking the shape of whatever she receives and thus giving birth to the visible world (50c).

Having now established his full cast of characters, Timaeus can finally get to the very beginning and describe the state of things "before" creation—the primordial chaos. "My verdict," he (re)begins, "can be stated as follows. There were, before the world came into existence, being, *khôra*, and becoming, three *distinct* realities" (52d, emphasis added).[29] This distinctness, we come to learn, is precisely what makes the scene chaotic. Since nothing could "become" when becoming was disconnected from the *khôra* that brings being into being, "chaos" is nothing more and nothing less than the stuff of the cosmos in nonrelation.[30]

Timaeus illustrates this chaotic state by saying that the precosmic *khôra* contained no objects—only "traces" or "vestiges" (*ichnê*) of the four elements. Because these forces (*dunameis*) had no proper existence yet, they "swayed unevenly," moving through *khôra* in a thoroughly disorderly fashion (52e). In keeping with his description of chaos as distinction, Timaeus explains that what made this movement "disordered" was its continual *separation* of forces. Rather than being drawn together, the precosmic forces were shaken "like the contents of a winnowing basket" into "different regions of space" (52e–53a).[31] Similar forces clustered together, held apart from what was different, and as such *they did not yet exist*. This vision of chaos begins to make sense when one considers, for example, the processes of evaporation, rarefaction, and combustion: water, air, fire, and earth exist only insofar as they are transformed into one another.[32] Clustering around nothing but themselves "before" the world was made, these "traces" could not become the full-fledged elements whose interrelations compose the cosmos. The work of creation is therefore to bring things into existence by *relating them*—by mixing together that which the chaos keeps separate.[33]

But here we should remember that, for Timaeus, this creative mixing is not simply made in accordance with the "invisible realm" of the Forms; rather, the invisible realm becomes *part of the mix* ("from the indivisible, eternally unchanging Existence and the divisible, changing Existence . . . he mixed a third" [35a]). If "chaos" names the unrelated plurality of the "three distinct realities" (being, *khôra*, and becoming), then "cosmos" names their interrelation—the mixture of mixtures that worlds the world. As such, it is only in relation that any of these "realities" can be said to *be* at all. At this point, the Platonist will surely object that the realm of the

Forms exists independently of *khôra* and the visible world; by definition, the Forms are unconstituted by the material realm of change. And, indeed, the Forms may "exist" on their own in some timeless time and placeless place (How would we ever know?). But when it comes to the birth of the cosmos, the Forms can *come into existence* only by means of the matter with which they are mixed.[34] In this sense, the Forms rely as fully upon *khôra* and her imprints as the imprints rely upon *khôra* and the Forms, and as *khôra* relies upon the Forms and imprints. Insofar as the world is composed only as the mixing of these "realities," the invisible is not merely a "model," the visible not merely a resource, and *khôra* not merely their receptacle. Rather, each of them is woven into the (mixed) fabric of the cosmos, becoming itself only as part of this melee. And so a world is born in a movement, not from difference to sameness, but from unrelated differences to their related mix—from plurality, one might add, to multiplicity.

Perhaps needless to say, however, this reading of Plato did not become the standard one.

Reflecting Singularity: The Aristotelian Cosmos

The Oneness of the World: Aristotle's De caelo

"Are we right then to speak of one universe," Timaeus wonders aloud, "or would it be more correct to speak of a plurality or infinity?" (31a). The question, of course, is rhetorical; for Timaeus, the cosmos has to be singular in order to reflect the singularity of its model. This "model" is not the demiurge who shapes it, but the eternal Form of the "perfect living creature"—an ideal being that comprises all other ideal creatures and that for this reason must be unique. To be sure, insofar as this singular Form does contain all living creatures, it is also a vast multiplicity, as is the world that is modeled on it and made by means of it (35a). It is for this reason that Michel Serres says the Timaean cosmos is, in spite of itself, "a mix of mixes"—a "negotiation" of the same and the different, the eternal and the changing, the chaotic and the cosmic.[35] Even though Plato often opens such floodgates onto multiplicity, however, he closes them as well—particularly when it comes to the question of cosmic eternity. We will recall that, for Plato, the oneness of the cosmos ensures the permanence of the cosmos; if there were other worlds bouncing around in some vast Atomist

void, then one of them might collide with ours and destroy it.[36] It is for
this reason that Timaeus insists that (1) the god used up all the precosmic
material in constructing the cosmos and that (2) "our universe was and is
and will continue to be his only creation" (29e, 32c, 31a). In short, if the
world is unique, then it is invulnerable to destruction or change—as long
as the god maintains his "sovereign bond" to preserve it (41a–b). Here, as
in the biblical tradition, oneness equals power—and both are secured by
the sovereignty of God.

Plato's concern over cosmic imperishability is reaffirmed in the works
of Aristotle, who in the *De caelo* (*On the Heavens*) likewise declares the
world to be "exempt from decay."[37] Unlike his teacher, however, Aristotle
seeks to establish this permanence through natural law rather than through
godly benevolence. In fact, he insists that the Platonic creator actually undermines the permanence of the cosmos—simply by *creating* the world.
Reading the Timaean narrative literally, Aristotle is baffled by what he
sees as the asymmetry of Platonic cosmology: as he explains in the *De caelo*,
the world can be either (1) generated and corruptible or (2) ungenerated and
incorruptible, but it cannot possibly be both generated and incorruptible at
the same time (279b19–80a11). Any world that comes into existence must
also go out of existence; in short, a universe with a temporal beginning
cannot have an eternal future. ("This," Aristotle writes parenthetically, "is
held in the *Timaeus*, where Plato says that the heaven, though it was generated, will none the less exist for the rest of time" [280a30].) Of the two
options that Aristotle has set forth (either generated and corruptible or
ungenerated and incorruptible), he will opt for the second, which leads
him to insist that the eternally existing world must be *unique*. "If the
world is one," he explains, "it is impossible that it should be, as a whole,
first generated and then destroyed. . . . If, on the other hand, the worlds
are infinite in number the view is more plausible" (280a24–28). Aristotle
here is echoing Timaeus's concern over the Atomists, whose teachings are
addressed in chapter 2. Against these thinkers, who posited an infinite number of *kosmoi*, Aristotle's task is to prove that there cannot be more than
one of them, much less an infinite number of them. And, once again, the
endurance of our cosmos is at stake: we can be sure the world will last
forever only if it is ungenerated, and we can be sure it is ungenerated only
if we know it is the only one.[38]

Aristotle establishes the singularity of the world by means of two distinct proofs. The first, found in the *De caelo*, is based on his theory of

"natural motion,"[39] which states that the elements of the universe move either "by nature" or "by constraint," according to their properties. For example, inherently heavy things, such as earth, move downward by nature and upward by constraint. Inherently light things, such as fire and air, move upward by nature and downward by constraint.[40] Each of the elements therefore occupies its own realm in the cosmos, which Aristotle demonstrates must be spherical.[41] Because earth is the heaviest element, it moves "naturally" toward the center of this universe. As the lightest element, fire moves toward circumference of the universe, and air and water move between earth and fire. Unlike Timaeus, then, Aristotle does not describe the cosmos as a product of elemental mixing. Rather, his cosmos looks more like Timaeus's "chaos," with each element assigned to a separate realm: "earth is enclosed by water, water by air [and] air by fire," so that the whole of "the heaven" is a sort of spherical nesting doll, eternally bounded by a rotating circle of stars (287a31). It is this vision that Ptolemy (ca. 90–168) will consolidate in the second century C.E. as the "geocentric" model of the universe, with the earth at the center, the sun and planets moving in concentric circles around it, and a ring of "fixed stars" that rotates around the earth every twenty-four hours (figure 1.1).

Having established the shape and arrangement of the cosmos, Aristotle goes on to demonstrate that this particular nesting doll must be the only one. He begins with the (poorly substantiated) premise that any hypothetical "other world" would have to be composed of the same substances as ours; otherwise, it could not be called a "world" (*ouranos*) at all (276a30–b10). This would mean that the elements of that other world would move the same way as ours: the hypothetical other-earth would move toward the center of the world, and the hypothetical other-fire would move toward its periphery. "*This, however, is impossible,*" Aristotle reasons, because in moving "down" toward the center of its own world, the other-earth would be moving "up" toward the periphery of our world. Likewise, in moving up toward its own periphery, the other-fire would be moving down toward our center. In short, both elements would be moving both up and down—naturally and unnaturally—at the same time (276b12–17, emphasis added).[42] But insofar as "moving downward" constitutes the essence of earth as such, the other-earth's upward motion with respect to our cosmos would make it not-earth. The same would go for fire; a downward-moving fire would not be fire at all. Thus "either we must refuse to admit the identical nature of the simple bodies in the various universes," or we have to admit that there is only one center

FIGURE 1.1 The Ptolemaic universe. (From Peter Apian, *Cosmographia* [Antwerp, 1524])

and one periphery, which is to say only one world. And because Aristotle (believes he) has already demonstrated that other worlds would have to be just like our world, "it follows that there cannot be more worlds than one" (276b21–22).

This one world, moreover, must be spatially limited, for, as we can see, the fixed stars rotate around the earth once a day. Because they always return to the same place, they cannot extend out to infinity; as Aristotle puts it, "a body which moves in a circle must necessarily be finite" (271b26). The "body" of the fixed stars therefore constitutes "the extreme circumference of the whole," a firm boundary outside of which nothing else exists.

33

Like Timaeus, then, Aristotle concludes that there is nothing beyond the world to threaten its eternity *or* its singularity; rather, "this heaven of ours is one and unique and complete" (279a10).

"A Somewhat Confused Interpolation": Many-Oneness in the Metaphysics

A very different proof of cosmic singularity can be found in Aristotle's *Metaphysics*. At first glance, this demonstration seems more straightforward than the argument from natural motion—if just as ridden with unsubstantiated premises. In the widely read Book Lambda of the *Metaphysics*, Aristotle stipulates that everything that moves must have a mover. But, he reasons, most causes of motion are subject to change or destruction—wind, for example, or hands or feet or anything with a body. So if the world as a whole is unchanging and indestructible (one of the unsubstantiated premises), then it must have some source of motion that is eternal and unchanging.[43] This source must itself have no "magnitude," or material extension, lest it be subject to division and change (1073a). Aristotle calls such an unalterable, disembodied force at the base of things the "prime mover," "primary essence," or "principle" (*archê*). It is this "principle" that anchors his second proof of the oneness of the universe, which here as elsewhere he calls "the heaven" (*ouranos*). The proof comes near the end of chapter 8 of the *Metaphysics* and proceeds as follows:

1. If there were many heavens, as there are many men, then
2. The principles for each heaven will be one in form but many in number.
3. But everything that is numerically plural has matter. . . .
4. But the primary essence does not have matter. . . . So
5. The unmovable first mover must be both formally and numerically single. So
6. The permanent and continuous object of movement must also be single. So
7. There is one single heaven only. (1074a)

Again, the argument seems straightforward: because the eternal source of motion is immaterial, it must be singular. And because it is singular, the world it moves must also be singular. Therefore, "there is one single heaven only."

The matter would seem to be settled, except this compact little proof comes directly on the heels of a very different argument. Earlier in the same short chapter, Aristotle begins with the premises that (1) what is moved must have a mover and that (2) this mover must itself be unmoved. To this, he adds the further premise that (3) "a single movement must be produced by something single" (1073a). So far, so good; this argument is what will allow him in a few pages to establish the singularity of the cosmos based on the singularity of its mover. But then Aristotle reminds us that this "single movement" applies not only to "the whole universe," but to the "planetary courses" as well (1073a). And because each of the planets (or "stars")[44] is a substance, "it is clearly necessary that the number of substances eternal in their nature and intrinsically unmovable (and without magnitude . . .) should equal that of the movements of the stars" (1073a). What has happened here is that without justifying the leap, Aristotle has attributed to each planet the "single movement" of premise (3), declaring that there must be as many "single movers" as there are planetary courses. Then, through a perfectly inscrutable calculus, he goes on to reveal the number of courses to be either fifty-five or forty-seven, depending on how many spheres one attributes to the sun and the moon. Therefore, Aristotle concludes that the number of prime movers must likewise be either fifty-five or forty-seven, but then adds, "We will leave to more rigorous thinkers than ourselves the proof of all this!" (1074a).

Even the most rigorous of thinkers, however, would be incapable of offering such a proof—or even of understanding the next move the *Metaphysics* makes. Immediately after concluding the number of prime movers to be either fifty-five or forty-seven, Aristotle launches into the seven-step proof elaborated earlier, which shows that the world is one because its prime mover must be one. It is at this point that, in the words of translator-commentator Hugh Lawson-Tancred, "we pass from the sublime to the ridiculous."[45] Aristotle never reconciles these proofs, nor does he adjudicate between them, so we are left at the end of chapter 8 uncertain as to which of them to take seriously. If we were to accept the text at its face value, we would be able conclude only that there are a varying number of prime movers, depending on which elements of the cosmos they are said to be moving. One, forty-seven, fifty-five . . . it all depends on the way we count. Yet to borrow a Peripatetic turn of phrase, this is clearly absurd: if the prime mover is eternally unchanged, then its number cannot shift with the whims of human argumentation. So what is going on here?

In the Loeb edition of the *Metaphysics*, translator Hugh Tredennick suggests that the problem might just be one of poor editing. Chapter 8 seems to superimpose two strands of Aristotle's writing: the chronologically "earlier" text asserting the singularity of the prime mover, and the "later" text numbering them at either forty-seven or fifty-five.[46] Even if these texts are so superimposed, however, this explanation does not account for Aristotle's failure at this "later" stage either to revise or to retract his assertions about the oneness of the cosmos. For if there are numerous prime movers, then according to Aristotle's own logic there are two possibilities: either the number of the world does *not* reflect the number of the power that moves it, so that the world can still be singular despite its plurality of movers, or the number of worlds *does* reflect the number of movers, in which case there are either forty-seven or fifty-five worlds—one for each mover. And if each planet can be said to compose a world, then Aristotle would have to conclude that each planet is in effect its own "earth," with its own center and periphery. But, of course, such a conclusion would violate the *De caelo's* theory of natural motion by positing multiple centers and multiple peripheries—which is to say no center and no periphery. But as the *De Caelo* itself might respond, "This, however, is impossible" (276b12).[47]

Faced with the deeply incompatible arguments of this section of the *Metaphysics*, Lawson-Tancred throws up his interpretive hands: "The chapter is on any account vexed. Although attempts have been made to show its compatibility with the rest of [Book] Lambda, the consensus of opinion is that it is a somewhat confused interpolation."[48] But it is precisely this vexed and confused interpolation that makes the chapter so fascinating. What is most compelling, I would suggest, is the extent to which the chapter both reveals and covers over what one might call singularity's *constitutive plurality*. In seeking to secure the absolute oneness of the cosmos, Aristotle stumbles upon its manyness and then tries to unfind what he has found. At first, this manyness is internal: in the process of establishing the eternity of the (singular) heaven, Aristotle recognizes that there are "other eternal courses, to wit those of the planets" (1073a). So the eternally singular cosmos *contains* an eternal plurality within it. But the very eternity of this plurality suggests that each planetary course must itself be "something single," which means that each must have its own mover (1073a). It is at this point that plurality breaks out of its singular containment, for if each planet has a mover, then each planet must surely be a world in its own right. This would mean that our ball of earth, water, air, and fire would be a singular element within a broader plurality of other (singular) *kosmoi*. For some reason, however, Aristotle finds this conclusion so troubling that he not only recoils from it but recoils from it badly. Messily, *illogically*, he concludes simply by restating that the cosmos must be singular because the prime mover is singular, and in this way he (thinks he) puts the messy multiplicity to rest.

But, of course, the repressed never quite stays put. One sentence after the proof's conclusion ("There is one single heaven only" [1074a]), Aristotle slips into a prosaic reflection on "tradition"—in particular, on the ancients' "mythical" belief "that the stars are gods, and that the divine embraces the whole of nature" (1074b). Although he does not go on to say much of anything about this belief—simply that we know about it because it has been preserved through time—the notion that "the stars are gods" reopens the line of thinking that he closed off in his formal proofs. For here, in this mythological postscript, we finally find the possibility that each of the planets/stars might correspond to its own mover and thus to its own world. Here, too, we can glimpse one way through the clash that Aristotle has unwittingly staged between the singular and the plural. He tells us the ancients believed "that the stars are gods, *and that the divine embraces [periechei] the whole of nature*" (1074b, emphasis added). That is to say, the divine is many (as numerous as the stars), and the divine is also one (embracing, encompassing, surrounding all of nature). The divine, one might say, is singular-plural: *multiple*. And what if the same were true of the world? Might it be that worlds are as singular-plural as the gods that get them going? This is more than Aristotle is willing to contemplate, but then again, it is his text that brought us here.

Just like the *Timaeus*, the *Metaphysics* thus insists on the oneness of the cosmos, only to run up against an irreducible multiplicity. If Plato's response to this multiplicity is a complex "negotiation," then Aristotle's might be seen as more of an allergic reaction.[49] In both cases, however, these tenuous openings have been collapsed by "tradition": the received reading has become simply that Plato and Aristotle believed the cosmos to be singular.

This Western enshrinement of cosmic singularity is largely the function of Ptolemy's second-century consolidation of the "classical" cosmos: a geocentric set of circles bound by a ring of "fixed stars" (see figure 1.1). As Apian's illustration and many other late-medieval diagrams attest, Ptolemy's model was largely unrivaled until the mid-sixteenth century, when Copernicus posited his heliocentric model of the universe (figure 1.2).[50] But even Nicolaus Copernicus, whose cosmology would turn the Christian

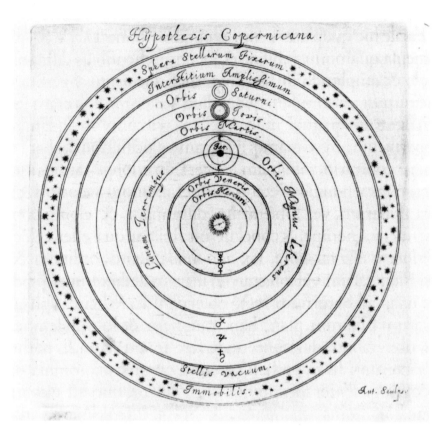

FIGURE 1.2 The Copernican universe. (From Johannes Hevelius, *Selenographia: Sive, lunae descriptio* [1647]. Reproduced by permission of the Master and Fellows of Trinity College, Cambridge)

West upside down in the hands of Galileo Galilei, still assumed that the cosmos was bounded by a ring of fixed stars, which is to say he assumed it was both finite and singular.[51] Galileo refused to weigh in on the matter, suggesting that astronomy confine itself to the one world that it could see. If we were following the most straightforward path through the history of science, we would then move to Johannes Kepler, whose seventeenth-century discovery of elliptical orbits vastly improved the predictive power of heliocentrism, yet who insisted that "this world of ours" (1) has a definite boundary and (2) "does not belong to an undifferentiated swarm of others."[52] From Kepler's reaffirmation of a finite and singular cosmos, it would then be a quick step to Isaac Newton's "absolute" space and time, which arguably undermined his (eventual) belief in an infinite universe.[53] And at the dawn of the twentieth century, Albert Einstein would cling to Newton's cosmic fixity *even as he abolished it*, establishing space-time as a plastic set of shifting relationships, but nevertheless believing the universe as a whole to be eternal and unchanging and to consist of nothing more than our Milky Way.

From Plato to Einstein, then, the dominant assumption has been that our cosmos "was and is and will continue to be" the only one.[54] Beneath this linear reign of singularity, however, there seethes a far more complicated story—a series of bold openings, frightened foreclosures, and radical retrievals of cosmic multiplicity—that makes the modern turn to the multiverse look a bit less sudden and certainly less unprecedented.

2

ANCIENT OPENINGS OF MULTIPLICITY

No thing is ever by divine power produced from nothing.

LUCRETIUS, *DE RERUM NATURA*

Accident and Infinity: Atomist Configurations of the *Kosmoi*

"Innumerable Worlds" in the Early Atomists

The theory that there might be a plurality of worlds traces back to the Atomist philosophers of the fifth century B.C.E.[1] Against the Ionians' single-material principles (water, air, fire, the indefinite), Leucippus and his pupil Democritus argued that the world is composed not of one basic element, but of microscopic, indivisible bits of matter called "atoms," from the adjective *atomos* (uncuttable).[2] These atoms move eternally in a void (*kenon*) and at some indeterminate time and place collide haphazardly to form a vortex (*dine*) that is "cut off from the unlimited."[3] In this vortex, the atoms "jostle against each other," "circling round in every possible way," and become "entangled" (*periplekomenon*) with one another.[4] These entanglements gradually form a circular (not spherical) cosmos, with the heavy materials at the center, the light materials ablaze at the periphery, and a shell or membrane (*hymen*) encasing the whole.[5]

Yet this "whole," for the Atomists, is not the only whole. As Diogenes Laertius, the third-century C.E. biographer of the philosophers, explains it, a cosmos is just a small, bounded part of "the all" (*to pan*), which includes both atoms and void and which Leucippus "declares . . . to be unlimited [*apeiron*]."[6] If the all is unlimited, then there are an unlimited number of atoms in existence and an unlimited amount of space within which they move. This means that the random atomic collisions that generate the cosmos must also generate other *kosmoi*—even an infinite number of them.[7] Classicists G. S. Kirk and J. E. Raven affirm that Leucippus and Democritus are "the first to whom we can with absolute certainty attribute the odd concept of innumerable worlds,"[8] and it is clearly these philosophers against whom Plato and Aristotle seek most energetically to guard their singular cosmologies. Interestingly, Plato never mentions either philosopher by name, an omission that Diogenes, at least, attributes to his inability to refute them properly.[9] Aristotle, by contrast, frames a great deal of the *De caelo* as an explicit rebuttal of "Democritus and Leucippus," and throughout his corpus argues against each of their tenets—from the theory of particulate matter to the existence of a void to the existence of an actual infinity to the existence of multiple worlds.[10] But whether they name their adversaries explicitly or not, we will recall that in both the *Timaeus* and the *De caelo*, the oneness of the cosmos establishes its imperishability. If there is nothing outside this world, then nothing can threaten its eternal endurance, whereas a cosmos that is one of many is expendable and would likely be destroyed by whatever exists beyond it.[11] Untroubled by this prospect, the earliest Atomists affirmed that *kosmoi* are destroyed regularly; as Hippolytus explains in the *Refutations*, "Some are growing, some in their prime, some waning; here they come to be, there they fail. They perish by colliding with each other."[12] Such a collision causes each world to unravel into its constituent atoms, and from these atoms new worlds eventually emerge.

This highly un-Platonic, un-Aristotelian view was developed a century later by Epicurus (341–270 B.C.E), who grounded philosophy in the Atomism of Democritus and Leucippus, even as he claimed to have no teachers.[13] Writing under the cosmogonic dominance of the *Timaeus* in particular, Epicurus sought a different explanation for the order of the universe than the benevolence of a god and the perfection of a transcendent model.[14] He did not teach that there were *no* gods—merely that to be divine meant to be free from all cares. The gods, then, would have no reason to be involved with the realm of mortals, whether as its creators, sustainers, or occasional punishers. With divinity off the table as a creative force, Epicurus found his alternative in the Atomist principles of infinity and accident: given infinite time, infinite substance, and infinite space, any material configuration that can emerge, will. There is thus no need to appeal to an intelligent cosmic designer; this world of ours, which seems so finely tuned and harmoniously composed, was bound to arise at some point.[15]

In fact, Epicurus argued, this world must have arisen elsewhere, even an infinite number of times, in time or space.[16] He was not the first to make this claim; Democritus had also taught that there are an infinite number of worlds just like ours.[17] Perhaps thanks to Aristotle's criticism,[18] however, Epicurus believed that Democritus had failed to secure this teaching mathematically. According to Aristotle, Democritus had said that worlds were composed by means of an infinite number of atoms *with an infinite number of shapes*.[19] Although these premises secured the infinity of worlds, however, they did not secure the *repetition* of worlds, for if there were an infinite number of types of atoms, then there would potentially be an infinite number of types of worlds. This would mean that no world would necessarily be the same as any other, and we would still be left asking why *this* world is so well suited to life. Epicurus therefore asserted that although there are an infinite *number* of atoms, there are a finite number of *types* of atoms.[20] This means that any given configuration is bound not only to occur, but also to recur—infinitely. So this world of ours, perfect and unique as it may seem, is just one of those things that happens from time to time.

In this manner, Epicurus can be said to have performed a Copernican revolution *avant la lettre* (and well beyond it): not only is our cosmos not the only one, he argued, but it is not even the only one of its kind. And it certainly will not last forever; like Democritus, Epicurus taught that the world will eventually be destroyed in relation to the material outside it. But unlike his predecessor, he did not say that this destruction would come about through the collision of different worlds. More like the pre-Socratic philosophers, he conceived of the world as an organism whose demise is merely the end of its life cycle. When a world begins, Epicurus taught, it takes matter into itself from the infinite space outside it. Like any organism, it will continue to grow as long as the amount it takes in exceeds the amount it gives off.[21] Midway through its life, a world will reach a point of equilibrium, after which "there is an excess of loss over gain and the Cosmos becomes weaker and weaker, losing not so much bulk as intrinsic compactness, until it finally breaks down" and unravels into the atomic whirl around it.[22]

Beyond the Fiery Ramparts: Lucretius's De rerum natura

Although the earliest sources of this Atomist cosmology survive only in fragments, it has been preserved and elaborated most fully in Lucretius's *De rerum natura*. Written in the first century B.C.E., this text brandishes the teachings of Epicurus against Platonic-popular creationism, on the one hand, and Aristotelian physics, on the other.[23] Lucretius (ca. 99–55 B.C.E.) assails the former from the very beginning of the treatise, lauding Epicurus as the first man to laugh at the gods' thunderbolts and their "menacing roar," to free himself from the "crushing weight of superstition [*religione*]" and leave it "trampled underfoot."[24] Lucretius goes on to say that it was precisely this triumph that allowed Epicurus to travel "beyond the fiery ramparts of the world" and out to the universe itself (1.73–74).[25] Once Epicurus was free from the tyranny of the gods, he could posit and contemplate an infinite number of *kosmoi* outside our starry cosmic shell.

In Epicurus's footsteps, Lucretius resolves to clear away all theistic cosmogonies in order to uncover the infinite universe, teeming with worlds.[26] For Lucretius, seeing the cosmos as a divine creation is unacceptable not only because it enshrines a singular cosmos where a vast multiplicity should be, but also because it commits a slew of other crimes—which, for the sake of convenience, can be grouped into four. First, creation narratives keep people in perpetual *fear* of the creators. This fear prompts them to engage in "foul" rituals in order to win the gods' favor or escape their wrath—practices that Lucretius sees as both humiliating to humanity and insulting to the gods. Second, creation stories assume narcissistically that the gods care about human affairs. But, as Epicurus taught, the gods exist in blissful, eternal contemplation; they would have no reason to disturb this serenity by engaging in cosmic affairs (1.62–70). Third, creationist cosmologies begin from the false premise that our world is particularly well suited for human flourishing—even that, as Plato insists, it is perfectly constructed for this purpose. Following Epicurus, Lucretius counters that the universe was clearly *not* designed "for us" or for any purpose at all, considering that it is "so profoundly flawed."[27] As evidence, he cites the uninhabitability of vast regions of the earth, the recalcitrance of the land that produces our crops, and the excesses of sun, frost, and wind that destroy them (5.200–17). Implicitly against Plato, Lucretius argues furthermore that the elements do not exist in perfect balance with one another; rather, fire wants to consume earth, water wants to extinguish fire, and air wants to dry up water in an agonistic tango of what Friedrich Solmsen calls "ceaseless cosmic strife."[28] "Besides," Lucretius continues, "why does nature feed and increase the frightful tribes of wild beasts, enemies of the human race, by land and sea? Why do the seasons of the year bring disease? Why does untimely death stalk about?" (5.218–21, translation altered slightly). Considering what David Hume's Demea will call "the perpetual war . . . kindled amongst all living creatures,"[29] it is the height of both ignorance and selfishness to insist that the world was in any way designed "for us."

Finally, Lucretius claims that in addition to being cosmologically, psychologically, theologically, and intellectually misguided, theistic cosmogonies are *morally* treacherous. He sets forth this view at the outset of the poem in response to the anticipated objection that if people no longer fear the gods, human society will devolve into chaos. Lucretius assures us that such concerns are unfounded. After all, it is not "impious" philosophy that leads one down "a path of crime. . . . On the contrary, more often it is . . . *religion* which has brought forth criminal and impious deeds" (1.82–84, translation altered slightly, emphasis added). As a prime example of religion's immorality, Lucretius cites Agamemnon's mythic sacrifice of Iphigenia. Persuaded by priests that the gods demanded his daughter to ensure his safe passage to Troy, Agamemnon brought her to be "slaughtered by her father's hand" (1.99, translation altered). The story, Lucretius suggests, speaks for itself. A man killed his innocent daughter for the sake of favorable sailing conditions: "so potent was religion in persuading to evil deeds" (1.101).

There is, therefore, no reason to believe the world is a deliberate product of divine craftsmanship. It is far sounder, Lucretius suggests, to contemplate nature as having arisen "of her own accord" (1.1092). The world is not the work of the gods, but of primordial particles colliding by chance, combining and recombining "through infinite time" (5.190–91). Like the proverbial typewriting monkeys who eventually hammer out Shakespeare, these atoms have an infinite amount of time, energy, and materials with which to collide—making it "no wonder that they fell into such arrangements . . . as this sum of things now shows" (5.193–95, translation altered slightly). The principles of infinity and accident therefore have just as much explanatory power as any theistic cosmogony and bring the added benefit of liberating humanity from its service to a fleet of inscrutable gods. "If you hold fast to these convictions," Lucretius promises, "nature is seen to be free at once and rid of her proud masters [*dominis privata superbis*]" (2.1090–91). And with the proud masters out of the way, we, too, can travel with Epicurus beyond this world to contemplate a vast universe of worlds, for the very processes that produce this imperfect cosmos are bound to produce an infinite number of others.

Mindful of the strangeness of this position, Lucretius sets forth the Epicurean doctrine of cosmic multiplicity through a series of progressive proofs and illustrations. His first step is to demonstrate—more against Aristotle than against Plato—that the universe, literally "all that is," cannot possibly be finite. "Let us grant for the moment that the universe [*omne quod est*] were finite," Lucretius begins. "Suppose someone proceed[s] to the very end and throws a spear" at what he believes to be its boundary (1.968–70, translation altered).[30] Two possibilities ensue. Either something stops the spear, or nothing does. If nothing stops it, then the universe has no boundary. If something stops it, then there is a boundary, which is to say there is a "beyond" into which the spear could have traveled had it not been stopped. In other words, Lucretius is suggesting that if there is a boundary, then there is necessarily something outside that boundary. So either way, the universe must be infinite: "Wherever you place your extremest edge, I shall ask about the spear" (1.980, translation altered slightly).

Lucretius deepens this proof of infinity by appealing to the mutual reliance of matter and void. Early in the *De rerum natura*, he argues (also against Aristotle) that if the void does not exist, then matter would not exist because atoms cannot move without empty space between them (1.369, 1.430–31). This means, furthermore, that the void (*inane*) is not external to matter (*corpus, materia*), but is mixed, intermingled, tangled up with things (*admixtum rebus*)—irreducibly constitutive of all that is (1.369). Insofar as void and matter are so tangled, they provide further testimony to the infinity of the universe. For if the void is inextricably *admixtum* with the stuff of the world, then "matter must be bounded by void and . . . void bounded by matter." Neither can provide an absolute edge to the cosmos because each must be surrounded by the other. "By this alternation," Lucretius concludes, "nature renders the universe infinite [*ut sic alternis infinite omnia reddat*]" (1.1009–11, translation altered).

Having established the endless entanglement of matter and void, Lucretius has laid the argumentative infrastructure that he will need to prove the existence of many worlds (*terrae, mundi*) within this infinite universe (*omnium, omne quod est, orbis*).[31] He begins this proof from the principle of infinity, suggesting that because space is infinite and particles of matter (the word he uses for atoms is *semina* [seeds]) are "innumerable," it is highly unlikely that this world could be the only one (2.1055–57). If one admits that there is matter outside our world, he reasons, it would be very strange to suggest that those seeds just sit still "out there" without forming anything. Rather, they most likely obey the same natural principles as the rest of the matter in the universe, "knocking together by chance, clashed in all sorts of ways, heedless, without aim, without intention, until at length those combin[e] which . . . could become . . . the beginning of mighty things, of earth and sea and sky" (2.1058–63). In short, Lucretius summarizes, "when abundant matter is ready, when space is to hand, and no thing and no cause hinders, *things must assuredly be done* and completed" (2.1067–69, emphasis added). In this manner, he provides an early articulation of what Arthur Lovejoy will in the 1970s call "the principle of plenitude": thanks to the infinite bounty of nature, anything that can be created will be created.[32] Lucretius seems to find this principle self-evident because immediately after stating it, he intensifies his tone and concludes: if it is the case that nature behaves uniformly, then "*you are bound* to confess that there are other worlds in other parts of the universe" (2.1074–75, translation altered, emphasis added).

Just in case his audience remains unconvinced, Lucretius follows this argument with evidence from experience. Whether we assent to the principles of atoms and void or not, we all have doubtless come to realize that "there is no thing in the sum of things that is unique" (2.1077–78, translation altered slightly). Not a single animal, man, or "dumb scaly fish" (2.1083) can be called (with apologies to William James's crab) "itself, itself alone";[33] rather, each of them is one of many others just like it. "Therefore," Lucretius explains, "you must in like manner confess for sky and earth, for sun, moon, sea and all else that exists, that they are not unique, but rather of number innumerable"—and here he subtly changes tack—"since there is a deepset *limit of life* equally awaiting them, and they are as much made of a perishable body as any kind here on earth" (2.1084–89, emphasis added). Worlds are ultimately innumerable because, like everything else, they are members of a species, and, like everything else, they are perishable. And so at the end of this series of arguments, we have run up against Plato's and Aristotle's primary concern: to concede the plurality of *kosmoi* is in fact to concede their impermanence. Worlds decay just as surely as they come into being, living for a time and then unraveling again into the chaotic flux from which they arose.

As in every cosmogony, however, the nature of this chaotic flux is not easy to pin down. In the *Timaeus*, we recall, a full description of chaos really comes halfway through the story of the formation of the cosmos, when our speaker interrupts himself to start his story again. Out of no-where, he suddenly says that he must add a "third thing" to his previous two and heads back to chaos: "[T]here were, before the world came into existence, being, *khôra*, and becoming, three distinct realities."[34] But the moment Timaeus adds *khôra* to the story, he drops the demiurge, leaving us with two fairly incompatible accounts of the state of things before the cosmos. If we shift to another mythic context, the book of Genesis also begins twice, its first primordial scene a triumvirate of God (Elohim), breath (*ruach*), and deep (*tehom*), and its second an empty earth, a stream of water, and the hands of YHWH.[35] And in a strikingly similar fashion, the *De rerum natura* also gives us two different accounts of the chaos be-fore and beyond the cosmos.

The first thing Lucretius has to say about this chaos is that it is not nothing. At the time of his writing, the formal doctrine of "creation out of nothing" (*creatio ex nihilo*) had not yet emerged—it would not do so for another two hundred years, when some of the early church fathers ad-opted it in order to affirm the sovereignty of God against the so-called Gnostics.[36] Put briefly, the fathers considered a God who could create out of nothing to be more powerful than a God (like Plato's demiurge) who created out of a preexistent material. After all, to say that God creates out of *something* is to imply that God depends on something or is limited by something or at the very least has to negotiate with something in order to create the world. God's creation out of "nothing," then, establishes God as the supreme (and sole) creative force in the universe.[37] But, again, this logic had not yet been formally articulated in Lucretius's time; in fact, Jesus of Nazareth had not yet been born. Yet the antiauthoritarian Lucre-tius seems nevertheless to anticipate the argument, making "creation out of nothing" his first target. In fact, he builds his entire cosmology on re-futing it: "The first principle of our study we will derive from this: that no thing is ever by divine power produced from nothing [*nullam rem e nilo gigni divinitus umquam*]" (1.149–50).

The primary reason Lucretius gives for affirming this "first principle" is that if something could come out of nothing, anything could come out of anything: men from the sea, fish from the earth, "and birds could hatch from the sky" (1.159–62). He takes this argument directly from Epicurus's *Letter to Herodotus*: "To begin with, nothing comes into being out of what is non-existent. For in that case anything would have arisen out of any-thing."[38] As the examples of Plato and the Ionians attest, however, neither Lucretius nor Epicurus was the first to hold this principle: the philosophers of antiquity were unanimous in holding that nothing comes from nothing.[39] So Lucretius is effectively beginning from what he would regard as an uncontroversial position, suggesting that once we truly understand what we already know, the rest of his argument will follow easily. In particular, excising nothingness as a cosmic starting point will excise any sort of sovereign God who might create out of it. As Lucretius predicts, "[W]hen we shall perceive that nothing can be created from nothing [*nil posse creari de nilo*], then we shall at once more correctly understand . . . the manner in which everything is done without the working of the gods" (1.155–58). If nothing is ever created from nothing, then there is no need for a creator to produce something. Eternal, inchoate matter is perfectly capable of calling itself to order—and of dismantling and rearranging any order it may establish.

Epicurus follows Leucippus and Democritus in teaching that cosmic order begins to emerge by means of a vortex (*dine*). From the primordial chaos, atoms form a turbulent swirl that gradually produces compounds, elements, and worlds.[40] The question that all these thinkers leave unanswered is: What produces the vortex itself?[41] How do the atoms move *before* they organize themselves into a generative whirl, and how does this whirl come about in the first place? In other words, what is chaos, and how does it produce the vortex that makes the cosmos?

Lucretius does describe these primordial operations, but as cosmogonists often do, he tells two different stories about them. One strand of the *De rerum* figures chaos as what Michel Serres calls a "cloud": a "stormy combat of atoms" colliding at random in every possible way.[42] This is the description of chaos usually attributed to Leucippus and Democritus, and Lucretius employs it in a number of places throughout the text. For example, immediately after his proof of cosmic infinity in book 1, he describes the state of things before "the first beginning" as involving a great number of seeds, "shifted in many ways . . . harried . . . trying every kind of motion and combination" (1.1024–26). Similarly, after reminding us of the principle of infinity in book 2, he says that the innumerable seeds of the universe had been, from eternity, "knocking together by chance, clashed in all sorts of ways, heedless, without aim, without intention" (2.1059–60). In both of these passages, the primordial scene is one of absolutely random motions and collisions, which, given enough time and material, eventually generated the earth, sky, sea, and their inhabitants by the sheer force of accident (1.1026–28, 2.1060–62; compare 5.416–31).

The work of this "cloudlike" chaos finds a fuller and slightly different description in book 5, which offers an extended explanation of the cosmogonic process. Before the world was made, Lucretius tells us, there was "a sort of strange storm [*tempestas*], all kinds of beginnings [*omnigenis e principiis*] gathered together into a mass, while their discord, exciting war amongst them, made a confusion of intervals, courses, connexions, weights, blows, meetings, motions, because, on account of their different shapes and varying figures, not all when joined together could remain or so make the appropriate motions together" (5.432–42).

From this eternal tempest and warlike confusion, an ordered world began to take shape. But rather than skipping straight from collisions to cosmos, as in the previous passages, Lucretius adds an intermediate step here: a gathering together of similar atoms. In the first place, he explains, there was a chaos of stormy indistinction: "In the next place parts began to separate, *like things to join with like*, and to parcel out the world, to put its members in place and to arrange its great parts—that is, to set apart high heaven from earth, and to make the sea spread with its water set apart in a place of its own, apart from the pure fires of ether set in their own place" (5.42–48, emphasis added).

The way this worked, Lucretius explains, was that the particles of earth, "being heavy and entangled [*perplexa*]," formed a fabric. Through the tiny holes of this earth-fabric, the lighter elements slipped out and up, like they do on those mornings "when the lakes and the ever-flowing streams exhale a mist [*nebula*], and the very earth seems sometimes to smoke . . . clouds with body now cohering weave a texture under the sky [*corpore concreto subtexunt nubile caelum*]" (5.450–65).

The notion of elements gathering together "like to like" echoes a fragment from Democritus in which atoms are sorted out through a rotational movement, as when a "sieve is moved around [*dinon*] [and] lentils are sorted and ranged with lentils, barley with barley, and wheat with wheat."[43] This passage is probably the one that lies behind Serres's suggestion in *The Birth of Physics* that Lucretius's cloud-chaos ultimately generates the cosmos by means of "the Democritean *dinos*." "The vortex (*tourbillon*) is thus the pre-order of things," Serres writes, "their nature, in the sense of nativity. Order upon disorder, whatever the disorder may be; the vortex arises."[44] Strangely, however, Lucretius does not mention the vortex in relation to the cosmic cloud—nor does he ascribe any sort of rotational movement to it. Rather, we have a storm of collisions out of which "like things" entangle themselves into a textile of earth and then the interstitial rising up of mist, sea, and cloud. Serres detects a vortical motion behind or beneath this primordial separation and entanglement, but it seems to me that the clues in the text are sparse.

The Democritean vortex haunts Lucretius more noisily in his description of another type of chaos, which appears briefly between his more numerous descriptions of the primordial storm. Having very recently described the original state of things as an "everlasting conflict [of] struggling, fighting, battling in troops without any pause, driven about with frequent meetings and partings" (2.118–20), Lucretius suddenly gives a very different account of things. "The first bodies," he writes, "are . . . carried downwards by their own weight in a straight line through the void [*rectum per inane feruntur*]" (2.216–17). Serres focuses much of his marvelous reading of Lucretius around this brief burst of linear chaos, which he calls the "laminar flow" or the "laminar cascade."[45] He borrows the term from fluid mechanics: a flow is said to be "laminar" if it is perfectly streamlined—that is, if all its particles move in the same direction, with no diagonal or perpendicular movements.[46] This primordial cascade, unlike its stormy counterpart, is not a site of atomic collisions and perpetual war; to the contrary, it is nothing but ceaseless, parallel flow—just an ideal of atoms in perfect motion—until, "at uncertain times and places, *they swerve a little [depellere paulum]* from their course, just so much as you might call a change of motion" (2.218–20, translation altered slightly, emphasis added). With this tiny swerve, everything changes; things begin to take place. As Serres explains it, this slight inclination (*clinamen*) produces "the minimum angle of formation of a vortex": it is the tiny deviation that draws the sea around it into an atomic whirl.[47] And as each of the ancient Atomists taught, this whirl alone produces everything and anything that is. As Lucretius concludes, "[I]f they were not apt to incline, all would fall downwards like raindrops through the profound void, no collision would take place and no blow would be caused amongst the first-beginnings: thus nature would never have produced anything" (2.221–24). What this means, Serres ventures, is that the real atomic unit is not the atom, but the *clinamen*: because nothing happens without the swerve, the swerve is "more atomic, so to speak, than the atom."[48]

The Lucretian text thus presents us with two very different descriptions of the world before the world: the omnidirectional "stochastic cloud" and the unidirectional "laminar flow."[49] It might be tempting to try to call this difference to order: to suggest, for example, that the laminar chaos precedes the stormy chaos so that atoms would move from an initially parallel pouring-out, through the slight angle of the *clinamen*, into the tempest's warlike collisions, which produce the cosmogonic vortex. But it seems important to note that Lucretius leaves these two chaoses stubbornly unconnected and unreconciled: the storm does not seem to produce a vortex, and the laminar flow that *does* produce a vortex does not give rise to the storm's atomic war. In fact, even though Serres wants to connect each of these primordia to the vortex, he, too, resists arranging them into a single story: "Now there are indeed two kinds of chaos, the cloud and the pitcher. In the first image, multiple aleatory collisions within the infinite void of space send disordered atoms moving in *all* directions. In the second image, against the second background, encounters and collisions are not possible, and the laminar atoms move only in *one* direction."[50] The two are irreducible to each other—even, in a sense, opposed—but Serres suggests that they are nevertheless communicating complementary things about the noise beneath all information or meaning (*sens*). Chaos "either dissipates in all directions, or flows in one direction. There is no sense when everything has the same sense. There is no sense when everything is in all senses."[51] The cloud and the pitcher are therefore "two thresholds of disorder"—mayhem, on the one hand, and absolute uniformity, on the other. In other words, as Empedocles knew, *both pure difference and pure identity are chaos* (see chap. 1, sec. "So Let Us Begin Again . . .").

In a sense, then, we are quite close to the *Timaeus*: if both pure difference and pure identity are chaotic states, then cosmos can only come about in and as their interdetermination ("again with the Same and the Different he made . . . compounds intermediate . . . and taking these three he mixed them into a single unity").[52] But, of course, there are two irreducible differences between Lucretius and Timaeus: the creator-god and the singular cosmos. Whereas Timaeus secures the latter by means of the former, Lucretius abolishes the former to dismantle the latter: an infinity of worlds does away with the sovereign god who secures the oneness of the world.

This is not to say that the infinity of worlds does away with divinity altogether; to the contrary, Lucretius calls on Venus at the beginning of the poem to guide his hand and put an end to earthly strife (1.1–55).[53] And toward the end of the poem, he infamously calls Epicurus "a god" for having overcome fear and suffering through the tranquility of wisdom (5.8–12). As we will recall, this tranquility was the result of Epicurus's having rid the cosmos of its "proud masters" (*dominis superba*, 2.2091). But even Epicurus never said that the gods did not exist; he merely said they did not govern the affairs of the universe.[54] In other words, Lucretius is not abandoning the notion of divinity as such; rather, his rejection of the world's singularity—like his preemptive rejection of creation out of nothing—indicates that he is abandoning divinity understood as domination, as *sovereignty*. If, as thinkers as varied as Augustine and Ludwig Feuerbach have argued, there is a reciprocal relationship between human beings and their visions of the divine—if, to put it crudely, people worship gods who look like them—then the Atomists' refusal of divine sovereignty should have profound consequences for human sovereignty as well. We have already heard Lucretius criticize religion for instilling violent tendencies in its adherents (1.82–101), suggesting that those who are free from the "proud masters" will be less likely to make themselves masters of others. And sure enough, centuries later, Plutarch will report that the mere thought of Atomist cosmology left as sovereign a character as Alexander the Great unhinged: "[H]aving heard Anaxarchus on the infinity of *kosmoi*, Alexander wept and, when his companions asked what was the matter, he said, 'Is it not worthy of tears that, when there are infinitely many *kosmoi*, we are not yet masters of one?' "[55]

To summarize, insofar as Lucretius refuses cosmic singularity, dismantles both divine and human sovereignty, and hinges all of it on a preemptive refutation of the idea of creation out of nothing, one might venture that he faces and even celebrates the multiplicity that his Platonic, Aristotelian, and eventually Christian counterparts would glimpse but then cover over. Indeed, multiplicity surfaces in a number of different ways in Lucretius's work. First, it establishes the cosmos as such, beginning with a vast plurality of "singular" seeds (1.584, 1.609) whose interactions come to establish each world through a misty entanglement (or vortical whirlwind) of sameness and difference. A cosmos is therefore *constitutively* multiple and as such is never simply itself. Second, as Judith Butler has argued, anything constituted by relation is "undone" by it as well—and everything is constituted by multiplicity.[56] Thus, for Lucretius, the "dance of atoms" that worlds the world will likewise one day unmake it, unraveling it back into a chaos of material from which new worlds will form. Third, these new worlds compose what one might call this world's "external" multiplicity—its being one of many worlds. But this manyness is not a strict plurality: precisely because each world is the product of and the material for other worlds, each of them is constitutively bound up with an untold number of others. And because the same processes of birth and death compose and undo all things, Lucretius's unbounded universe, in its infinite plurality, can also be said to be one. So multitude and mortality, like atoms and void, encase each other infinitely. The multiplicity that forms the world also leaves it vulnerable to destruction, and its destructibility makes it part of a (loosely united) multiplicity.

But there are multiple ways for worlds to be multiple. For the earliest Atomists, what I am calling "external" cosmic multiplicity was figured primarily in *spatial* terms: our world is one of an infinite number of worlds that exist simultaneously. It will be destroyed by a kind of spatial calamity, when it collides with another world out there. Following Epicurus's revisions of Democritus, Lucretius likewise configures the world as one of a spatially coexistent many, but he also places a great deal of emphasis on the *temporal* multiplicity of worlds. As Epicurus taught, our world will most likely not be destroyed by a collision in space, but by the gradual wreck of time. It will grow old like any other organism, lose its vitality, and die. And this is where the Epicurean "first principle" is of central importance. To say that nothing comes from nothing is also to say that nothing *goes* to nothing (1.216). Just as the cosmos did not spring out of *nihil*, it will never be annihilated. Rather, as Lucretius says the moment he mentions the matter of beginnings, nature brings all things back into the "first bodies" (*corpora prima*) that compose them (1.60; compare 1.249), and these bodies go on to make new worlds. Worlds are "one of many" both spatially and temporally: an infinite number of *mundi* exist throughout infinite space, and each is destroyed and recombined into new worlds throughout infinite time.

This emphasis on cosmic rebirth among Democritus's successors is most likely a function of the rise of Stoic cosmology in the late fourth century B.C.E., which figures multiplicity in strictly temporal terms. Like its Epicurean rival, Stoic cosmology will be demonized, repressed, and mostly forgotten by the Western philotheological tradition that it nonetheless haunts. We will explore these various hauntings in chapters 3 and 4, but first we turn to the Stoics themselves.

Fire and the Phoenix: Stoic Configurations of the *Kosmoi*

The Stoic school was born in 300 B.C.E. when Zeno of Citium began to lecture at the *stoa poikile* (painted colonnade) in Athens. Ethically speaking, the teachings of Zeno were quite close to those of Epicurus, who had founded his Garden only six years earlier. Although the two men were "implacably hostile to each other," both taught that the highest ethical good was a "calm imperturbability and the living of the simple life."[57] Moreover, both taught that a person could cultivate this imperturbability only by understanding nature and living in accordance with it. That having been said, Zeno and Epicurus disagreed vehemently over the *nature* of nature and set forth incompatible doctrines concerning everything from the composition of matter to the nature of the gods to the infinity of worlds.

Against the Epicureans, the Stoics taught that there is no such thing as a smallest unit of matter, or atom. Rather, they followed Aristotle in arguing that matter is continuous, which is to say infinitely divisible. If there are no atoms, then there is no empty space in which they must move—no void "tangled up with things." The only space that one might call "empty" would therefore lie outside the realm of things altogether, and so some commentators explain the Stoic universe as a bounded, continuous cosmos, surrounded by an infinite void.[58] As historical linguist Michael Lapidge points out, however, it is not clear whether the Stoics spoke of "void" (*kenon*) in the positive or privative sense; that is, although they have traditionally been interpreted as having posited *a* nothing outside the cosmos, they might just as plausibly have meant there is simply *nothing* outside the cosmos.[59] Either way, the absence of any extracosmic material means there is nothing "out there" that might form other worlds. For the Stoics, ours is the only world in the universe.

Even as the Stoics asserted the singularity of the cosmos, however, they rejected the imperishability that has traditionally secured it: "[T]he world, they say, is *one* and *finite*."[60] Against Plato and Aristotle alike, Zeno taught that insofar as the world has a beginning, it must have an end. Moreover, anything whose "parts are perishable is perishable as a whole. . . . Therefore the world is doomed to perish."[61] As we will recall, Epicurus maintained a similar view on similar grounds. But whereas Epicurus the Atomist attributed the world's demise to an excess of excretion over absorption, Zeno the Stoic said that it will end when the sun dries up the earth and consumes the cosmos in flames.[62] From these flames, a new universe will be born, live for a time, and then be set on fire again—and the process will repeat eternally. Over against the Atomists' spatial multiplicity, then, the Stoics offered a *temporal* multiplicity: there is only one world, but it is destroyed and re-created throughout infinite time, like the mythic Phoenix, out of fire.

This is by far the most distinct feature of early Stoic cosmology: the periodic destruction and regeneration of the universe, a process called *ekpyro-*

sis (literally, "out of fire"). The idea was not without precedent; nearly two centuries earlier, the philosopher Empedocles had taught that the cosmos undergoes periodic regeneration under the influence of "love" (*philia*), which draws all the elements together, and "strife" (*neikos*), which separates them from one another.[63] But the Stoics described their cosmic cycles "more naturalistically or mechanically" (not to mention more dramatically) as the result of condensation and rarefaction, dual processes through which fire would eventually consume and re-create the cosmos.[64]

Unfortunately, this Phoenix universe in its ancient formulation remains only in a few scattered fragments and some anti-Stoic treatises.[65] Moreover, the middle and later Stoics abandoned *ekpyrosis* entirely, along with cosmology in general. This means that there is nothing like a "Stoic Lucretius" when it comes to cosmology—no Greek *or* Roman source that has preserved these teachings in any depth. Even Cicero, whose *On the Nature of the Gods* is often cited in relation to *ekpyrosis*, mentions the doctrine only in a qualified, almost embarrassed hurry. His Stoic character Balbus takes only a moment in a lengthy exposition of his own philosophy to say, "[T]he philosophers of our school believe that in the end it will come about (though Panataeus is said to have thought it doubtful) that the whole universe will be consumed in flame: because when all the water is dried up, there will be no source from which air can be derived and nothing but fire will be left. From this divine fire a new universe will then be born and rise again in splendor. But I must not dwell too long upon the system of the stars and planets."[66] This strange little passage leaves one wondering: Why such haste? Why does Cicero find *ekpyrosis* so repellant as to announce it as doubtful to begin with, describe it as quickly as possible, and then change the subject as soon as he can? And why is there no lengthy engagement of this idea among *any* of the Roman Stoics?[67]

In *The Myth of the Eternal Return*, historian of religions Mircea Eliade suggests that the Romans shied away from *ekpyrosis* because it promised the inevitable demise of the "empire without end." In the same book, Eliade argues that every sociopolitical ritual repeats the "cosmogonic act," or creation of the universe. If this is the case, he argues, then building the Roman Empire might be seen as an effort to take the reigns of the universe away from the Stoic god. As Eliade ventures, the Roman emperor attempted to "liberate history from the law of cosmic cycles" by enacting these cycles himself—by unmaking and remaking the world with each invasion.[68] To be sure, the doctrine of *ekpyrosis* would undermine the integrity of this imperial project, which is to say the very sovereignty of the sovereign. Just as Alexander the Great wept over the possibility of infinite *kosmoi*, Caesar Augustus would be far less august if a cosmic conflagration threatened to wipe out his "eternal city" and start the whole process again.[69] Furthermore, as we shall see momentarily, it is not just late Stoic philosophy that excises *ekpyrosis*; Augustine of Hippo (354–430) will find the doctrine so repugnant that he will foreclose any Christian consideration of it, and this foreclosure will hold throughout sixteen subsequent centuries of theological reflection. But what exactly did ekpyrotic cosmology entail, and why was it so unanimously rejected?

According to Diogenes Laertius, the early Stoics taught that "the whole world is a living being [*zoon*], endowed with soul and reason."[70] This means that the Stoic world is governed providentially from within; reason pervades the cosmos, just as the soul pervades the human body. This "reason" (*nous* or logos) is often also called "fate" or "god" and suffuses the material cosmos as its "rational, perfect . . . providential" spirit.[71] Unlike the rational, perfect, and providential spirit of the *Timaeus*, however, Stoic spirit is irreducibly corporeal: the source texts figure it variously as breath (*pneuma*), ether (*aither*), or fire (*pyr*).[72] In short, the material world contains everything it needs for its own generation and governance within it. There are no ideal Forms regulating the universe, nor is there a First Mover setting it in motion from beyond the cosmic fray. Rather, the principles of creation, animation, providence, and eventual destruction are *immanent* to this one cosmos. In this sense, the early Stoics can be called "monists": there is no disembodied realm hovering above or beyond the corporeal realm of this only-world.

That having been said, the Stoics often figured the corporeal *components* of this only-world in dualistic terms. Like the rest of the ancient philosophers, they taught that "nothing comes from nothing," so the cosmos must have emerged from some kind of primordial material. Because they held that matter is continuous rather than particulate, however, they rejected Epicurus's aboriginal atoms and void, returning instead to the pre-Socratic notion that the world emerges from an initial substance, into which it will eventually dissolve. Yet according to classicist Michael Lapidge, the Stoics had also learned from both Plato and Aristotle that "genesis could only take place from the intersection of opposite forces."[73] So they found a cosmic compromise, teaching that the primordial substance was not one, but two. "[The Stoics] hold that there are two principles [*archai*] in the universe," writes Diogenes, "the active principle and the passive."[74]

The active principle is often associated with the cosmic breath or fire we have just glimpsed—that is to say, with reason, providence, and the god. The passive principle, by contrast, is a wet and "formless material" (*apoios hyle*) that is "shaped or 'qualified' by the active principle into a universe."[75] Some Zeno fragments offer a highly gendered gloss of this duality, calling *hyle* the cool "female secretion" that mixes with the hot pneumatic sperm to generate the world.[76] Chrysippus of Soli goes further, in one fragment likening *hyle* to Hera and *pneuma* to the sperm of Zeus[77] and in another illustrating the cosmogonic process with reference to a painting of Hera fellating her husband.[78] In an effort to salvage early Stoicism from such unsavory dualisms, Lapidge argues that the active and passive principles are technically inseparable—just "two aspects" of the same creative substance. Unfortunately, he explains, some Stoic teachers "forgot" this inseparability and "resorted to biological terminology," thereby compromising the dignity and complexity of their own thought.[79]

Classicist David Hahm chooses not to pay much attention to the fellatio fragment, arguing instead that Chrysippus avoids Zeno's dualism problem by locating the primal substance back behind the distinctions of active/passive, male/female, and spirit/matter. According to Hahm, the creative force for Chrysippus is not the two *archai*, but pure fire (*pyr*), which is also called the "god" (*theos*) and which itself gives rise to the opposing forces that assemble the universe.[80] In this sense, the Stoics can be seen as hearkening back to Heraclitus, for whom the principle of the world "was and is and shall be: an everliving *fire*."[81] At the same time, Hahm concedes that many Stoic fragments do confine this allegedly primordial fire to the "active principle" alone, which they say operates upon an equi-primordial (and watery) "passive" principle.[82] There is therefore a tension in early Stoic thought between the monism of fire and the duality of the *archai*; it is not clear whether the world begins in oneness or in manyness.[83] Of course, one might say that this very interpretive difficulty speaks to a certain pluri-singularity at the beginning; as Jean-Baptiste Gourinat explains, "these two principles are two different bodies, even if they are always mixed together, and constitute *by mixture* a unified body."[84] Here we might recall Michel Serres's description of the Timaean cosmos: "The world is a harmony, the world is a mix, the world is unitary . . . but it is only mixed with mixtures."[85] One significant difference between these cosmogonies, however, is that whereas the demiurge remains outside the world he creates, the Stoic divinity is inextricably part of the mix; as Gourinat reminds us, "God himself only exists in matter."[86] The Stoic creator exists without remainder in and as the world's fire-watery mix, which is neither one nor two, but the relation between them—an aboriginal many-one that tangles together not just spirit and matter, but also world and god.

Insofar as the Stoic divinity is irreducibly *in* the cosmos it governs, it is markedly different from the Platonic demiurge (who transcends the cosmos he assembles) and from the Epicurean gods (who float beyond the worlds they ignore)—not to mention from the transcendent Abrahamic sovereign who creates and governs the universe. In some accounts, the Stoic god is said to be the breath or life force of the world, "pervad[ing] all that is in the air, all animals and plants, and also the earth itself, as a principle of cohesion."[87] In others, the god is said to *be* the cosmos.[88] Therefore, although later Stoics would adopt a "more Platonic," proto-Christian vision of a governor god outside the universe, one can call the early Stoics "thoroughgoing pantheists."[89] They believe the divinity to be utterly bound up with—and sometimes identical to—the world it creates and regulates. As such, Diogenes Laertius tells us that the god can be called by many names "according to its various powers": Dia, Zeus, Athena, Hera, Hephaestus, Poseidon, and Demeter.[90] The gods, then, are every bit as pluri-singular/female–male as the hydropyric cosmos they inhabit. They are, in short, both providentially and ontologically bound up with the life of the worlds they create. When we bring this insight to bear on the doctrine of *ekpyrosis*, we therefore begin to see that the pluri-singular god both oversees *and undergoes* the fiery cycles of birth and destruction that constitute the life of the cosmos.

Ekpyrosis is certainly the strangest and least integrated of all the Stoic teachings. Where could they have gotten the idea that the world periodically consumes itself in flames? And what does it have to do with the rest of their cosmology—let alone their ethics? According to Lapidge, there are three explanations of the doctrine scattered throughout the Greek source texts, but they are mutually incompatible, and none of them is well elaborated. The most common story is that the sun and stars will eventually dry up all the moisture in the world, setting the whole thing on fire. As Zeno writes, "[T]he sun is *fire*—shall it not, then, burn up what it has?"[91] Another says that when the planets return to the positions they occupied at the moment of creation, the universe will be engulfed in flames and renewed.[92] And a third account suggests simply that the god, figured as pure fire, "keeps on increasing until he absorbs himself into himself."[93]

Whatever the cause of its ignition, the cosmos is said to burn gloriously for an unspecified amount of time, after which there emerges a watery mass upon which the fire again operates to re-create the cosmos.[94] Unsurprisingly, there are numerous and conflicting accounts of the source of this watery mass: Does the fire produce it? Does the fire *become* it? Or does the fire just spare a little bit of water from the world before the blaze? This difficulty aside, there is a more basic question that all three versions leave unanswered: Why would the universe undergo these cycles at all? This question is at least double-sided: in terms of physics, one is left asking how the creative principle of the universe suddenly becomes a destructive one; and in terms of theology, one wonders why a "perfect," "providential" god would annihilate the universe it creates—much less a universe it *pervades*.[95]

Jaap Mansfield offers a sustained resolution to both prongs of this problem by explaining that for Zeno the conflagration is not a tragic or violent occurrence; rather, it is the "best possible state of affairs."[96] Edward Adams similarly writes that the conflagration "is not the death of the cosmos but its acme" because when everything is fire, everything is god.[97] If this is the case, then to the physical question one can answer that the "destructive" fire does not oppose the "creative" fire at all. Rather, both are stages in the life of the same substance, which like everything else in the cosmos undergoes periodic cycles of death and rebirth. To the theological question of why the god would unmake its handiwork, one can say, first, that the god has no choice but is bound to the laws of physics that it both governs and pervades, and, second, that the conflagration is not annihilation. It is, rather, a divine assimilation of all things, through which the life of the world is renewed.

But perhaps the best-known vindication of these cycles emerges through the teaching of cosmic repetition. Because the divine is rational and perfect, and because it permeates every corner of the cosmos, the early Stoics insisted that this world must be the best possible world. And because it is the best possible world, *it can only be* the way that it is.[98] This means that every cosmos emerging like a phoenix from the flames will look just like the one that burned before it—identical down to the smallest detail. "There will be another Plato and another Socrates," says Zeno,[99] another Caesar Augustus and another Barack Obama, another me who writes this book and another you who reads it, forever and ever. Some of Zeno's followers allowed for minuscule variations among successive *kosmoi*—a freckle, perhaps, on a face that had been freckle free the previous time around.[100] But for all intents and purposes, each world will be exactly the same as its predecessors and descendants because the cosmos is perfect as it is.

Once More, with Feeling

Stoic cosmology is, of course, where Nietzsche's Zarathustra gets his infamous idea of "the eternal return." In *Ecce Homo*, Nietzsche speculates that "the unconditional and infinitely repeated circular course of all things *might* in the end have been taught already by Heraclitus. At least the Stoics have traces of it, and the Stoics inherited almost all of their principal notions from Heraclitus."[101] In the eternal return, we therefore have a nineteenth-century repetition of the ancient idea of repetition, whose genealogy seems to bend time back on itself. "In the end," Nietzsche writes—which presumably means at some point in the future—it "might" turn out to "*have been*" the case that the ancient Heraclitus taught the eternal recurrence of all things, which is now proclaimed by Zarathustra (a futural repetition of the ancient Persian prophet). At the very least, Nietzsche says, this untimely idea recalls the infinite cycles of Stoic cosmology, which were clearly haunted by Heraclitus. Despite this Stoic heritage, however, Alexander Nehemas suggests that it is not clear how seriously Nietzsche took the eternal return as a cosmology. As far as Nehemas can see, Nietzsche was far less concerned to assert this circular repetition as a physical fact than he was to offer it as a psychological possibility.[102]

It is in the spirit of possibility, for example, that *The Gay Science* confronts its reader with a dizzying "what if": "What if some day or night a demon were to steal after you into your loneliest loneliness and say to you: 'This life as you now live it and have lived it, you will have to live once more and innumerable times more.'" The demon goes on to specify that living "this life" again would amount to reliving every joy and every sorrow "and everything unutterably small or great"—everything exactly as it has been, forever. What, Nietzsche asks, would you do? "Would you not throw yourself down and gnash your teeth and curse the demon who spoke thus?" Or would you instead say to the demon, "You are a god and never have I heard anything more divine"?[103] For Nietzsche, the *possibility* of the eternal return asks us what it would take to opt for the latter and call this demon a god. What would it take for you to affirm your life so radically that you would be able not only to accept but to *will* the whole thing back again—your own life, with all its stupidities and triumphs—exactly as it has been, once more and "innumerable times more"?[104]

For Nehemas, "what it would take" does not necessarily entail cosmology, but it certainly entails ontology, which in turn opens onto ethics. This interpretation will therefore help to round out our investigation of the Stoics, for as centuries of commentators have noted, the ekpyrotic eternal return does seem to make for "a rather awkward ethics."[105] If nothing can be other than the way it is, what room does this leave for decision, freedom, or change? Does the Stoic vision of "living in accordance with nature" amount simply to resigning oneself to the world as it is? To thinking of oneself, as Zeno is said to have said, "as a dog tied to the back of a cart"?[106] Following Nehemas's lead, I would like to suggest that by presenting the eternal return as a psychological hypothesis rather than a physical law, Nietzsche allows us to think differently about what it would mean to live in relation to this teaching—which is to say, to live as a set of relations.

Throughout his corpus, Nietzsche insists that just as there is no lightning without its flash, there is "no 'being' behind doing," no "thing-in-itself," "no subject . . . beyond . . . its characteristics and effects, it experiences and actions."[107] Applied to the human "subject," this means that there is no "I" without the specific events that compose "my" life—however trivial or accidental these events might be. Trivia and accidents constitute me to such an extent that if any of them were different, *I would not be me*. In response to Nietzsche's demon, then, it would make no sense to try to affirm some parts of my existence without the others—to wish, for example, that "I" could come back next time being able to draw—because changing anything about me would amount to abolishing me *tout court*. By a similar logic, Nehemas reminds us that "every event in the world is inextricably connected with every other," so that if anything in the world were different, everything in the world would be different.[108] Positively stated, if in response to the demon you could wish anything to be the same, you would have to wish everything to be the same. Thus asks Zarathustra: "Have you ever said Yes to a single joy? O my friends, then you have said Yes too to all woe. All things are entangled, ensnared, enamored: if ever you wanted one thing twice, if ever you said, 'You please me, happiness! Abide, moment!' then you wanted all back. All anew, all eternally, all entangled, ensnared, enamored—oh, then you *loved* the world."[109]

If it is the case that the eternal return is above all an ontoethical possibility, then to *will* the identical recurrence of all things is not simply to resign oneself to whatever happens, but to recognize the world as fundamentally "entangled and ensnared"—more radically, to *love* it as such. This affirmation is itself the space of freedom: we can choose to love the entangled world or not. Moreover, this affirmation constitutes a *difference* in the eternal return of the same. For a world that is willed and loved in its interdetermined insanities is different from the "same" world passively endured or misunderstood as the sum of discrete entities. Perhaps this is what Gilles Deleuze means when he writes that "it is not the 'same' or the 'one' which comes back in the eternal return but return is itself the one which ought to belong to diversity and to that which differs."[110] To "want it all back" is to affirm the pluri-singular inextricability of cause and effect, spirit and matter, god and world—and this very affirmation makes "it" other than itself.

Perhaps surprisingly, a conclusion approaching this one can be drawn from the Roman emperor and Stoic philosopher Marcus Aurelius Antoninus (121–180). Although he thought cosmology a waste of time and never mentioned *ekpyrosis* or cosmic repetition,[111] he retained enough of this ancient-modern idea to affirm its entangled ontology. "All things are interwoven [*implexa*] with each other," he writes in the *Meditations*. "Everything is coordinated, everything works together in giving form to the one universe. The world-order is a unity made up of multiplicity: God is one, pervading all things."[112]

Half a century later, the Christian theologian Origen of Alexandria (185–254) reverses these priorities, retrieving Stoic cosmology while dropping its pantheizing ontology. In *De principiis*, Origen tells us that a cyclic cosmology is the best way around that perennial, "impious[,] and absurd" question of what God was doing "before" the creation of the world (Was he just lounging around, waiting for the right time to start creating? What was God the God *of* if there was no world? If God is eternally God, then mustn't the world be eternal as well?).[113] Origen's answer is that the world is not coeternal with God, but that God has nevertheless always been in action *as God* because there have been worlds before our own and will be worlds after it. Origen finds ample support for this position in Scripture, citing the Psalms' conviction that "the heavens will perish" (Psalm 102:26), Matthew's notion that "heaven and earth shall pass away" (Matthew 24:35), and, above all, the Isaianic promise that "there will be new heavens and a new earth, which I will make to abide in my sight" (Isaiah 65:17).[114] Origen also cites Ecclesiastes, who reaffirms the existence of multiple worlds when he asks, "What is that which hath been? Even that which shall be. And what has been created? Even this which is to be created: and there is nothing altogether new under the sun. Who shall speak and declare, Lo, this is new? It has already been in the ages which have been before us. "[115]

In short, the soundest way to secure the eternal divinity of God without asserting the eternity of the world is to affirm that this world is not the only one—there have been and will be more. How many? Origen confesses that he does not know—"although, if anyone can tell it, I would gladly learn."[116]

Although Origen affirms the existence of multiple worlds, he is careful to make a number of qualifications to keep the doctrine from slipping into paganism. First, he maintains that although God does create a succession of worlds, he creates each of them not out of the stuff of any previous world, but rather ex nihilo.[117] Second, Origen insists throughout the treatise that God is not "in any degree corporeal."[118] This means that when Deuteronomy 4:24 calls God "a consuming fire," for example, it is speaking metaphorically: God cleanses us from all moral impurities, but God (unlike the Stoic divinity) is not literally *fire*—or any other physical substance at all.[119] Third, Origen argues that the different worlds that God creates cannot be identical to one another because souls are driven by free will and so are not bound to any predetermined cyclic course. Because souls can "direct the course of their actions" as they see fit, different worlds must be different from one another—some better and some worse—with different souls populating each one.[120] And finally, although there can be said to be a plurality of worlds, "it is not . . . to be supposed that several worlds existed at once, but that, after the end of this present world, others will take their beginning."[121] In other words, the scriptural "ages of ages" are incompatible with the spatial riot of the Epicureans, but they fit perfectly well with the temporal multiplicity of the Stoics. A few crucial differences notwithstanding, Origen's cosmos looks very much like that of the Stoics: "I am of the opinion," he writes, "that the whole world ought to be regarded as some huge and immense animal, which is kept together by the power and reason of God as by one soul."[122] Just like the Stoics, Origen sees the world as a living being and an interrelated, God-permeated one at that—except his God is immaterial, and therefore transcendent to the creation that God suffuses.

But even Origen's highly qualified reconciliation of Christian theology with Stoic cosmology never quite takes root—thanks in large part to Augustine's sustained invective against the idea in *The City of God* (413–426).[123] Writing as bishop of Hippo in the wake of the Sack of Rome by the Visigoths in 410, Augustine is concerned throughout this book to assert the sovereignty of God over that of any earthly ruler. When it comes to cosmic origins, Augustine's concern for sovereignty leads him to assert God's creation out of nothing against two principal adversaries whose positions we have already glimpsed. The first are those who argue that the world must be eternal if God is eternally God. The second are those who argue, against the first, that God is always God because this world is just one of many.[124] In other words, Augustine takes on Origen's opponents and Origen himself, rejecting both the problem the former presents and the solution the latter provides. He does away swiftly with those who say that the world is eternal, arguing simply that nothing can be co-eternal with a singular God.[125] The many-worlders, however, prove a trickier group of opponents, and Augustine's efforts to refute them throws his argument into strange disarray.

Augustine tells us that of the philosophers who assert a plurality of worlds, some posit an infinite number of perishable worlds, whereas others believe that there is only one world, which undergoes "an infinite series of dissolutions and restorations."[126] In other words, there are the followers of Democritus and Epicurus (who posit what I am calling a spatial multiplicity) and the followers of Zeno (who posit a temporal multiplicity). After this brief reference, however, Augustine drops the Epicureans altogether, directing the entirety of his argument against the Stoics. "The Physicists," he begins, referring to the Greek Stoics, "assert . . . an unceasing sequence of ages, passing away and coming again in revolution." Although he mentions the Platonic interpretation of these ages as internal to the imperishable world (as in the periodic destruction of cities through floods and fires),[127] his real concern is with the Stoic teaching that the world *itself* "disappears and reappears" throughout an infinite succession of ages. In particular, he zeroes in on the baffling notion that each world will be filled with people and cities and events "which appear as new, but which in fact have been in the past and will be in the future."[128] This, of course, is the doctrine of eternal return—Zeno's idea that there will be another Plato in another Athens who teaches another Aristotle, "time after time . . . in innumerable centuries in the future."[129] Augustine seems also to be familiar with Seneca's meditation on this teaching, which imagines that after the coming conflagration "the souls of the blest, who have partaken of immortality . . . shall be changed back into our former elements" to live their lives once more, as they were.[130] And the idea nearly drives the bishop crazy. "They are utterly unable to rescue the soul from this merry-go-round," he charges, "even when it has attained wisdom."[131]

For under such conditions, each soul must come back each time to live and die again, the same way as before, without improvement or escape.

As it turns out, then, Augustine's chief argument against the Stoics is not really an argument so much as an extended expression of discomfort:

> It is intolerable for devout ears to hear the opinion expressed that after passing through this life with all its great calamities (if indeed it is to be called life, when it is really a death . . .) that after all these heavy and fearful ills have at last been expiated and ended by true religion and wisdom and we have arrived at the sight of God and reached our bliss in the contemplation of immaterial light . . . that we reach this bliss only to be compelled to abandon it, to be cast down from that eternity, that truth, that felicity, to be involved again in hellish mortality, in shameful stupidity, in detestable miseries, where God is lost, where truth is hated, where happiness is sought in unclean wickedness, and to hear that this is to happen again and again, as it has happened before, endlessly, at periodic intervals, as the ages pass in succession.[132]

"Who could listen to this?" Augustine asks. "Who could believe it? Who could even tolerate it?"[133] The idea is so intolerable, in fact, that it prompts even this inveterate seeker of truth to suggest, "[I]f it were true it would be more prudent to *suppress the truth*, nay, wiser to be in ignorance—*I am trying to find words to express what I feel*."[134] What Augustine feels is something like a revolted despair, and this despair produces a far dimmer view of this world than he usually sets forth. Just one book earlier, he had taken pains to defend the inherent goodness of creation against the "delirious raving of the Manichaeans."[135] But faced with the possibility that this creation might return, all he can see is death, fear, misery, and untruth ("*tartareae mortalitati, turpi stultitiae, miseriis exsecrabilibus*").[136] One imagines Augustine here thinking back to his own "pestilential" youth, to those wasted years of hating truth and seeking happiness "not in God but in his creatures."[137] One can almost see Nietzsche's demon steal into Augustine's "loneliest loneliness" to taunt him with the possibility of doing it all again. The raving Manichaeans, the shameful sex, the death of his mother again, forever—it is as though Augustine pictures an infinity of his adolescent selves throwing pears at an infinity of hapless pigs, and the last thing the man can do is affirm the Stoic perfection of creation. Rather, he throws himself down and gnashes his teeth and curses the demon: "God forbid [*absit*] that what the philosophers threaten should be true," he exclaims.[138]

God forbid that the soul, having finally escaped this "hellish" life into beatific existence with God, might have to do it all over again. For life thus construed would be an "unremitting oscillation between false bliss and genuine misery."[139]

As for Origen, whom Augustine otherwise considers a "learned and experienced" theologian,[140] his erroneous endorsement of Stoic circularity stems, Augustine explains, from a misunderstanding of Ecclesiastes. For when "Solomon" says that what has been is the same as what will be, or that "there is nothing new under the sun," he is likely just referring to "successive generations, departing and arriving." Or perhaps, Augustine continues, he is referring to categories rather than to individuals: to men, trees, and plants in general rather than to *this* Plato or *that* hibiscus. Or, he flounders, maybe Solomon is referring to the eternal mind of God, which contains all things so completely that, strictly speaking, there is nothing new. Augustine admits that the meaning of the passage remains unclear, but whatever it may be, "heaven forbid [*absit*] that correct faith should believe that those words of Solomon refer to those periodic revolutions of the Physicists."[141] Again, Augustine has less of an argument here than a wish—a plea, even, that this not be the way things are. He substantiates this plea by appealing to "the promises of God," but of course he chooses his promises carefully, taking them not from Isaiah ("there will be new heavens and a new earth"), but from the letters of Paul. He focuses on two in particular: first, that "Christ died once for all for our sins" (Romans 6:9) and, second, that "we shall be with the Lord forever" (1 Thessalonians 4:17). Both of these assurances, we might note, preserve the eternal integrity of the singular subject, whether human or divine. Like Christ, we humans will live and die only *once* before dwelling with God forever. And so "heaven forbid," Augustine repeats, that Christ should be said to die in more worlds than one.[142] Aside from subjecting Our Lord to perpetual violence, this act of unsettling his singularity would unsettle our own, dooming us to endless earthly repetition. And this repetition, to be frank, seems to be Augustine's worst fear of all: "heaven forbid" that we be forbidden from heaven, dwelling with God only to lose him again and condemned to these mortal cycles forever.

Augustine will ultimately decide that these two Pauline promises—the uniqueness of the Christ event and the *permanent* immortality of the human soul—amount to nothing less than geometric salvation. They turn the Stoics' hellish circles into a "straight and right" path from creation to eternal life with God.[143] "So let us keep to our straight way, which is

Christ," he concludes, "and turn our minds from the absurd futility of this circular route of the impious." In Christ, we shall never find happiness only to lose it. Rather, "we shall keep it always, in assured security, and no unhappiness can interrupt it."[144] And now that he has finally secured our eternal security, Augustine can return to the problem of creation. Beginning from the certainty that this world is not eternal, and "seeing that those cycles of [the Physicists] have been hissed off the stage," he concludes that the only remaining possibility is that God created the world "in time," yet with no change in God's being or design.[145] If this sounds like a contradiction, he adds, then it is likely close to the truth, for "who could plumb this unplumbable depth of God's counsel, and scrutinize his inscrutable design?"[146] An absurd creation narrative, in other words, is better suited to an inscrutable God than a straightforward story would be.

Can this be the reason that Christian theology never quite came to terms with the Stoics' cycles—because St. Augustine "hiss[ed] them off the stage" in the fourth century? And did he do so because of the threat to divinity they posed or simply because he could not stomach the notion? Whatever the reason and with very few exceptions, the Western tradition did not even consider Stoic cosmology long enough to dismiss it until the early twentieth century, when astronomers and physicists begin to grapple with the cosmological implications of general relativity. And sure enough, fifteen centuries after Augustine's revolted dismissal of the Stoics, a British astrophysicist named Arthur Eddington will be hit by a similar wave of nausea. As Alexander Friedmann realizes in the 1920s, if the amount of matter in the universe were to exceed a critical density, it would draw the universe into a "big crunch," from which another big bang might bang, and the cycle of the cosmos could start anew.[147] Contemplating this possibility in 1935, Eddington (like Augustine) has no concrete data with which to refute it;[148] rather, he opposes the notion "from a moral standpoint," calling the idea of a cyclical universe "wholly retrograde. Must Sisyphus for ever roll his stone up the hill for it to roll down again every time it approaches the top? That was a description of Hell. If we have any conception of *progress* as a whole reaching deeper than the physical symbols of the external world, the way must, it would seem, lie in *escape* from the Wheel of things."[149]

As we read of the hellish, pagan circles overcome by the linear "way" of "progress," we might think this a passage straight out of *The City of God*. Indeed, Eddington's theological recapitulation deepens as he goes on to quote scripture: "Since when has the teaching that 'heaven and earth shall pass away' become ecclesiastically unorthodox?"[150] Leaving to one side the question of why Eddington is contemplating ecclesiastical orthodoxy in a technical paper on entropy and cosmic expansion, it is interesting to note that this quotation from Isaiah, which Eddington cites as evidence *against* the periodic renewal of the cosmos, is the same passage that Origen used to support it. So Eddington's scriptural argument is a flimsy one. But that is not really the point; for Eddington, the bottom line is simply that the idea of eternal recurrence is distasteful. "I am no Phoenix worshipper," he writes in *The Nature of the Physical World*. "It seems rather stupid to keep doing the same thing over and over again."[151] The whole idea seems to Eddington childish, inane, *annoying*—as if the cosmos itself were some unending hymn that all creatures started singing,

> not knowing what it was,
> And they'll continue singing it forever just because
> This is the song that doesn't end.[152]

Heaven forbid.

In this chapter, we have explored two different models of cosmic multiplicity. Against the Academic insistence that there can be only one, finite world, the Atomists posited an infinity of *kosmoi* haphazardly moving through infinite space. The Stoics, by contrast, posited one world destroyed and reborn throughout infinite time. Perhaps needless to say, neither model became the template for observational astronomy, Western philosophy, or Christian theology. As we have seen, neither Roman philosophy nor early Christian orthodoxy was at all hospitable to the early Stoics' endless cycles. The Atomists suffered even greater ridicule beginning in the first century B.C.E. and lasting at least through the early fifteenth century.[153] Throughout this period, a poorly understood "Epicureanism" was consistently demonized as both intellectually stagnant and ethically repulsive—failures that were attributed to its godlessness. This fairly unanimous rejection can be traced back to Cicero, on the one hand, whose "Epicurean" character in the *De natura deorum* is nothing short of imbecilic,[154] and to the early church fathers, on the other, who sought not only to refute Epicurean physics, but also to imagine the depths of sin to which such physics must have led. As Howard Jones explains in his history of Epicurean philosophy, "Epicurus and his latter-day disciples were credited with a colorful array of depravities and perversions . . . : swinish gluttony, drunkenness, fornication, adultery, homosexuality, sodomy, incest—[with] Theophilus, Clement, Pseudo-Clement, Ambrose, Epiphanius, Peter Chrysologus, Filastrius, and Augustine each contributing a little to the list."[155] The condemnation was so virulent and so exceptionless that by the year 410, Augustine could say of Epicureanism that "its ashes are so cold that not a single spark can be struck from them."[156]

Had Augustine given a little more credit to the Stoics, however, he might have taken more time to worry that even from ashes life might resurge. And, indeed, although the question of multiple worlds would be extinguished for centuries, it would come firing back again in a series of recurrences—not so much of the same, but of the different—a wave of repetitions that were nonetheless new.

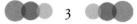 3

NAVIGATING THE INFINITE

My God, you are absolute infinity itself, which I perceive to be the infinite
end, but I am unable to grasp how an end without an end is an end.

NICHOLAS OF CUSA, *ON THE VISION OF GOD*

Each thing (the glass surface of a mirror, let us say) was infinite things,
because I could clearly see it from every point in the cosmos.

JORGE LUIS BORGES, "THE ALEPH"

Ending the Endless: Thomas Aquinas

From the ashes of the mid-thirteenth century arises a familiar question: Is
there one world, or are there many? Might there even be an infinite number of them? The author of the question is Thomas Aquinas (1225–1274),
and one almost wonders why he bothers to ask. Hadn't the matter been
put to rest by Plato, sealed by Aristotle, diagrammed by Ptolemy, and
Christianized by Augustine? Hadn't all the pluralizing dissenters been
"hissed off the stage" or consigned to dust and ashes centuries ago? And
yet here we find Thomas in the *Summa theologiae*, beginning as usual with
the position he will refute, saying: "[I]t would seem that there is not only
one world, but many."[1] To whom would it seem that this is the case? What
has changed since Augustine declared the matter closed nearly a millennium ago?

What has changed is, in one sense, a return of the "same," which is to
say a rediscovery of Aristotle. As is well known, most of Aristotle's works
had been lost to the Latin-speaking world between the third century B.C.E.
and the twelfth century C.E., when scores of his manuscripts were translated, debated, and made the object of lengthy scholastic commentaries.[2]
Among these newly recovered manuscripts was the *De caelo*, which was
translated from Arabic into Latin in 1170. As we saw in chapter 1, this text

insists against the Atomists that the cosmos must be unique, basing its claims on the principles of "natural motion." If another world existed, Aristotle reasons, then its earth would move unnaturally "up" with respect to our world, even as it moved naturally "down" with respect to its own. "This, however, is impossible," he says, because it is the property of earth to move down.[3] Elements cannot move both naturally and unnaturally at once, so "it follows that there cannot be more worlds than one."[4] With the rediscovery of the *De caelo,* medieval Europe thus possessed a seemingly definitive argument for the singularity of the cosmos, one that reaffirmed the teachings of Platonists and church fathers alike.

This agreement notwithstanding, Aristotle also held cosmological positions that contradicted the received teachings of medieval Europe—perhaps most problematically concerning the eternity of the cosmos.[5] Here we might recall Augustine's insistence that the world could not be eternal without compromising the singularity of God: if God alone is God, then nothing else can exist alongside him. Thus began centuries of Latinate efforts to evaluate Aristotelian cosmology in the light of Christian theology, with the universities' "secular masters" ready to adopt Aristotle wholesale, the Franciscans looking to abandon any position that seemed to contradict scripture, and the Dominican Thomas Aquinas working to reconcile the two.[6] In this era of intellectual fervor, even the most firmly entrenched doctrines were reopened for debate, including the doctrine of cosmic singularity. So even though the Epicureans themselves would not be given a fair hearing until the seventeenth century, their teachings on the plurality of worlds were tentatively engaged five hundred years earlier as the medieval West came to terms with the very philosopher who had rejected them.[7]

Among Aristotle's Christian interpreters in particular, the central cosmological concern was to uphold the sovereignty of God with respect to creation. This emphasis on sovereignty, in turn, reopened the question of cosmic plurality in the high Scholastic period. After all, one might ask, if God is omnipotent, then why would "he" limit himself to creating one world? It is in this sense that Thomas Aquinas concedes in Question 47 of the *Summa theologiae* that it might "seem" that there are many worlds. "For the same reason He created one," Thomas reasons, "He could create many, since His power is not limited to the creation of one world; but rather is infinite" (1.47.3).[8] Indeed, the infinity of God's power might even lead us to posit an *infinite* number of worlds.

Yet just as an eternal world would threaten God's singularity, Thomas seems concerned that infinite worlds might rival his infinity. Indeed, a standoff between material and divine infinity can be seen as early as Question 2 of the *Summa*, in which Thomas proves the existence of God based on the absurdity of an infinite causal regress. "Whatever is in motion must be put in motion by another," he argues in the first of his five proofs. "If that by which it is put in motion be itself put in motion, then this also must needs be put in motion by another, and that by another again. *But this cannot go on to infinity*, because then there would be no first mover, and, consequently, no other mover. . . . Therefore it is necessary to arrive at a first mover, put in motion by no other; and this everyone understands to be God" (1.2.3, emphasis added).

In the work of this proof, Thomas aligns "everyone's" God with Aristotle's prime mover: the extracosmic stopgap that prevents the causal march to infinity. God puts an end to worldly endlessness. How, then, could there be an endless number of worlds? Where is the place for a *first* mover if worlds extend backward eternally?

In short, the doctrine of a plurality of worlds threatens Thomas's whole theological infrastructure: if worlds have existed from eternity, then there is no starting point for God to occupy. In the *Summa*, Thomas therefore raises the possibility of a cosmic plurality, only to launch a multipronged attack against it. He calls briefly on the Gospel of John (" 'the world was made by Him,' where the world is named as *one*"), offers a brief paraphrase of the *De caelo*'s argument from natural motion, and appeals in passing to its neo-Timaean insistence that "the world is composed of the whole of its matter" (1.47.3, emphasis added).[9] In this manner, Thomas lines up the usual sources of authority against cosmic multiplicity: scripture, Plato, and Aristotle all seem to say no. But his chief strategic move in the face of this possibility is to refocus the question, shifting the measure of divine sovereignty away from brute force and toward singularity. An omnipotent God *could* make other worlds, Thomas imagines, but doing so would compromise his unity.

The argument proceeds as follows: all things come from God, and all things find their end (terminus) in God. This means that "[w]hatever things come from God, have relation of order to each other, and to God himself."[10] This "relation of order" denotes the hierarchy of creation—the Neoplatonic "Great Chain of Being" under which all things from angels to snails are peacefully, permanently, and vertically related to God and one another.[11] In an earlier question, Thomas calls on this ordered relation to demonstrate the oneness of God: the unity of creation, he argues, testifies to the unity of its creator (1.11.3). In the question at hand, the demonstration is simply reversed: because God is one, "it must be that all things should belong to one world" (1.47.3). Taking these questions together, we can see that the oneness of the cosmos is both a function and a sign of the oneness of God. The only way to "assert that many worlds exist" would therefore be to deny the "ordaining wisdom" of God altogether—to say that there is no providential order of things (1.47.3). Here Aquinas offers the example of "Democritus, who said that this world, besides an infinite number of other worlds, was made from a casual concourse of atoms" (1.47.3). To affirm an infinity of worlds is therefore to deny the involvement of God, for God is said to be the *end* of all creation. But insofar as "the infinite is opposed to the notion of end" (1.47.3), infinite worlds would mean the absence of end; there would be no single source, no final home, and no ordered passage from one to the other. Many worlds, in short, would mean no God, and "this reason proves that the world is one" (1.47.3).

Although the tone is far more somber, one can thus detect in this argument echoes of Augustine, who likewise rejected cosmic infinity because of its endlessness. A soul destined to recurring embodiment, he feared, would never find its rest in God.[12] To his credit, Thomas has arguably found firmer ground for his rejection than the Augustinian "heaven forbid": if all things have their end in one God, he reasons, then all things must belong to one world. Thomas may well have adapted this strategy from Book Lambda of Aristotle's *Metaphysics*, which asserts that the cosmos must be singular because its source of motion is singular.[13] As we will recall, however, Aristotle's proof undermines itself even before it is concluded, producing either forty-seven or fifty-five prime movers in the process of trying to secure one and opening in spite of itself onto just as many worlds. Thomas's argument similarly can be made to tremble at the very point that it hinges the number of the cosmos on the number of God. For however closely Thomas may align them, his God differs from the prime mover in not being simply one. His God is rather three-in-one, an eternal interrelation of identity and difference. So even if God *is* the end of all things, God is an "end" that is both one and many: multiple. Might the things of creation not therefore occupy multiple worlds?

This might be a compelling possibility, but it would be unacceptable to Thomas for two reasons. First, he insists on the priority of identity over difference even within the Trinity, arguing in the previous article that "unity" pertains to the Godhead and "multiplicity" only to creation.[14] God might *contain* plurality, and God certainly produces plurality, but God is primarily *one*.[15] Second, Thomas assumes that if numerous worlds were to exist, they would bear no relation to one another, constituting nothing more than a haphazard plurality à la Democritus. But if we push on the first assumption, then the second moves as well: if the Christian God is eternally triune, then God is not single first and plural afterward, but *eternally pluri-singular*. What, then, would prevent such a God from creating multiple worlds that are nonetheless interrelated? If the number of creation really mirrors the number of God, then wouldn't an entangled multiplicity of worlds reflect God's many-oneness more fully than a single world would?

It is perhaps for this reason that Thomas shifted his strategy the next and last time he addressed the question of multiple worlds. Two years before the end of his life, he wrote a detailed commentary on the *De caelo* in which he hinges the oneness of the cosmos not on the oneness of God, but on the *omnipotence* of God. Although it might seem that an omnipotent God would create as many worlds as possible, Thomas counters that "it takes more power to make one perfect [individual] than to make several imperfect."[16] A "perfect" individual world would be one that includes all beings within it, and (back to the *Timaeus* again) a world that contains all beings within it would have to be singular. It therefore does more justice to the omnipotence of God to say the world is one than to suggest it might be one of many.

As it turned out, however, neither of Thomas's approaches succeeded in putting cosmic multiplicity to rest. A mere three years after Thomas's death in 1274, Bishop Etienne Tempier of Paris issued a list of 219 heretical Aristotelian teachings, among which was Condemnation 34: "Quod prima causa non posset plures mundos facere."[17] Anyone who taught that "the first Cause cannot make many worlds" would henceforth be excommunicated for undermining the absolute power of God. And so the Scholastics of the late thirteenth century and the fourteenth century could not rest with Thomas's Christianized repetitions of Aristotle. More important, they could not rest with Aristotle himself—and would even have to face the possibility that his nesting-doll cosmology was *wrong*. In the long run, then, these ecclesiastical prohibitions opened a surprising space of intellectual freedom, one that eventually led to the wholesale abandonment of Aristotelian physics in the late seventeenth century.[18] However coercive the Condemnations of 1277 may have been, they eventually prompted a shift in thinking so radical that Pierre Duhem calls them "the birth certificate of modern physics."[19]

But, of course, the first three hundred years of "modern physics" never went so far as to teach that there were multiple worlds. Neither, as it turned out, did the Scholastics after 1277. Rather, reading Condemnation 34 as closely as possible, they found a number of ways to argue that although God *could* create worlds other than this one, he never *would*. For the sake of clarity, these strategies can be grouped into two. The authors one might call "voluntarist" held to Aristotelian physics even as they accepted the bishop's chastisement, arguing that although the laws of nature preclude the existence of more than one world, an omnipotent God could decide to override the laws of nature if he so wanted.[20] The "naturalist" authors, by contrast, used the Condemnations as an opportunity to undermine the very principles of Aristotelian physics. By attacking the *De caelo*'s two proofs of cosmic singularity, they argued that other worlds could exist in full *accordance* with the laws of nature. For example, they maintained, other worlds might be composed of different elements, with different sorts of motion.[21] Or even if the materials were the same, another world's "earth" and "fire" might move down and up with respect to that world alone, preventing any conflict between "natural" and "unnatural" motion.[22] Finally, they argued, it is senseless to say that all the matter in existence has been used up on this world because an omnipotent God can always make more. In a strange turn of events, then, the very teaching that Lucretius found inimical to the plurality of worlds came back in the late medieval period to support it: if God created one world ex nihilo, then God could create any number of them ex nihilo. "In order to establish this position," wrote Richard of Middleton (1249–1302), "one can invoke the sentence of Lord Etienne . . . he has excommunicated those who teach that God could not have created several worlds."[23] We would therefore do well to teach that he could have.

For all their daring flirtations, however, none of these authors dared to assert the *existence* of other worlds. Rather, at some point in each of their arguments, one finds the sort of "standard disclaimer" issued by Nicole Oresme (ca. 1320/25–1382) in *Le livre du ciel et du monde* (1377): although God in his omnipotence could create and care for numerous worlds (Oresme was particularly taken with the possibility that there might be smaller worlds embedded concentrically within our world, which might itself be embedded within larger ones), "there never has been nor will there be more than one world."[24] Voluntarists and naturalists alike, the late Scholastics exhibited what Steven Dick calls a "uniquely medieval mixture of boldness and conservatism":[25] they went to extraordinary lengths to defend the possibility of other worlds but would not even contemplate the actuality of those worlds. The upshot of this medieval mixture was that although these authors called into question almost all of Aristotle's cosmological *principles*, they left his cosmic geography in place. At the close of the fourteenth century, Europe still imagined the world as a set of concentric circles with earth at the center; rings of water, air, and fire surrounding it; and a halo of "fixed stars" moving calmly around the circumference once a day. These stars were held to be the Primum Mobile, or "first moved" of the cosmos. Set in motion by the prime mover itself, the fixed stars then conferred movement on the lesser cosmic bodies within their bounds.

This motive gradation allowed the Aristotelian cosmos to be mapped onto the Neoplatonic "Chain of Being" in which physical position was thought to coincide with spiritual rank. Of this worldview, Ernst Cassirer explains that "the higher an element stands in the cosmic stepladder, the closer it is to the unmoved mover of the world, and the purer and more complete its nature."[26] The realm of the stars, made of an incorruptible "fifth essence" (*quinta essentia*), was thought to be nearest to God, whereas the corruptible earth was farthest away; here we might recall Dante's journey from the inferno at the center of the earth, up the purgatorial mountain, to the stars at the gates of paradise (figure 3.1).[27] On the earth itself, the beings that participate most fully in divine intellect are ranked above the others—hence, the superiority of angels to humans, humans to animals, men to women, and reason to the passions. And so cosmology recapitulates theology: as Thomas insisted, the order of the universe mediates the singular God down through the hierarchical ranks of the singular cosmos. This means that if any of these terms were to be challenged, the rest would have to change as well. Any real departure from Aristotle's tidy circles would need to reconsider the singularity of God, the singularity of the cosmos, *and* its hierarchical arrangement. It is therefore striking that this departure initially came from within the Christian theological tradition itself. The thinker who genuinely abandoned Aristotelian cosmology was not Nicolaus Copernicus, who put the sun at the center of the universe in 1543, but Nicholas of Cusa, who declared a hundred years earlier that the universe had no center at all.

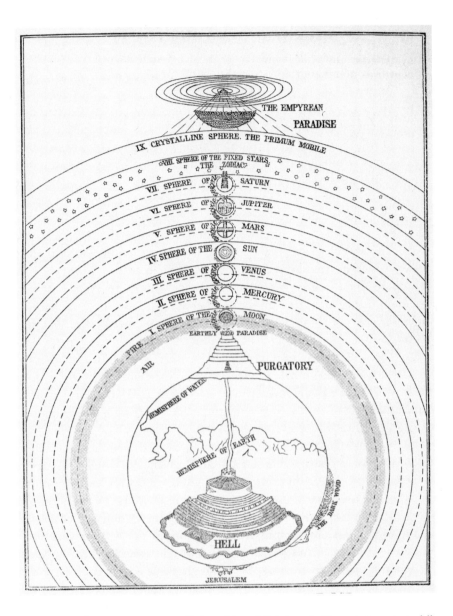

FIGURE 3.1 The Dantean universe. (Adapted from Michelangelo Caetani, *La materia della "Divina Commedia" di Dante Alighierie* [Monte Cassino, 1855])

End Without End: Nicholas of Cusa

"The Earth Is Not the Center of the World"

Picture yourself on a boat, Cusa tells us, sailing through a vast ocean. Unless you can see the shore recede behind you or the waters rush beneath you, you will think you are at rest no matter how fast you may be moving. Indeed, even if you do gaze down at the waters flowing by, you may at first perceive that they are moving while you are standing still. So it is with our position in the universe. Although the earth moves through a vast expanse of space, we perceive ourselves to be at rest in the middle of the world because we lack an unmoved point of reference. The same holds for every other cosmic body: everything in the universe moves in imperfect circles around its neighbors.[28] And yet precisely because nothing is at rest, he says in *On Learned Ignorance*, "it always appears to every observer, whether on the earth, the sun, or another star, that one is . . . at an immovable center of things and that all else is being moved" (2.12.162). For Cusa, nothing is at the center of the universe, which means that everything is at the center—from its own perspective. Even those stars that we see at the outer edge of the universe occupy the center of creation from their own vantage point and see us at the outskirts. Thus there is neither center nor circumference, but a shifting series of cosmic configurations depending on the observer's point of view.

Nicholas of Cusa (1401–1464) has long been touted as a surprising forerunner of modern cosmology,[29] and, given this brief meditation alone, it is easy to see why. Not only did Cusa remove the earth from the center of the universe and set it in motion, but he also set it in motion in orbits that are not quite circular. He did the same for all other celestial bodies as well, and, astonishingly, this led him to an early articulation of the theory of special relativity. In 1905, Einstein would write that "electrodynamics [and] mechanics possess no properties corresponding to the idea of absolute rest"; rather, rest and motion are wholly relative to the standpoint of the observer.[30] Cusa had considered this proliferation of vantage points four and a half centuries earlier in *De docta ignorantia* (1440), concluding, moreover, that all the "stellar regions" beyond our own were most likely "inhabited"—although by what sorts of species, he could not say. The furthest he would go was to imagine that the inhabitants of the sun were likely to be "more solar, bright, illuminated, and intellectual, even more spiritual than those on the moon, who are more lunar, and than those on the earth, who are more weighty." For the most part, however, Cusa maintained that the inhabitants of distant cosmic bodies "remain completely unknown" (2.12.172). A mobile earth, noncircular orbits, the relativity of motion, extraterrestrial life—each of these postulates can be seen as strikingly protomodern. Nevertheless, the development for which Cusa has become best known is his systematic destruction of Aristotle's cosmic nesting dolls. As Alexandre Koyré explains, "[I]t was Nicholas of Cusa, the last great philosopher of the dying Middle Ages, who first rejected the medieval cosmos-conception and to whom, as often as not, is ascribed the merit, or the crime, of having asserted the infinity of the universe."[31]

Of course, this radical idea was not exactly new. As we might recall, Lucretius argued for the infinity of the universe by entreating us to hurl a spear at whatever we might think to be its boundary. If nothing stops the spear, then there is no boundary; if something stops the spear, then there is something beyond boundary. Either way, the boundary is not a boundary, and the universe must be infinite.[32] Cusa offers a similar line of reasoning with far less fanfare in *De docta ignorantia*, saying that "the universe is limitless [*interminatum*], for nothing actually greater than it, in relation to which it would be limited, can be given" (2.1.97). In other words, because the universe is all that *is*, there cannot be anything outside it to bind it. When one considers, furthermore, that Cusa was the first Western philosopher since Lucretius to assert the existence of extraterrestrial beings and that he was sixteen years old when the papal secretary Poggio Bracciolini rediscovered Lucretius's *De rerum natura*,[33] one might even be tempted to see Cusa's cosmology as some sort of resurrected Atomism. Yet as Steven Dick has cautioned, Cusa's physics is more clearly a reaction against Aristotle than it is a retrieval of Epicurus. Furthermore, although Cusa was at least nominally aware of Epicurean cosmology, he never once cites Lucretius and probably never read the *De rerum natura*.[34] Perhaps even more important, however, aligning Cusa too closely with any of the Atomists would cause us to miss the distinctive character of his cosmology. If the Atomist worlds are an interrelated many and the Stoics' an ever-recurring one, the Cusan universe can be regarded as either one or many or both of them at once—depending on how you look at it.

Nowhere and Everywhere

The first crucial distinction to draw between Cusa and the Atomists is that the Cusan universe is not exactly infinite. It is not finite because, as Cusa states in *De docta ignorantia*, "it lacks boundaries within which it is enclosed" (2.11.156), but neither is it infinite because it is not "from itself." "Nothing is from itself except the simply maximum," which is to say God, who alone is absolutely infinite (2.2.98). This distinction clearly constitutes another major difference from Lucretius, whose gods are finite beings that have nothing to do with the creation or governance of the world. Lucretian worlds, we will recall, can bring themselves to order through the stormlike collisions or vortical swirl of atoms. The Cusan universe, by contrast, cannot get itself together; the source of its being lies beyond itself in God. Because "the universe embraces all things that are not God" (2.1.97), it cannot, strictly speaking, be called infinite. And yet, again, it has no bounds.

To account for this finite sort of infinity, Cusa borrows some terminology from Thomas Aquinas, who in *Disputed Questions on the Power of God* (1265–1266) distinguishes between the "negative infinite which simply has no limit" and the "privative infinite . . . which should have limits naturally but which lacks them."[35] Cusa attributes the former to God and the latter to the universe: God has his reason for being within himself and is thus "negatively" infinite, whereas the universe has its reason for being outside itself and is thus "privatively" infinite.[36] In Cusa's own language in *De docta ignorantia*, God is the "absolute" infinite, whereas the universe is a "contracted" infinite—a concrete and, for that reason, restricted infinite (2.4.113). But this very difference between God and the universe constitutes their inexorable relation: in its contracted infinity, the universe exists as a created reflection of God. Like God, it has no limits; like God (and like Timaeus's "perfect living creature"), it contains everything that is as well as the seeds of what might yet be. "It is as if the Creator had spoken, 'Let it be made,'" writes Cusa, "and because God, who is eternity itself, could not be made, that was made which could be made, which would be as much like God as possible" (2.2.104). Emerging from the very being of God, the universe is the expression (*explicatio*) of the divine enfolding (*complicatio*): as Elizabeth Brient puts it, "a concrete likeness of God unfolded in the diversity and multiplicity of space and time."[37]

This likeness is perhaps nowhere more apparent than in the dizzying geometry of *De docta ignorantia*. As we have already seen, the Cusan universe has neither center nor circumference; rather, it appears to have its center wherever an observer finds herself and its circumference as far as she can see. Our sense of the universe is thus irreducibly perspectival. And yet, Cusa promises, we can visualize the whole—provided we are willing to shatter our spatial sensibilities. "You must make use of your imagination as much as possible," he advises, "and enfold the center with the poles" (2.11.161). The result will be something like a sphere whose center coincides with its periphery. Only if you can picture such an unpicturable thing will you begin to "understand something about the motion of the universe" (2.11.161). Moreover, you will begin to understand the likeness between the universe and its creator, for insofar as God is both omnipresent and boundless, God can be thought of as *an infinite sphere, whose center is everywhere, whose circumference is nowhere*" (2.11.161, emphasis added).[38]

The image of an infinite sphere itself is not unique to Cusa: it had appeared in the work of Alain de Lille, St. Bonaventure, Thomas Aquinas, and Meister Eckhart (from whom Cusa most likely picked it up)[39] to describe the ineffable being of God. What *is* unique is that Cusa is applying what had been a theological metaphor to the creation itself, thereby rendering the universe just as incomprehensible as its creator. "Therefore enfold these different images," he entreats us, "so that the center is the zenith and vice versa, and then . . . you come to see that the world and its motion and shape *cannot be grasped*, for it will . . . have its center and circumference nowhere" (2.11.161, translation altered slightly, emphasis added). Yet even here, we should note, Cusa is careful not *quite* to identify the infinite sphere of the world with the infinite sphere of God (Jorge Luis Borges's Tzinacán will go a bit further four centuries later, conflating "the deity" with "the universe" in a mystical vision).[40] The difference, for Cusa, is that God is a sphere with its "center everywhere," whereas the universe is a sphere with its center *nowhere*.[41] Just as we saw the universe's difference from God secure its resemblance to God, we therefore now see the resemblance ratchet up the difference: both are infinite spheres, but God is omnicentric, whereas the universe has no center at all.

But then, again, can't the universe be said to have as many centers as there are positions within it, and is it not in this sense omnicentric? The issue once again boils down to perspective. The universe has no absolute center *within itself* because there is no body in the universe that is equidistant to each of its "poles." "Precise equidistance to different points cannot be found outside God," Cusa explains, which is to say that God alone is equally proximate to all parts of creation. But insofar as God is equally proximate to all parts of creation, the universe does indeed have a center: "*God* is the center of the earth," Cusa proclaims, "of all the spheres, and of all things that are in the world" (2.11.157, emphasis added). And so in this very particular sense, the center of the world is not nowhere, but everywhere—because God is everywhere. Even as Cusa asserts this principle, however, he adds a qualification—just in case we have missed the context: "[T]he world machine will have, *one might say* [*quasi*], its center everywhere and its circumference nowhere, for its circumference and its center is God, who is everywhere and nowhere" (2.12.162, emphasis added). So just as the universe is finite in one respect and infinite in another, its center is nowhere on its own, but everywhere in God. This quasi-omnicentrism establishes the universe as the strongest possible *imago dei*, a concrete expression of divine being itself. Yet we should note that this likeness holds only insofar as God *occupies* the center(s) of the very universe that resembles God. Cusa, in other words, is shattering the simple mirror game between God and the universe by folding God *into* God's own image, as its omnicentric center. The universe does not resemble a God who stands outside it; it resembles God only insofar as it embodies God everywhere, equally.

It is with this insight that Cusa truly demolishes the graduated cosmos of his Scholastic predecessors. God is not mediated down through the heavenly ranks to the lowly earth at its center; rather, God is directly present to every part of the boundless universe. As Cassirer explains it, "[T]here is no absolute above and below, and . . . no body is closer or farther from the divine, original source of being than any other; rather, each is 'immediate to God.'"[42] There is no privileged place in the universe, no distinction between the astral and sublunar spheres. And so the order of things is not a static hierarchy under an extracosmic divinity; instead, it is a dynamic holography in which God is fully and equally present to everything in creation. This radical indwelling is, for Cusa, what it means to make a world in the first place: "[C]reating," he ventures in *De docta ignorantia*, "seems to be not other than God's being all things" (2.2.101).

If all things exist as the image of God, then it is not the case that God is mediated by some things (intelligences, Reason, Man) to other things (matter, the passions, women, and nonhuman animals). Rather, Cusa writes, "God communicates without difference or envy" so that every creature becomes a "perfect" image of God: "[E]very creature is, as it were, a finite infinity or created god, so that it exists in the way in which this could best be" (2.2.104). Precisely because God immediately communicates Godself to every creature, however, every creature also mediates God to every other creature. Because God is in each thing as the being of each thing, everything mediates God to everything. And insofar as "everything" as such is the universe itself, Nicholas suggests that "God is in all things as if [*quasi*] by mediation of the universe" (2.5.117). "As if" by mediation of the universe, it is not just the case that God is in all things and all things are in God, but also that "*all things are in all things*" (1.15.118, emphasis added).

And yet . . . and yet—with Cusa, one must always say "and yet." All things are in all things, and yet all things are not identical. No created thing is identical to any other because, again, pure equality belongs to God alone. Therefore, although each thing has the same essence (*quidditas*) as all other things, each thing participates in that essence differently (1.17.48). Thus the things of creation are at once radically co-implicated and radically particular: everything shares the essence of God, yet every thing is irreducibly itself. By a similar logic, while God is equally present to all parts of the universe, God also remains distinct from the universe.[43] Creation is the expression of God, the contraction of God, the holographic dwelling place of God, and yet creation is not God. Cusa often expresses this difference as a matter of number: whereas God is unity itself, the universe is unity "contracted in plurality" (1.2.6).

At this point, it might be helpful to recall Thomas Aquinas's concern that a plurality of worlds would compromise the singularity of their creator. If there were more than one world, he argued, then they would be ordered to different ends. If there were an infinite number of worlds, then they would have no end at all. Because Thomas's God is the one end of all things, all things must occupy this one world. As we have just seen, however, Cusa departs radically from this theocosmology, figuring God not as an extracosmic point, but as the omnipresent center, and the world not as a hierarchical chain, but as a holographic web. Insofar as God can be called an end for Cusa, the notion of "end" is therefore not at all opposed to infinity. "My God," he exclaims in *De visione dei* (*On the Vision of God*, 1453), "you are absolute infinity itself, which I perceive to be the infinite end, but I am unable to grasp how an end without an end is an end."[44] Considering, then, that the "end" of "all things in all things" is an "end without end," how would Cusa answer the perennial question of whether there is one world or many?

A Perspectival Multiverse

At first blush, the answer seems fairly straightforward. If the earth is just another star, then there are a vast number of "earths," each of them inhabited and occupying the center of its own world. In this sense, Cusa's one universe can be said to be composed of a plurality of worlds—perhaps even an infinity of them, considering the universe is spatiotemporally boundless.[45] Cusa's cosmology would thus look like a Christianized Atomism: a staggering plurality of worlds, plus God, minus the unqualified "infinity."

And yet even if we put both Cusa's God and his contracted infinity to one side, the Cusan *mundi* differ significantly from the Atomist *kosmoi* when one considers the nature of their relation to one another. The interaction between worlds in Democritus, we might remember, was limited to their colliding with and destroying one another. Their atoms would then recombine to form new worlds, which would live until they crashed into other neighbors nearby. For Epicurus, this type of interaction disappears because cosmic destruction comes about through gradual decay rather than through collision. As we might remember, however, this very propensity to decay renders each world the product of other worlds and the material for others still, and in this sense Epicurean worlds bear the physical imprint of others. That having been said, Epicurean worlds never interact with one another while they are alive; rather, they drift by themselves through an infinite void, living for a time and then dying.

Cusa's "worlds," by contrast, not only interact with one another but compose one another—simultaneously. The crucial image of this real-time intercomposition can be found in one of Cusa's meditations on the sun. To support his claim that the earth is a star like any other, he speculates that the earth must look like a sun from far away. Likewise, the sun must not look bright to its own inhabitants, who probably see their dwelling place much as we see our own earth. After all, each heavenly body occupies the center of its own world, and here Cusa figures each "world" in surprisingly Aristotelian (but typically qualified) fashion. Just as our earth is surrounded by layers of water, air, and fire, he argues in *De docta ignorantia*, "the sun's body, on examination, is discovered to have, *as it were* [*quasi*], a more central earth, a fiery brightness, *as it were*, along its circumference, and in between, *as it were*, a watery cloud and brighter air" (2.12.164, emphasis added). The reason the sun looks so bright to us is that we are outside its "region of fire" and so cannot see its inner elemental regions. By that same principle, "if someone were outside [our] region of fire, this earth of ours would . . . appear to that individual as a bright star." Cusa goes on to speculate that "if the moon does not appear so bright to us, it is perhaps because we are within its watery region, *so to speak*" (2.12.165, emphasis added). The Cusan universe is therefore not a disconnected proliferation of worlds, but an overlapping, interconstituted set of them. There is no void between worlds—nor is there any clear separation between them. There is no "ours" floating along independently of "theirs." In fact, to push Cusa a bit, one might even say that each cosmic body inhabits many different worlds at the same time: our earth, for example, occupies the "center" of its own world, the midranges of others, and the peripheries of others still. As it were.

Then, again, it is precisely this inextricability among "worlds" that leads Cusa at one point to *deny* any cosmic plurality, calling creation "a single universal world [*unus mundus universalis*]" (2.12.172). As he explains it, every thing of creation participates in the being of God through the mediation of every other thing. Cusa might therefore say along with William James that "our 'multiverse' still makes a 'universe'" because the order of these "strung-along" and interdetermined worlds is one in its multicentricity.[46] The way Cusa ultimately puts it is that "the creation as creation [*creatura ut creatura*] . . . cannot be called 'one,' since it descends from unity, nor 'many,' since it takes its being from the One, nor both 'one and many' conjointly" (2.2.100). Rather, like everything else in the Cusan landscape, the creation can be said to be one thing from one perspective and another from another: creation *as creation* is many, whereas creation in God is one. From the perspective of any world, there are many worlds; from the perspective of the one God, the world is one.

That having been said, the oneness of God is different than we might think, by virtue of the one God's also being three. As Cusa understands the doctrine of the Trinity, God is at once Unity, Equality, and Connection (Unitas, Aequalitas, Nexus), eternally generating unity from unity (as equal to itself) and holding this unity in unified relation to unity (as the connection between them) (1.7.20–21, 1.9.24). As such, God cannot be said to be one as opposed to many, identity as opposed to difference, or unity as opposed to multiplicity. God is, rather, the radical co-implication of all of these, a "oneness" that "precedes all opposition" (1.24.77). The unity of God is therefore not different from plurality, but a "unity to which neither otherness nor plurality nor multiplicity is opposed" (1.24.76).

To say that creation is "one . . . since it takes its being from the One" (2.2.100) is therefore *not* to say that creation is not also many, because the One *itself* is not "not many." This is the reason Cusa dedicates a full chapter to elaborating the Trinitarian operations of the universe itself: as the created expression of God, the universe generates all things through a triune dance of potentiality, actuality, and the connection between them (2.7.127–31). As the image of God, the universe is therefore both one and many, even though as creation it can be only *either* one or many (2.2.100). The answer to the question of the number of creation is therefore that there is one world and there are many worlds and that there is either one world that is not opposed to manyness or many worlds held together in oneness. Any way you look at it, the universe is neither simply one nor simply many, but a complex co-implication that shifts according to your vision. What Cusan cosmology comes down to, what it opens up, is something like a "perspectival multiverse."

Postscript

Infamously, the Roman Catholic Church would not share Cusa's vision of the perspectival multiverse. A mere century and a half later, the whole idea would be obliterated along with the former Dominican Giordano Bruno, who, as the story goes, was burned at the stake for professing the same cosmic ideas as Nicholas of Cusa.[47] We will turn to this story and its complications in a moment, but it is important to note that the Cusan controversy ignited immediately after the publication of *De docta ignorantia*, when the Heidelberg professor Johannes Wenck issued a lengthy accusation of heresy against its author.[48] Calling his refutation *De ignota litteratura* (*On Ignorant Erudition*, 1442–1443), Wenck rails against each of Cusa's positions—from his method of learned ignorance ("[H]ow . . . are we to understand incomprehensible things incomprehensibly?") to his departure from Thomas's Aristotelian God ("[T]his deduction destroys the Prime Mover!").[49] But each of these epistemological and theological errors stems, for Wenck, from Cusa's having destroyed the hierarchical Scholastic cosmos. By setting the earth in motion and relating all things immediately to God, Cusa allegedly "deifies everything, annihilates everything, and presents the annihilation as deification."[50] Wenck's stated concern is that Cusa is a pantheist, that he has destroyed the difference between the creator and creation. As we have seen, however, Cusa takes pains to maintain this distinction. The real problem seems to be that by flattening out the order of mediation, Cusa's cosmology threatens to do the same to ecclesiastical authority. After all, if all things participate directly in God, then what use are the professional ranks of bishops, priests, and deacons? This is part of the reason Cusa's ideas would become so inflammatory in 1600 as the Roman Catholic Church sought to regain control over the Protestant disaster, and it certainly lies behind Wenck's distaste for them. Whether in the seventeenth century or the fifteenth century, it is no surprise that an authoritarian doctor of the church might recoil from Cusan cosmology.

What is far more surprising is Cusa's *own* departure from his perspectival multiverse. In the works that follow the *De docta*, Cusa barely mentions cosmology at all, instead offering varied meditations on the soul's ascent to God.[51] Although these works refrain from addressing the constitution of the universe directly, they have striking, perhaps disappointing, cosmological implications. One year before his death, for example, Cusa published a treatise called *De venatione sapientia* (*The Hunt for Wisdom*, 1463), which likens the soul to a hunter and God to its prey.[52] Many of the familiar elements are here—from the quest to know an unknowable God to the language of contraction to the elaboration of the Trinity. There are some developments as well: Cusa has begun to call God "Non Aliud" (Not Other) to mark God's transcendence of identity and difference and "Possest" (Possibility to Be) to mark God's transcendence of actuality and possibility.[53] Perhaps most dramatically, *De venatione* introduces a brand-new category, which Cusa calls *posse fieri* (the possibility to become). Because God is being rather than becoming (a distinction that Cusa either did not make or just did not mention in *De docta*), Cusa explains that God must make the *possibility* to become before making anything that becomes. *Posse fieri* is therefore "the first and greatest of all creatures": the stuff out of which God then creates everything else.[54]

The difference, then, is that whereas *De docta* figures the universe as the created expression (*explicatio*) of God, *De venatione* inserts an ontological category between them. And, coincidentally, the universe that God creates out of *posse fieri* is no longer the destratified proliferation of perspectives that explicated the infinite sphere. Rather, it is a bizarrely re-Scholasticized cosmos. The *posse fieri* itself takes the form of the heavenly bodies, which Cusa now calls "intellectual natures" and which are made of an incorruptible substance at the circumference of the cosmos.[55] Our earth, made of finite, sublunar materials, has been carefully placed at its center.[56] And in a book that Cusa wrote in the same year, a character called "the Cardinal" tells the duke of Bavaria—with no complication or qualification, no "as it were" or "so to speak"—that because the world is almost perfectly round, "there will . . . only be one world."[57] So the late-Cusan cosmos becomes a single sphere with stars at the periphery and the earth at the center. After all that.

What was it that prompted Cusa to retreat so severely from his perspectival multiverse? Was it fear of other Wencks? Something about the *posse fieri*? Intellectual exhaustion in the face of a multitude that was nevertheless one? One will likely never know. And yet, as Cusa consoles us, "one will be the more learned, the more one knows that one is ignorant" (1.1.4).

Infinity Unbound: Giordano Bruno

Copernican Convulsions

The Wenck affair aside, Cusa's era will turn out to have been one of comparative intellectual freedom—one in which it was possible to assert the mobility of the earth, for example, without being sentenced to a lifetime of house arrest, as was Galileo Galilei, or to death, as was Giordano Bruno (1548–1600). The case of Bruno is particularly striking, for, as we shall see, his cosmology was far more radical than Galileo's and in large measure directly imported from Cusa. And although Cusa's *De docta ignorantia* generated some controversy, it was never officially interrogated or condemned; to the contrary, Cusa was made a cardinal eight years after its publication. Retrospectively, one can therefore locate his multiversal musings in a strange space of quiet before the convulsions of the sixteenth century, when a series of anthropological, scientific, and ecclesiastical upheavals prompted a protracted inquisitorial frenzy on the part of a besieged Christendom. It was in large part this frenzy that made impossible for Bruno in 1600 what had been possible for Cusa in 1440. In the context of the Copernican Revolution and of the Counter-Reformation, Bruno's neo-Cusan multiverse posed such a direct and coherent threat that it led to his being burned at the stake in Rome's Campo de' Fiori on an Ash Wednesday at the turn of the seventeenth century.

All this notwithstanding, historical differences alone cannot account for Cusa's and Bruno's wildly divergent fates. For one thing, the two men seem to have had vastly different personalities. That which is carefully qualified in Cusa is loudly proclaimed in Bruno, the philosopher from Nola who is said to have "preached" his boundless cosmology "with the fervor of an evangelist"[58] and who made use of anatomical jokes, elitist jabs, narcissistic flourishes, and a "rich repertory of Neapolitan curses" in order to do so.[59] Far more important, the teachings of the Cusan and the Nolan were not in fact the same. Bruno did import most of his cosmological principles from the man he called "the divine Cusanus,"[60] but he went on to make some subtle alterations that would have profound theological consequences—most significantly concerning the divinity of Christ, the doctrine of the Trinity, and the eternity of creation. Strictly speaking, it is these theological "errors" rather than the cosmological teachings themselves that eventually led to Bruno's execution. But as we shall see, Bruno's heretical theology was not unconnected to his multiversal cosmology; in fact, each was the product of the other.

Like Cusa, Bruno held that "the earth is not in the centre of the universe," insisting that it revolves around the sun.[61] Of course, Bruno's most immediate predecessor with respect to this principle was not Cusa, but Nicolaus Copernicus (1473–1543), who had set forth a heliocentric model of the universe in *De revolutionibus orbium coelestium* (*On the Revolutions of the Heavenly Spheres*), published just before his death. In the throes of the ensuing controversy—at this point, more academic than ecclesiastical— Bruno offered *The Ash Wednesday Supper* (1584) in part as a defense of Copernicus, "a man of deep, developed, diligent and mature genius," against his contemporary detractors: a mainly Oxonian throng of "Peripatetics who get angry and heated for Aristotle."[62] Aristotle, Bruno charged, had kept philosophy imprisoned for centuries within a bounded cosmos, a notion that would have baffled earlier teachers such as Democritus and Pythagoras. Bruno therefore interpreted the Copernican "revolution" as something more like a restoration—a return to the more authentic teachings of the pre-Socratics. In Bruno's characteristically baroque turn of phrase, Copernicus was "ordained by the gods to be the dawn which must precede the rising of the sun of the ancient and true philosophy, for so many centuries entombed in the dark caverns of blind, spiteful, arrogant and envious ignorance."[63] But we should note that Bruno's praise of Copernicus here is carefully circumscribed: Copernicus was the *precursor* to a revival of the ancients— the dawn before the sun rather than the sun itself. And the reason Copernicus remained confined to the dawn was twofold: his retention of a finite cosmos and his failure to philosophize.

For all the controversy that Copernicus's heliocentric model generated, Bruno charged that it did not depart all that radically from the geocentric model because it still retained a motionless center and periphery. Of course, the body at the center had changed—Copernicus put the sun where the earth had been—but the so-called fixed stars remained firmly in place, making the heliocentric model as much a series of nested circles as the geocentric model. Bruno offers his most memorable elaboration of this symmetry in a diagram in *The Ash Wednesday Supper*, which holds the Copernican system within the same stellar boundary as the Ptolemaic one (figure 3.2). In a sense, however, Copernicus's boundary is not quite the same as Ptolemy's, for in the geocentric model the fixed stars make one full rotation around the earth every twenty-four hours, whereas in the heliocentric model they stand still while the earth turns. The Copernican cosmos might therefore seem even *more* bounded than the Ptolemaic cosmos, with the stars truly "fixed" in the heavens. Ironically, however, this fixing of the fixed stars had the opposite effect, opening the heliocentric model out to infinity by undermining Aristotle's only demonstration of cosmic finitude.

This demonstration, we will recall, begins from the premise that "the body which moves in a circle must necessarily be finite."[64] Cosmologically, this means that if the stars at the edge of the universe move in a circle around the earth, the whole universe must be finite. Of course, this reasoning works as long as the earth is thought to occupy the center of the universe. But if the *sun* occupies the center of the universe, then the stars must stand motionless as the earth rotates on its axis. Because the stars of the heliocentric model do not move in a circle, they cannot assure us of the finitude of the cosmic "circumference." And if the circumference cannot be said to be finite, then it cannot really be said to be a circumference at all. The Copernican universe might, in other words, be infinite.

The first Copernican to ascribe infinity to the heliocentric universe was the English mathematician and astronomer Thomas Digges (ca. 1546–1595), whose translation of the *De revolutionibus* in an encyclopedia appendix (1576) features a diagram in which the stars, rather than being confined to a thin ring around the cosmos, are extended out indefinitely (figure 3.3). As we can see, there is still only one world at the center of Digges's infinite universe—only one solar system, surrounded by an endless, starry sky.[65] This conclusion, however "Copernican" it might have been, nevertheless had proved a bit too much for Copernicus himself to handle. The astronomer was certainly aware that an immobile circumference opened the door

FIGURE 3.2 The symmetry between the Ptolemaic and Copernican models of the universe. (From Giordano Bruno, *The Ash Wednesday Supper/La cena de le Ceneri* [1584], trans. Stanley L. Jaki [Paris: Mouton, 1975], 140, fig. 7. Reproduced by permission of Walter de Gruyter GmbH)

to an infinite universe, "for the strongest argument by which they try to establish that the universe is finite is its motion." Nevertheless, he preferred not to think too much about it, proposing to "leave the question whether the universe is finite or infinite for the natural philosophers to argue."[66] As far as Bruno was concerned, this was the real reason that Copernicus stopped short of overturning "vulgar" Aristotelian cosmology once and

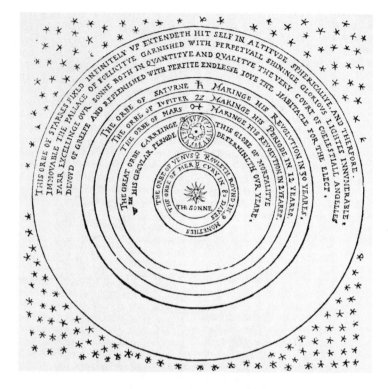

FIGURE 3.3 The caption to Thomas Digges's illustration of the Copernican universe reads: "This orb of stars fixed infinitely up extendeth itself in altitude spherically and therefore immovable, the palace of felicity garnished with perpetual shining glorious light innumerable far excelling our sun both in quantity and in quality, the very court of celestial angels devoid of grief and replenished with perfect endless joy, the [home] for the elect" (quoted in John D. Barrow, *The Infinite Book: A Short Guide to the Boundless, Timeless, and Endless* [New York: Vintage, 2005], 118). (From Thomas Digges, "A Perfit Description of the Caelestial Orbes," in Leonard Digges, *A Prognostication Everlasting* [London, 1576], fol. 43. Reproduced by permission of the Huntington Library, San Marino, California)

for all—because he was too much of a mathematician and not enough of a natural philosopher.[67]

Cosmological Provocations

If Bruno confined Copernicus to the "dawn" of the ancient-new cosmology, he associated his own teachings with the rising of its sun. It was "the

Nolan," declared the Nolan, who finally had the courage to do away with the sphere of the fixed stars and assert the cosmic infinity that Copernicanism implied. It was the Nolan who had "freed the human mind" by breaking the bars of its Aristotelian prison. "Now behold," says one of Bruno's characters of his own creator, "the man who has surmounted the air, penetrated the sky, wandered among stars, passed beyond the borders of the world, effaced the imaginary walls of the first, eighth, ninth, tenth spheres, and the many more you could add according to the tattlings of empty mathematicians and the blind vision of vulgar philosophers."[68]

If this hymn of praise sounds familiar, it is because Bruno had found a "living teacher" in Lucretius, whose work had been back in circulation for more than a century. We might recall that the *De rerum natura* lauded Epicurus as a god for having passed "beyond the fiery ramparts" of the cosmos and out to an infinite number of worlds.[69] In Bruno's work, the god becomes Bruno himself—the only man of his age to travel beyond the false confines of Aristotelian circles and proclaim that "the universe is of infinite size and the worlds therein without number."[70] Again, this conclusion was a direct consequence of the Copernican turn, but no one had yet had the courage to give it any thought. Such thought had been discouraged, above all, by the preface to the first edition of Copernicus's *De revolutionibus*, which announced that the book was of strictly mathematical importance and that it made no truth claims about the actual constitution of the universe.[71] Bruno was the first person to insist that this preface could not have been written by Copernicus himself but was "added by I know not what conceited ass."[72] Even though he could not name the ass who had written it (Johannes Kepler would eventually discover him to be the Lutheran theologian Andreas Osiander), Bruno's rejection of the preface cleared away the chief obstacle to philosophical reflection on the Copernican hypothesis. As Hilary Gatti has summarized the matter, "Bruno attempted to pilot a recalcitrant sixteenth-century public, convinced of the falsity of the Copernican hypothesis except within a strictly mathematical formulation of it, toward a realist acceptance of the heliocentric principle, together with much else that Copernicus himself would not have been prepared to accept."[73]

All told, the "much else" that Copernicus would have resisted amounts to a kind of Epicurean-inflected, pseudo-Cusan cosmology. Bruno posits an infinite universe filled with innumerable worlds that move in a "void." As a corrective to Aristotle, who denied the existence of an absolute vacuum, Bruno explains that the void is not nothing; it is simply the space in which worlds come to be.[74] Like Plato's *khôra*, Bruno's void is the "bosom in which the whole has its being." Of the infinite worlds within this infinite bosom, many "contain animals and inhabitants no less than our own earth [does],"[75] rendering our planet just one of countless others like it. It follows that neither the earth nor the sun nor any other cosmic body occupies the center of the universe; rather, as Cusa suggested, the "center" is wherever one happens to be in the universe. Thus, according to Bruno in *Cause, Principle, and Unity* (1584), "we may affirm that the universe is entirely center; or that the center of the universe is everywhere, and the circumference nowhere insofar as it is different from the center."[76]

Insofar as this infinite sphere has no fixed center or periphery, there is no distinction between the sublunar and astral spheres; "consequently, that beautiful order and ladder of nature is but a charming dream, an old wives' tale."[77] As many commentators have explained, the Brunian universe is therefore "homogeneous" rather than hierarchical; no place within it is either geographically or ontologically privileged.[78] That said, this "homogeneity" does not amount to indifference; elements of the cosmos are differentiated from one another, as they are in Cusa, by virtue of each thing's particular capacity to contain the whole. And insofar as this is the case, insofar as "this universe . . . is found entire in each of its parts," one can go so far as to say that "all things are in all things" *and* that "all things are in each thing."[79] It is as if every point in the cosmos were Borges's "aleph": "the place where, without admixture or confusion, all the places of the world, seen from every angle, coexist."[80] For Bruno, all things are both radically particular and radically co-implicated by means of the grand cosmic holography in which everything participates.

It may therefore strike us as strange that Bruno thought he had made such an unprecedented departure from the cosmological stupidity of his age. After all, each of his principles seems to have been lifted directly out of Cusa. Indeed, Bruno admits in *On the Infinite Universe and Worlds* (1584) that "the Cusan seems to have approached" the teachings he himself now proclaims. But the Cusan fell short, Bruno maintains, because he failed to distinguish between types of cosmic bodies.[81]

For Cusa, we will recall, the earth is a star like any other; to the inhabitants of the sun, the earth looks just as bright as the sun does to us. Conversely, every cosmic body is an "earth" at the center of its own world, surrounded by progressive rings of water, air, and fire. Bruno changes this model considerably by making what might seem to be a subtle distinction between "hot" and "cold" bodies, which is to say between stars and planets. Although all cosmic bodies move, these two classes of bodies move "differently," with the watery, cold ones orbiting the fiery, hot ones. This means that Bruno's worlds, unlike Cusa's, are not centered around any given cosmic body; rather, they are centered (in good Copernican fashion) around a *sun*. For Bruno, to say there are an infinite number of worlds is therefore to say there are "innumerable suns, and an infinite number of earths [that] revolve around those suns, just as the seven we can observe revolve around this sun which is close to us."[82] Worlds, in other words, are solar systems.

Unlike Cusa's worlds, then, Bruno's do not overlap: they look far more like Epicurean *kosmoi* than Cusan *mundi*. As Miguel Granada has explained, each Brunian world is "separated from the adjacent ones by a vast extension of space filled with pure air or ether," terms that Bruno used interchangeably with "void."[83] This separation allows Bruno to avoid the Aristotelian claim that the "earth" of a hypothetical other world would be drawn to "our" center as well as to its own; for Bruno, the distance between worlds is so great that cosmic bodies can only "move toward communication with convenient things," having no bearing on things far away.[84] Each system thus exists in a self-sustaining equilibrium, with its sun and planets moving "according to an intrinsic principle" that Bruno calls the "soul" of each cosmic body.[85] By attributing an animating principle to each planet and star, Bruno eliminates any need for a prime mover—not to mention for a "first moved" sphere of fixed stars that would confer movement on the rest of the cosmos. This is not to say that Brunian cosmology has no need for a divine principle, but that Bruno's God acts through *immanent* processes of motion, "giving the power to generate their own motion to an infinity of worlds."[86]

But, again, the bodies within Bruno's infinity of worlds move only in relation to the bodies closest to them. As they so move, Granada explains, they exchange particles of matter with one another, "the stars emitting light and heat and the planets emitting humid exhalations" so that the life of each system is constantly renewed.[87] And with this renewal, Bruno swerves away from Epicurus and back toward Plato. Securing the self-containment of each system allows Bruno to ensure their eternity, skirting the Democritean collision, the Epicurean demise, *and* the Stoic cataclysm all in the same gesture. Worlds will never come close enough to collide with one another; every earth constantly receives new life through its exchange with the sun; and the sun is so tempered by water that it will never consume the cosmos in flames. As they do in the *Timaeus*, then, the elements sustain a constant, harmonious relation that ensures the eternity of the creation they compose. The only significant difference is that Bruno's elemental dance takes place within an infinity of *kosmoi*, each of which is just as everlasting as the universal whole.

In *On the Infinite Universe and Worlds*, Bruno goes on to deploy this process of elemental exchange to rid himself of Cusa's one Aristotelian remnant: the elemental striation of *kosmoi*. Worlds are not layer cakes of earth, water, air, and fire, Bruno argues; they are "heterogeneous bodies" composed of a strikingly Timaean sort of mixing. Because Bruno's worlds inhale and exhale particles of matter, they are like animals, "great globes in which earth is no heavier than the other elements," but is intermingled with them. The same goes for all innerworldly things; although one element—fire, for example—may *predominate* in any given body, it exists in tensional dependence with all the others: "fire" is not fire without the "earth" it consumes and the air that contains it. So there can be no "outer" sphere of pure fire. We might say that with this destratification of worlds, Bruno radicalizes Cusa's antihierarchical gesture even further, showing even "the famous order of the elements . . . to be in vain."[88]

But Bruno's most dramatic cosmological departure from Cusa lies in his demolition of the boundary between absolute and contracted infinity. For Bruno, the universe is not "finitely" or "privatively" infinite, nor is it infinite from one perspective and finite from another. The Brunian universe is just infinite—*genuinely infinite*—both in power and in extension. To be fair, in *On the Infinite Universe* Bruno does make one fairly perfunctory qualification to this designation, conceding that the universe, unlike God, is not "all-comprehensive infinity" because it, unlike God, is "comprehensively in the whole but not comprehensively in those parts which we can distinguish within the whole." From this statement alone, we might think that Bruno is adopting Cusa's distinction between "absolute" and "contracted" infinity. But immediately after making this qualification, Bruno goes on to undermine it, pointing out that the "parts" of the universe cannot actually be called parts at all "since they pertain to an infinite whole."[89] If the universe is infinite, he insists, then each of its "parts" would have to be infinite. As Bruno explains in *Cause, Principle, and Unity*, this means that an infinite universe can only be "found entire in each of its parts"— just like God.[90] This, then, is the cosmological departure from Cusa that will open the theological floodgates: Bruno collapses the distinction between the boundless universe and its infinite God.

Theological Consequences

For Cusa, we will recall, the relationship between God and creation is secured only by their distinction; it is precisely because the universe is not God that it can be called God's created image. Whereas book II of *De docta ignorantia* explains this relationship in predominantly cosmological terms (the universe is the contracted expression of an uncontracted divinity; the universe is divine unity unfolded in the multiplicity of space and time), book III, which I have not yet addressed, shifts to a strictly theological account. In this last third of his treatise, Cusa sets forth an extended systematic Christology, proclaiming that the true bridge between the infinity of God and the finitude of creation is the incarnation of God in the person of Jesus Christ. Insofar as Christ is both infinite and finite, he enables creation to participate in that from which it is ontologically different, bringing into relation entities between which there would otherwise be "no proportion" (3.4.203–7).[91]

It is precisely this lack of proportion that Bruno abolishes when he attributes full infinity to creation: for Bruno, there is no irreducible difference between the universe and God. This does not mean that Bruno's universe *is* God—at least not exactly; it means that the universe issues forth *from* God and that it does so immediately, without the help of Christ, the *posse fieri*, or anything else. The Brunian universe is the direct, outward expression of an infinite God, and this is the reason it must itself be infinite. The universe is infinite because the creator holds nothing back, conferring on the universe everything God has, which is everything God is. The universe, in short, is not the contracted image of God, but the *unmediated outpouring of God*; it is what Hans Blumenberg has called "the essential undisguisedness of divinity itself."[92] There is therefore no need for a mediator—no need for a Christological bridge between God and the universe—because nothing has ever separated them to begin with. So Bruno's accusers were not exactly wrong when they charged him with denying the divinity of Christ;[93] Bruno must have thought it strange to make such a fuss over the manifestation of God in one man when God was already manifest in the whole universe. What seems to have been lost on the inquisitors, however, is that Bruno does not abolish Christology here so much as he cosmicizes it: insofar as the creation flows forth eternally from God, the universe is *itself* God's "Son," eternally begotten of the Father.[94] There is, in other words, no distinction between God's "inner" expression in the persons of the Trinity and God's "outer" expression in creation: creation is the "God from God" and "light from light" that orthodoxy calls the Word of God. Consequently, there is no gap between the eternity of God and the temporality of creation, nor is there a gap between the moment of creation and the event of incarnation. Rather, the universe is the outpouring of God from the beginning, and *creation itself is the incarnation.* The whole universe is God-in-the-universe.

And so to the old Scholastic question of whether an omnipotent God could create more than one world, Bruno answers that God cannot *not* do so. "If in the first efficient Cause there be infinite power," he reasons, "there must result a universe of infinite size and worlds infinite in number."[95] This statement is among the teachings that would run him into so much trouble with the Inquisition—not because the church denied that God *could* create a plurality of worlds (recall Bishop Tempier's Condemnations), but because it balked at Bruno's insistence that God *had* to do so. As the mathematician-theologian Marin Mersenne (1588–1648) would charge a quarter-century after the Nolan's execution, Bruno's doctrine of an eternal and ineluctable creation compromises the creator's free will, "reducing God to the rank of a natural and necessary agent."[96] The only way to avoid this indignity, Mersenne insisted, would be to return to that late-medieval solution and teach, as Granada puts it, that "God can do infinite things which he nevertheless does not will to do,"[97] that God restrains himself from creating the infinite universe of infinite worlds that he very well could have created. To suggest otherwise—to say, for example, that God is bound to create everything God can create—would be to suggest that God cannot quite control Godself: that the universe pours out of its creator eternally, even womanishly, and there is nothing God can do about it.

Although Bruno would not live long enough to refute this argument in person, his thoughts on the matter are just the opposite. As far as Bruno can see, the insult to God's omnipotence comes not from an infinite universe, but from a finite one: it is a derogation of God's creative power, he insists, to teach that God creates anything less than God might create. This teaching is also a derogation of divine goodness, for why would a benevolent God refrain from creating as many good things as possible? In fact, as Bruno's mouthpiece Philotheo points out in *On the Infinite Universe and Worlds,* to say that God creates anything less than an infinite universe is effectively to say that God creates nothing at all, "for every finite thing is nothing in relation to the infinite." The infinite God is therefore "glorified not in one, but in countless suns; not in a single earth, a single world, but in a thousand thousand, I say in an infinity of worlds."[98] Finally, contra Mersenne, Bruno considers the insult to divine masculinity to come from a restricted creation rather than from an infinite one. "Why do you desire that [divinity] should remain grudgingly sterile," Philotheo asks, "rather than extend itself as a *father*, fecund, ornate, and beautiful?"[99]

As is evident from this string of adjectives alone, however, Bruno's account of divine fatherhood is remarkably queer, unsettling boundaries of sexual and ontological difference alike. The language of "ornate fecundity" aside, we might note that the passage we have just read hinges divine fatherhood on God's being "extended" into the infinite universe. To call the universe finite, Bruno says, is tantamount to calling God "sterile." This view squares with what we already know of Bruno's theology: God creates eternally, and so there is no "God" without the creation that expresses God exhaustively. This means there are no strictly intratrinitarian relations—no "Father" or "Son" gazing at each other before or beyond the universe. Rather, the "only-begotten Son" is creation itself. This means that God can be said to be "Father" only in relation to the infinite universe and worlds that God creates. The fatherhood of God, in other words, depends on creation's materiality and multiplicity—terms that have traditionally been coded as feminine.

Bruno tackles this coding most directly in *Cause, Principle, and Unity* by putting it in the mouth of his least intelligent and most Aristotelian interlocutor, Poliinnio. During a conversation about the nature of "primary matter," Poliinnio explains that Aristotle "compares it to the female sex—that sex, I mean, which is intractable, frail, capricious, cowardly, feeble, vile, ignoble, base, despicable, slovenly, unworthy, deceitful, harmful, abusive"—the list, in true Brunian fashion, goes on.[100] The connection between matter and women lies, for Poliinnio, in the notion of "appetite": just as women's desire brings about the ruin of men, matter's "desire" renders corrupt and perishable the form that would otherwise be perfect and eternal. And so two pages later, Poliinnio concludes that "one must condemn the appetite of both women and matter, which is the cause of all evil, all affliction, defect, ruin and corruption" (246). Although the character Gervasio spends a good amount of time ridiculing this position ("this person has given me a pain in the head with his comparison between matter and woman"), it is the twin characters Philotheo and Theophilo, "reliable reporter[s] of the

Nolan philosophy," who set out to break the whole network of associations on which the position relies (85–86, 101).

They accomplish this by means of a two-part strategy that might be called proto-deconstructive. The first part gets under way in the first dialogue, when Philotheo reverses the ancient "table of opposites" that the *Metaphysics* attributes to Pythagoras. According to Aristotle, the Pythagoreans divide all cosmic principles as follows:

limited	unlimited
even	odd
one	many
right	left
male	*female*
still	moving
straight	bent
light	darkness
good	bad
square	oblong[101]

As the feminist philosopher Grace Jantzen has glossed this infamous table, "[T]he left-hand column was orderly, good, and masculine, while the right-hand column was chaotic, bad, and feminine. In the thought of Plato . . . the items in the left-hand column were further associated with mind and spirit, and the right-hand column with body and matter. Women, who were seen in terms of their reproductive function, were thus conceptually linked with matter and with the chaotic and evil, while men were linked with reason, spirit, and form."[102]

Bruno's Philotheo begins to unravel this whole two-column scheme, with all its attendant associations, by aligning the characteristics that are usually coded "female" with maleness and vice versa. To make matters more confusing, he introduces this set of upside-down associations by ascribing them to the Aristotelian Poliinnio. Looking to reason Poliinnio out of his "slanderous rage" toward women, Philotheo tells him, "You hold, on the one hand, the body, masculine, to be your friend, and the soul, feminine, your enemy. On the one hand, you have chaos, masculine, and on the other, organization, feminine" (32). Of course, apart from the "friend" and "enemy" designations, this is precisely the *opposite* of what Poliinnio holds, for if, as he believes, "prime matter" is female, then the body and chaos must be female as well—in accordance with the usual alignment of attributes. Nevertheless, Philotheo continues unhindered, perhaps hoping to confuse his rather sluggish opponent. Running down all the usual attributes, but backward, he lists on the "masculine" side "error, on the other [feminine side], truth; here, imperfection, there, perfection; here, hell, there, happiness; on this side, Poliinnio the pedant, on the other side, Poliinnia the Muse. In short, all the vices, imperfections, and crimes are masculine, and all the virtues, merits, and goodnesses are feminine [*Finalmente tutti vitij, mancementi, et delitti son maschi: et tutti le virtudi, eccellenze, et bontadi son femine*]" (32).

Once Philotheo has inverted these associations, Theophilo goes on to subvert them, crossing so many categories that the old table collapses altogether. For example, Philotheo has rather cheekily asserted that materiality is masculine (and therefore corruptible, changeable, etc.), whereas immateriality is feminine (and therefore incorruptible, unchangeable, etc.) (32). As we may notice, this particular assertion may rupture the traditional associations of materiality with femininity and of femininity with corruption, but it preserves the traditional association of *materiality* with corruption. This last link will be broken in the fourth dialogue, when Theophilo explores the relationship between matter and form. Making an uncharacteristic and formidable departure from Plato, Theophilo argues that form (much like God) does not exist in some extracosmic realm before it gives shape to matter. On the contrary, form without matter cannot exist at all, "for when form is separated from matter it ceases to exist, as is not the case with matter" (86). For example, when a clay pot falls off a shelf, the form of the pot ceases to exist, but the material of the clay remains. Or when a log is thrown on a fire, the form of the log is lost, but the matter is transformed into particles of smoke, air, and ash. Form, in other words, comes and goes. But matter is eternal.

It is therefore not the case that matter "desires" form in purportedly feminine fashion, waiting for it to bring her to life. Rather, "since matter clearly preserves form, *form must desire matter* in order to perpetuate itself, and not the other way around" (86, emphasis added). Thinking in line with Philotheo's initial reversals, we might be prompted at this juncture to think that Theophilo is associating matter with masculinity and form with the femininity that desires it—all the while leaving in place the incorruptibility of the masculine and the corruptibility of the feminine. Yet Theophilo makes one more move than this, saying that insofar as matter is the constant substratum beneath all forms, matter must in a sense "contain" all forms, producing them from within itself and then reassimilating them as things are transformed into other things. Matter, as he explains it, "unfolds [*esplica*] what it possesses enfolded [*implicato*]" (84).

This enfolding-unfolding, we will note, is a direct citation of Cusa's *implicatio* and *explicatio*, the glaring difference being that Bruno is attributing to *matter* what his predecessor attributed to God. Matter, for Bruno, has the infinite capacity to implicate and explicate all that is. And strikingly, it is only in the process of divinizing this eternal material principle that he finally gives it a gender: "[T]his matter which unfolds what it possesses enfolded must, therefore, be called a divine and excellent parent, generator and mother [*genetrice et madre*] of natural things" (84). This is the point at which the "table" finally collapses: matter is not feminine (Philotheo), and matter is feminine (Theophilo), depending on what "feminine" means. But either way we look at it, matter is eternal, so if we want to "relegate" materiality to femininity, then we end up elevating femininity to eternity in the process. And if we want to save eternity for masculinity, then we must saddle masculinity with materiality—and the distinctions recollapse.

In terms of cosmogony, the bottom line for Bruno is that matter eternally generates an infinity of forms from within itself, thereby producing and sustaining the infinite worlds of the infinite universe. Needless to say, then, Bruno does not follow Cusa in reaffirming the doctrine of *creatio ex nihilo*. Insofar as creation is the direct and necessary expression of God, there is no "God" in excess of the material universe—no self-contained divinity that suddenly decides to make something—nor is there ever "nothing" for "him" to create "out of." Rather, what we have is an infinite God who eternally generates an infinite creation—a God who cannot, in fact, do otherwise. And although this account may feminize and materialize God, although it may compromise the absolute power of God's will, Bruno mentions in passing that his teachings are no different from those of "Moses himself." When "Moses" wrote " 'Let the earth bring forth its animals, let the waters bring forth living creatures[,]' it is as if he had said: 'Let matter bring them forth.' " And when he wrote "that the efficient intellect (which he calls spirit) 'brooded on the waters,' " he meant that the intellect "gave the waters a procreative power and produced from them the natural species" (83). "God," then, is both matter and intellect, both father and mother—the total coincidence of opposites from *and as* which the universe eternally unfolds.

A Multimodal Multiverse

As the unmediated outpouring of God, the universe must also be said to be the coincidence of opposites, and it is in this sense that Bruno's infinite worlds can be said not only to be many, but also to be one. Although this theme is a familiar one, Bruno's many-oneness is subtly different from Cusa's. For Cusa, we will recall, the universe "in itself" is *either* one or many, whereas the universe "in God" is both. Because Bruno makes no such distinction, his multiverse is always both one and many, always at the same time, just like God. Yet how does this many-oneness play itself out cosmologically? From what we have already seen, the "manyness" of the universe is clear: there are an infinite number of suns surrounded by an infinite number of earths, each of which has its own principle of motion within it. But how can these infinite worlds also be said to be "one"? If each body has its "own" soul, and if worlds are separated from one another by a vast expanse of interstellar space, and if each of them lives forever, then do they not constitute a more unrelated plurality than the Atomist *kosmoi*? Just a random scattering of cosmic individuals?

No, replies the Nolan; "the universe is one [*l'universo é uno*]" (87). Bruno accounts for this cosmic unity in two different ways. First, he posits a "world soul" that permeates the whole of matter to such an extent that neither exists without the other, a corporeal spirit much like the ancient Stoic *pneuma*. So although all cosmic bodies have their "own," individual souls, they are also just varying expressions of the one "immense spirit, [which] under different relations and according to different degrees, fills and contains the whole" (6). Moreover, whereas each individual soul moves its own body, the world soul as a whole is at rest. After all, the entirety of an infinite universe cannot rotate (because circular motion implies finitude), and it cannot move rectilinearly (because it has nowhere outside itself to go). So the world soul is the "infinite, immobile" spirit substance of all things, the stuff that lies around, between, and within the infinite worlds of the infinite universe and makes the whole thing one (11).

Bruno's second argument for the unity of the universe stems from its infinity. If it is true, as he insists in *Cause, Principle, and Unity*, that "in the infinite there are no parts," then every bit of the universe contains the entirety of it. In other words, the multitude of things in an infinite universe must, in essence, be one, to such an extent that "all that we see of diversity and difference is nothing but diverse aspects of one and the same substance" (11). At this point, considering the omnipresence of the world soul and the indivisibility of the infinite, Bruno's cosmology looks nothing like a hyper-Atomist multitude. It suddenly looks far more like a hyper-Stoic monism, with all the diversity of the universe subsumed under a more primordial, more ultimate oneness. "In the multitude I find no joy," Bruno writes in *On the Infinite Universe*; "it is Unity that enchants me."[103]

Thanks in large part to a scholarly overvaluation and misunderstanding of this utterance,[104] Bruno has traditionally been interpreted as a thoroughgoing monist: a neomystic (or protoromantic) who assimilates all diversity into an undifferentiated one.[105] But it is crucial to recall that, for Bruno, nothing exists without everything. So just as there is no form without matter, no matter without spirit, and no God without creation, there is no "one" without the many. This is not to say that the many is the outward expression of the one or even that it is the created reflection of the one; for Bruno, the many *composes the one as such.* This, in turn, is not to say that the many is therefore prior to the one or that it makes the one many. Rather, the many makes the one *multiple.* To use Bruno's language in *Cause, Principle, and Unity*, "[I]n the infinite and immobile one, which is substance and being, if there is multiplicity . . . [it] does not, thereby, cause being to be more than one, but to be *multi-modal, multiform, and multi-figured*" (90, emphasis added). This, then, is the peculiar ontology of Brunian cosmology: the universe is neither monistic nor dualistic nor pluralistic, nor is it one from one perspective and many from another. Rather, the universe is irreducibly "multimodal"—one *by virtue of* its infinite multiplicity, with the whole reflected holographically in each of its partless parts.

Postscript

When the inquisitors read Bruno the verdict that called for his execution, he is said to have replied, "You may be more afraid to bring that sentence against me than I am to accept it."[106] This response has become part of a vast stock of Brunian lore and tends to be interpreted as a confirmation of any number of his heretical teachings. Bruno is said to have been unafraid of death because he believed in the transmigration of souls or because he "denied that sins deserve punishment" or because he denied the divinity of Christ and so did not fear he would stand before him in judgment.[107] But one might also interpret this final outburst in cosmological terms as a confession of faith in his multimodal universe. If "the universe is in all things and all things are in the universe," then this must also mean that "we [are] in it and it in us." "We" are, in this sense, as eternal as the universe itself. And so the endless sea of matter, constantly unfolding the forms she enfolds, will find something else to do with us—even with a body that has been thrown like wood on a fire: "See, then, how our spirit should not be afflicted, how there is nothing that should frighten us: for [this] unity is stable in its oneness and so remains forever."[108]

 4

MEASURING THE IMMEASURABLE

What a wonderful and amazing scheme have we here of the magnificent
vastness of the universe!

CHRISTIAAN HUYGENS, *COSMOTHEOROS*

'Tis all in pieces, all coherence gone.

JOHN DONNE, "AN ANATOMIE OF THE WORLD"

From Infinity to Pluralism:
Seventeenth-Century Cosmologies

An Extraterrestrial Age

It will perhaps come as no surprise that even the elimination of its most
vocal proponent failed to do away with the cosmology of an infinite universe and its infinite worlds. To be sure, Giordano Bruno's execution in
1600 encouraged a degree of caution for a number of decades, especially
among Catholic scholars on the Continent, but for the most part the seventeenth century witnessed a veritable explosion of treatises concerning
the existence of other worlds.[1] This explosion can be attributed in part to
the exceedingly public nature of Bruno's trial and execution, which arguably circulated Bruno's heretical cosmology more widely than it would
have been otherwise and which lent to it a certain harrowing caché.[2] Even
more significantly, this fascination with the possibility of other worlds was
fomented by the publication in 1610 of Galileo's telescopic discoveries.[3] In
them lay the observational proof that the earth was just one of many planets
orbiting the sun. Karl Guthrie therefore credits Galileo Galilei (1564–1642)
with having lit the fire of seventeenth-century "pluralism,"[4] explaining that
"the lay public was not sufficiently convinced by Copernicus's theory until

Galileo had actually *seen* through his telescope in 1609/10 that the surface of the moon was extremely similar to that of the Earth; that Jupiter was circled not by one moon (like the Earth) but by four, which could only shine on Jupiter's inhabitants; and that the stars of the Milky Way were suns."[5]

Galileo's success in this regard is clear from the Inquisition's only having condemned "Copernicanism" six years after the publication of Galileo's *Sidereus nuncius* (1610). After the emergence of his *Dialogue Concerning the Two Chief World Systems* (1632), the Inquisition then tried and convicted Galileo himself, placing him under lifelong house arrest and his writings on the burgeoning Index of Prohibited Books.[6] In spite of the church's best efforts, however, this sentencing only intensified the anxious excitement stirred up by Bruno's fate, an excitement that seems to have been unfazed by Galileo's abject apology and recantation ("I neither did hold nor do hold as true the condemned opinion of the earth's motion and the sun's stability").[7] Apology notwithstanding, the *Dialogue* had clearly demonstrated the superiority of the heliocentric model, leaving geocentrism all but indefensible. Hence the infamous lore of Galileo's having recanted his recantation with a softly muttered "eppur si muove": "and yet it moves."[8]

Not incidentally, it was this era that also staged a spirited retrieval of the long-demonized Atomists. With Aristotelian cosmology definitively thrown in the Copernico-Galilean dustbin, the intellectuals of the late Renaissance and baroque periods turned to Lucretius in particular to provide a heliocentrically viable alternative. The resulting neo-Atomist movement, which seems to have had its earliest roots in England's Northumberland Circle,[9] finds its best-known elaboration in the *Syntagma* (1647) of Pierre Gassendi (1592–1655), a philosopher, astronomer, and Roman Catholic priest who sought to reconcile Epicurus with Christianity.[10] His most significant move in this regard was to deny the eternity and infinity of atoms, asserting instead that God had created a finite number of them out of nothing, set them in motion, and used them to form the planets and suns of the universe.[11] Strikingly, Gassendi also rejected the Atomist doctrine of a plurality of worlds, calling it both incompatible with scripture and beyond the reach of observational verification.[12] In line with Marin Mersenne and the late Scholastics, Gassendi argued that an infinite cause need not produce an infinite effect and that God's omnipotence is reflected not in the volume of his creation, but in his ability to bring it out of nothing.[13]

Even as Gassendi insisted that "this world is the only one," however, he postulated that the other planets in our solar system are most likely inhabited, with the more perfect creatures living close to the sun and the less perfect creatures living farther from it.[14] At this point, then, we run into a bit of a semantic difficulty concerning the definition of *world*. Gassendi denies emphatically that there are other *kosmoi*, let alone an infinite number of them, but he is happy to assent to the existence of any number of inhabited planets, and many other thinkers (Nicholas of Cusa, for one) would call an inhabited planet a "world." But this fuzziness is not merely definitional; rather, it serves as an index of a significant conceptual shift among the pluralists of the seventeenth century. During this period, even the scholars who *did* affirm a "plurality of worlds" were affirming not a plurality of self-contained systems strewn throughout infinite space, but a plurality of inhabited planets orbiting a plurality of suns.[15] "Worlds," in other words, became the specific bodies that could be seen through a telescope.

The invention of the telescope therefore had the duplicitous effect of both extending and narrowing the cosmological vision of western Europe. As it became possible to see all the way out to the craters on the moon and the satellites of Jupiter, the question of "other worlds" was pulled within those observational limits. As a consequence, the meditations on "other worlds" became far more concrete than they had ever been, if no less speculative. Rather than wondering whether worlds were perishable or imperishable or whether one earth might in principle be drawn to the center of multiple *kosmoi*, the seventeenth-century pluralists wondered whether there was life on the moon or on Jupiter or Mars and what it might look like. This is not to say that such questions had never been asked before—here we might recall Cusa's brief but spirited catalog of different cosmic residents[16]—but is simply to say that these residents received new and intensified focus in the seventeenth century. Observationally speaking, this newfound fascination with extraterrestrials was indeed conditioned by Galileo's construction of an "astronomically useful telescope."[17] But philosophically and even theologically, it had been conditioned a century and a half earlier with Europe's "discovery" of the Americas.

In his inexhaustible *City of God* (413–426), Augustine devotes one brief chapter to "the fabled 'antipodes,' men, that is, who [are said to] live on the other side of the earth." From the outset, he insists that "there is no rational ground" for believing that such people exist because scripture tells us that all humanity came from one man, and "it would be too ridiculous to suggest that some men might have sailed from our side of the earth to the other."[18] Although debates on this issue would reopen during the twelfth and thirteenth centuries, the church's official position for more than a thousand years after Augustine was that "antipodeans" did not exist.[19] This teaching was called into question rather dramatically after the voyages of Columbus, and it was finally abolished in 1573 when Pope Paul III declared that the natives of the lands encountered in these voyages could not be enslaved (and could be converted) because they were, in fact, "human."[20] The so-called antipodean debate therefore provided a blueprint for the seventeenth-century pluralists: If humans existed on "the other side" of this planet, what would stop them from existing on other planets as well? This line of reasoning took flight following Tommaso Campanella's *Apologia pro Galileo* (1622), which dubs Galileo "the new Columbus" and suggests that it would be as ridiculous to deny the existence of people on other planets as it would be to deny the existence of people in the Americas.[21]

And sure enough, the unfolding extraterrestrial debate opened a number of anthropological and theological questions that were remarkably similar to those concerning non-European humans. To what extent could the inhabitants of other planets be considered "rational"? Were they less or more intelligent than Europeans?[22] Were their bodies like "ours" or different?[23] Were they darker, lighter, taller, shorter?[24] Were they descendants of Adam or members of some other species entirely?[25] Had they fallen when Adam fell, or had they somehow remained sinless? Did Christ save Venusians when he died on a cross on the earth, or would he have to jump from planet to planet to be born again, teach for a while, and die at the hands of some alien Pontius Pilate?[26] Could a Lunarian be baptized with the water in a crater on the moon?[27] In short, seventeenth-century pluralism both reflected and enacted a massive anxiety over the place of "European man" not only on the globe but throughout the universe, the vast expanse of which threatened to render him insignificant.

A distraught John Donne gave voice to the state of things in "An Anatomie of the World" (1611):

And new Philosophy calls all in doubt,
The element of fire is quite put out;
The sun is lost, and th'earth, and no mans wit
Can well direct him where to looke for it.
And freely men confesse that this world's spent,
When in the Planets and the Firmament
They seeke so many new; they see that this
Is crumbled out againe to his Atomies.

'Tis all in peeces, all cohaerence gone;
All just supply and all Relation.[28]

For Donne, the Atomist revival was shattering the cosmos, "crumbling" the heavens into particles, and leaving humanity without any solid point of orientation: "The sun is lost," he mourns, "*and* th'earth." A similar anxiety surfaces half a century later in Blaise Pascal's *Pensées*. Contemplating humanity's position between the endlessness revealed by the telescope and the endlessness revealed by the microscope, Pascal confesses, "The eternal silence of these infinite spaces terrifies me."[29] In the French, this infamous utterance is followed immediately by the exclamation, "How many kingdoms know us not!" Here we might remember Alexander the Great weeping when he learned of "the infinity of *kosmoi*" or Thomas Aquinas worrying that an infinite world would compromise the integrity of an infinite God (see chap. 2, sec. "Accident and Infinity," and chap. 3, sec. "Ending the Endless"). For better or worse, cosmic infinity always seems to undermine sovereignty—whether human or divine.

"The World and Its Prisons": The Curious Case of Johannes Kepler

Provoked by the telescopic unfolding of "these infinite spaces," the cosmological writings of the seventeenth century display a dizzying combination of anthropological fascination, existential revulsion, scientific restraint, and theological innovation. Such a combination is perhaps nowhere more apparent than in the work of Johannes Kepler (1571–1630), who is best known for his three laws of planetary motion. The first of these laws is that planets orbit the sun in ellipses rather than circles. This discovery improved the predictive power of Copernicanism dramatically, establishing Kepler as something of a heliocentric hero. His position on cosmic plurality, however, was complicated. On the one hand, he shared his pluralist contemporaries' conviction that the earth was not the only inhabited planet in the solar system; he even wrote a fictional account of a journey to the moon called *Somnium*, which imagined and cataloged numerous species of lunar creatures.[30] On the other hand, Kepler refused to speculate about life beyond our solar system or even about *solar systems* beyond our solar system, insisting that our sun occupied the center of a finite universe, bounded by a ring of fixed stars.

For this reason, Kepler was disinclined to trust "the speculations of the Cardinal of Cusa" and was downright horrified by the "dreadful philosophy" of "the unfortunate [Giordano] Bruno," who had "made the world so infinite that [he posited] as many worlds as there are fixed stars."[31] Faced with such a prospect, Kepler went so far as to express sympathy for Aristotle, who may have been wrong about the position of the earth, but who was right, he said, to demolish the Atomists' execrable infinity of worlds. Anticipating the sentiments expressed by Donne and Pascal, Kepler wrote that "the very cogitation [of infinite worlds] carries with it I don't know what secret, hidden horror; indeed one finds oneself wandering in this immensity, to which are denied limits and center and therefore all determinate places."[32] In order to close down the "immensity" that Bruno had reopened, Kepler suggested that one could always appeal to scripture, which never mentions any worlds other than our own. Yet he cautioned that, in essence, "the question we are discussing is not a dogmatic one," and so the problem of Bruno's "immeasurabilities" would have to be "dealt with . . . by scientific reasoning."[33]

Kepler executed this scientific reasoning along two major lines. First, he attacked the cosmic homogeneity that Bruno's infinite universe implied. As we will recall from Cusa and Bruno alike, an unbounded universe has no absolute center or periphery. Rather, every place looks like every other place, such that the inhabitants of any planet would see themselves at the center of the universe, surrounded by a dense ring of stars. But Kepler countered that observation had shown this not to be the case: of all the stars in the sky, he argued, our sun appears to be brightest, and therefore (according to his own interpretation of Galileo's findings) it is the largest. Because all other stars are smaller than our sun, the view from anywhere else in the universe would be totally different from ours.[34] We must therefore live in an immense cavity in the sky: a unique clearing in which a few planets orbit this one brilliant sun, surrounded by a shell of stars. And because these stars cannot be said to be suns—because, in fact, "the body of our Sun is brighter beyond measure than all the fixed stars together"— Kepler was able to conclude that "this world of ours does not belong to an undifferentiated swarm of countless others."[35] In effect, he countered Bruno's neo-Epicureanism with a modulated dose of early Stoicism: the world for Kepler was a singular, bounded whole in the midst of a giant void. The modulation, of course, was that this universe would neither disappear in fire nor reemerge from it; both temporally and spatially, the Keplerian cosmos must be unique.

Kepler's second approach to reigning in Bruno's cosmology was simply to insist that astronomy must limit itself to observable phenomena. An infinity of other worlds is in principle unobservable, he maintained, and therefore of no scientific use.[36] In *De stella nova* (1606), Kepler therefore expressed the fervent wish that "Astronomy herself" might cure "this madness of the philosophers" by limiting "her" imagination to the visible and the measurable. Only proper scientific restraint could bring an errant philosophy "back within the bounds of the world and its prisons. . . . Surely," he cautioned, "it is not good to wander through that infinity."[37] Nevertheless, by allowing observation to maintain the walls of the cosmic "prison," Kepler also allowed observation to move or demolish those walls. He knew this, of course, and had been particularly worried that four recently discovered "planets" might be seen to orbit one of the fixed stars— in which case the fixed stars might indeed be "suns" in their own right.

To Kepler's relief, this was not the case: in 1610, Galileo reported that all the new "planets" he had seen were moons of Jupiter. Just a few months later, an exuberant Kepler exclaimed in a written response to Galileo, "I rejoice that I am to some extent restored to life by your work. If you had discovered any planets revolving around one of the fixed stars, there would be waiting for me chains and a prison among Bruno's innumerabilities." Perhaps realizing how inapt this metaphor was (after all, the *Stella nova* had commended the "prisons" of the *bounded* cosmos), Kepler immediately reversed it: "I should rather say, exile to [Bruno's] infinite space. Therefore, by reporting these four planets revolve, not around one of the fixed stars, but around the planet Jupiter, you have *for the present* freed me from the great fear which gripped me as soon as I heard about your book."[38] For the present, Kepler was free to be confined within a singular and finite cosmos. For the present, he would not have to think about anything he could not observe.

Well, not quite "anything." As Karl Guthrie has shown, Kepler added an appendix to his cosmic travelogue *Somnium* well after he had learned about Galileo's discoveries.[39] And although Galileo himself had considered "false and damnable the view of those who would put inhabitants on Jupiter, Venus, and the Moon,"[40] Kepler used the occasion of Galileo's observations to offer even more details about the many varieties of Lunarians—whom no one had ever seen. Even more puzzlingly, the young Kepler in particular indulged in some remarkably constructive theologizing, finding in his rebounded cosmos a perfect image of the Trinity. As the central and brightest body in the universe, as the source of its light and life, the sun was for Kepler the created image of the Father. The fixed stars on the periphery could be said to embody the Son. And the interstellar space between them, "the relationship between the point and the circumference," reflected the Holy Spirit.[41] As it was for both Bruno and Cusa, then, the universe for Kepler was the created expression of God Godself. Yet thanks to an intricate implication of existential panic, ecclesiastical bullying, and scientific rigor, Kepler fled the infinity of Bruno *and* the boundlessness of Cusa, contracting even the contracted universe back into a stubbornly finite cosmos.

His trinitarian vision notwithstanding, what we begin to see in Kepler is a new disciplinary separation between scientific and humanist approaches to cosmology. As Steven Dick has shown, the question of a plurality of worlds began to move in the mid-seventeenth century out of the realm of astronomy and "into the domain of philosophical explorations, both secular and religious."[42] Strikingly, the "religious" branch of this philosophical inquiry generated far less drama in the second half of the seventeenth century than it had during the time of Copernicus, Bruno, and Galileo. With a few vocal exceptions, it simply became commonplace to assert that a universe filled with inhabited cosmic bodies was more befitting of an omnipotent God than one lonely planet or solar system would be.[43]

Chief among these vocal exceptions was Walter Charleton (1619–1707), whose neo-Epicurean *Physiologia* (1654) not only denies any plurality of worlds but even suggests that "*Democritus* and *Epicurus*" themselves never taught such a thing, "proposing it as a necessary Hypothesis" rather than as a cosmic reality.[44] Charleton raises a number of objections to this doctrine in the *Physiologia*, but they can for all intents and purposes be grouped into two: first, he finds the notion theologically offensive, and, second, he thinks it an affront to reason. Everyone agrees, Charleton begins, that God *could* have created worlds apart from this one. But to claim that God *has* done such a thing would be downright blasphemous. To say, for example, that there are more atoms in creation than were needed to compose this world would be to imply that God does not know how "to proportion the quantity of his materials to the model or platform of his structure." And to say that there is an *infinite* "Residue" of such material would be to imply that God's creation is somehow "equal to Himself" and, consequently, that God has "an Imperfection in his Nature."[45]

We have already heard both of these arguments in defense of the singularity of God. The line of attack more unique to Charleton, on which he spends far more time, concerns the *unreasonableness* of the doctrine of a plurality of worlds—a charge that will be leveled against modern multiverse theorists (see conclusion). According to Charleton, it simply makes no sense to posit other worlds beyond our own; there is no reason to do such a thing. He admits that both cosmic pluralists and cosmic monists lack observational evidence that might support their positions; there is no way to *see* that the world is either unique or one of many. But Charleton finds this inability all the more reason to affirm the "Unity of the world"; in the absence of evidence, he argues, the monists ground their position "upon that stable criterion, our *Sense*," and the pluralists "found theirs upon the fragil [*sic*] reed of wild Imagination."[46] To make matters worse, the pluralists spend so much time attending to invisible worlds that they fail to understand the one world that they *can* see. Charleton finds it painfully ironic that "there were whole Schools of *Philosophers*, who fiercely contended for a *Plurality of Worlds*, when, indeed, their Understandings came so short of conquering all the obvious Difficulties of this one, that even the grass they trod on, and the smallest of Insects, a Handworm, must put their Curiosity to a stand." To anyone who might feel seduced into cosmic pluralism, Charleton therefore prescribes a "Remedy." Contemplate the handworm, he suggests, or a dandelion or the motion of leaves in the wind: "The most inquisitive may find Difficulties more than enough within the Little World of their own Nature"; what is more, their efforts "will prove more advantageous to the acquisition of Science, then [*sic*] the most acute metaphysical Discourse."[47]

In Charleton, then, we see a remarkable concordance between a particularly Scholastic style of conservative theology and an early modern scientific commitment to observation and common sense. When it came to the question of a plurality of worlds, the new scientists followed a very similar approach, resolving along with Kepler to "avoid the march to the infinite."[48] In the meantime, a new breed of Brunian theologians and philosophers joyfully entertained what they regarded as an unproblematic proposition: an infinite God can be glorified only in an infinite number of worlds. Strikingly, then, "sacred" and "secular" attitudes toward cosmic multiplicity swapped places in the seventeenth century: the new theology began to revel in the pluriversal glory of God, whereas "Astronomy herself" remained confined—from the second half of the seventeenth century to the second half of the twentieth—to the one world that astronomers could measure and see. Nevertheless, the growing philotheopoetic body of cosmic pluralism was not without scientific grounding. Even as the infinite worlds of "mad philosophy" were reviled by figures such as Kepler and Charleton, they found rational support in the cosmology of René Descartes.

The Ambivalent Pluralist: René Descartes

In the British Isles and on the Continent alike, the latter half of the seventeenth century produced an increasing number of "pluralist" treatises, the majority of which were to some degree based on the cosmology of René Descartes (1596–1650).[49] By all accounts, this Cartesian-inspired pluralism reached its apex in the flamboyant prose of Bernard le Bovier de Fontenelle (1657–1757), whose immensely popular *Conversations on the Plurality of Worlds* (1686) proclaims an infinite number of solar systems inhabited by a staggering variety of creatures.[50] These conversations, addressed to a "Marquise de G***," are presented in plain (if melodramatic) prose so that even women might understand them ("I only ask of the ladies," Fontenelle writes in the preface, "for this whole System of Philosophy, the same amount of concentration that must be given to the *Princess of Clèves*").[51] The Dutch astronomer Christiaan Huygens (1629–1695) offers a similarly delightful journey through multiply inhabited planets in the popular, posthumously published *Cosmotheoros* (1698).[52] Having made his way from the earth to the other planets and moons in our solar system and then to the (most likely infinite) multitude of orbited suns beyond it, Huygens exclaims, "What a wonderful and amazing scheme have we here of the magnificent vastness of the universe! So many suns, so many earths, and every one of them stocked with so many herbs, trees, and animals, and adorned with so many seas and mountains! And how must our wonder and admiration be increased when we consider the prodigious distance and multitude of the stars!"[53] Like Fontenelle, Huygens grounds his ecstatic journey through the ranks of an infinite universe in Descartes's "vortex" theory.[54] What is strange about calling either of these works "Cartesian," however, is that Descartes himself either avoided or denied each of his own theory's most thrilling cosmic implications. He would never say whether he thought there might be life on other planets,[55] he refused to call the universe "infinite,"[56] and he said quite explicitly that "there cannot be a plurality of worlds."[57] Nevertheless, his universal theory of "vortices" seemed to indicate otherwise.

In a letter to none other than the anti-Brunian Marin Mersenne, Descartes wrote, "I have resolved to explain all the phenomena of nature, that is, all of physics."[58] The year was 1629, Descartes was thirty-three years old, and he had just begun working on a book he would come to call *Le monde, ou Traité de la lumière* (*The World, or a Treatise on Light*). This book, which sets forth a full heliocentric cosmology *and* cosmogony (Stephen Dick calls it the "first complete physical system . . . since Aristotle"),[59] took Descartes far longer to complete than he had expected it to. As he explained to Mersenne, he was plagued by concerns not only that his physics might be wrong, but also that his theology might be poorly received—especially his conviction that once matter is created, it goes on to form a world with no further help from God.[60] Descartes therefore resolved to publish the treatise anonymously, "principally because of theology, which has been so subjected to Aristotle that it is almost impossible to set out another philosophy without its appearing at first contrary to faith."[61]

But just as Descartes had finally completed the manuscript in 1633, he learned that Galileo had been placed under house arrest, his books condemned and burned. In a letter to Mersenne that never reached its destination, Descartes wrote, "[T]his has so astonished me that I am almost resolved to burn all my papers, or at least not to let anyone see them."[62] Although Descartes had foreseen the controversy that his cosmology might generate concerning God's function in the universe, it seems that he had not known how staunchly opposed the church remained to heliocentrism, on which his little treatise depended entirely: "I confess that if [the motion of the earth] is false, all the foundations of my philosophy are also. . . . It is so linked to all parts of my treatise that I cannot detach it without rendering the rest completely defective."[63] And so the manuscript was suppressed until after Descartes's death in 1650. Nevertheless, he managed to scatter some of its central ideas throughout the formally dry, almost neo-Scholastic *Principia philosophiae* (*Principles of Philosophy*, 1644), thereby circulating an attenuated version of his vortex theory while he was still alive.

Descartes's first major cosmological move in the *Principia* is to sidestep the Brunian controversy over infinity by returning to the Cusan solution. "We should never enter into arguments about the infinite [*numquam disputandum esse de infinito*]," he writes; "things in which we observe no limits—such as the extension of the world, the division of the parts of matter [*partium materiae*], the number of the stars, and so on—should instead be regarded as indefinite [*pro indefinitis habenda*]."[64] Descartes insists on this distinction not only in the *Principia*, but also in the controversial *Le monde* and throughout his written correspondence—most notably with the enthusiastic English pluralist Henry More, who tried at length to convince him to abandon it.[65] Depending on the context, Descartes offers one of two different reasons for maintaining that the world is indefinite rather than infinite. The first is theological: in short, it allows Descartes "to reserve the term 'infinite' for God alone";[66] here he reminds us of Thomas's distinction between a limitlessness that is strictly "privative" and one that is fully "negative" (although Descartes somewhat confusingly calls the negative infinity "positive"). The second reason is epistemological: to call the world and its matter "indefinite" is to say that *we do not know* how to ascribe limits to them. "Whether they are *simpliciter* infinite or not," he wrote to More, "I confess not to know. I only know that I do not discern in them any end."[67] This may sound like a kind of Cusan *ignorantia*, but Descartes has no desire to call the universe (or anything else) unknowable. Rather, this admission of nonknowledge becomes for Descartes the *condition* of knowledge: it is only when we admit that we do not know about infinity that we will refrain from getting lost in it, stand on determinate ground, and gain certainty about everything else.

It is this epistemological strategy that Descartes is employing in *Le monde* when he invites us to imagine with him the creation of a world. A signal of how long it has been since western Europe has had a proper cosmogony, Descartes prefaces his own by hearkening back to Plato. "In order . . . to make the length of this discourse less boring for you," he begins, "I want to wrap part of it in the cloak of a fable [*dans l'invention d'une fable*]."[68] Much like Timaeus, Descartes offers us a "likely story." In this newer fable, however, the storyteller is also the demiurge. "For a short time," he entreats us, "allow your thought to wander beyond this world to view another, wholly new one, which *I shall cause to be born* [*que je feray naistre*] before you in imaginary space" (49, translation altered, emphasis added). Lest we get lost in this imaginary space, however, Descartes is careful to circumscribe its limits: "In order that infinity not impede us and not embarrass us, let us not try to go all the way to the end; let us enter only so far in that we can lose from view all the creatures that God made five or six thousand years ago and [stop] there in some fixed place [*en quelque lieu déterminé*]" (51). We seem, then, to be back behind the biblical creation, with nothing more than Descartes, ourselves, and a finite, "fixed place."

For the sake of having a place to start, we can think of this "fixed place" as place itself: the place in which the world will come to be. Here, again, Descartes reminds us in *Le monde* that it will do no good to think of this place as infinite, proposing instead that we "purposely [*à dessein*] restrict it to a determinate space that is no greater, say, than the distance between the earth and the principal stars of the firmament" (51). And as we have already seen, restricting ourselves to a determinate space is, for Descartes, the condition for our knowing anything about the world at all. In fact, he suggests, because "we" are making this cosmos in the first place, we can just declare that "there is absolutely nothing" about it "that anyone cannot know as perfectly as possible" (51–53). So we have a very large but perfectly comprehensible place, and this place constitutes the most primitive state of the cosmos. Crucially, however, this state is not empty. Rather, our determinate place is filled with "*the matter that God shall have created . . . out to an indefinite distance*" (51, emphasis added).

This little passage is doing an extraordinary number of things. First, it suddenly reintroduces God, who is absent from the beginning of the story and then is excised as soon as he is mentioned. In fact, the story begins and rebegins with nothing more than Descartes, our imaginations, and a fixed place. As the story develops, however, Descartes concedes that it cannot begin at the beginning because it must assume that God "shall have created" matter already along with the space that it occupies. Theologically, this concession allows Descartes to suggest (although he never actually says it) that God created matter out of nothing. Literarily, it allows him to begin right where Timaeus begins: with a demiurge (in this case, Descartes) who finds a disordered material already there. And, epistemologically, it saves Descartes from preemptive ruin by giving him some precosmic material, for, in the words of two twenty-first-century cosmologists, "there are no rigorous physical principles that dictate how to go from 'nothing' to 'something.'"[69]

Second, this passage repudiates the Atomist void, indicating that there is no "place" without matter—no time when space will not already have been filled. This is to say, as Descartes explains in the *Principia*, that "there is no real distinction between space . . . and the corporeal substance contained in it."[70] For even if we were to remove a stone from a place, we would not thereby remove the extension of the place itself, and anything extended is material. In terms of his cosmogony, then, Descartes is beginning not with a void—the very notion of which he calls "a contradiction" insofar as there is no difference between matter and space[71]—but with a "plenum," a space so "filled" with matter that it is matter itself.

Third, with no void, there can be no atoms moving within it. This is another point that Descartes clarifies in the *Principia*: no unit of matter is indivisible (*a-tomos*). "For if there were any atoms," he reasons, "then no matter how small we imagined them to be, they would necessarily have to be extended; and hence we could in our thought divide each of them into two or more smaller parts."[72] In other words, insofar as "an indivisible unit of matter" is already the product of our imagination, the very fact that we can imagine dividing it amounts to its divisibility. Besides, Descartes adds, even if there existed a particle so small that no human could divide it, God could divide it, and so it would be in principle divisible.[73] As it was for Aristotle, then, matter for Descartes is infinitely divisible— or as he scrupulously phrases it, "indefinitely divisible."[74] By rejecting atoms as well as the infinite, however, Descartes ironically allows himself to configure matter in units that are genuinely discrete. As we may recall from Bruno in particular, nothing infinite can be divided into real parts; rather, any "part" of the infinite must itself be infinite. Hence Bruno's (and Cusa's and even Lucretius's) total holography: if creation is infinite, then "all things are in all things"—the universe in a grain of sand or, for that matter, in an atom. Such a riot of in(ter)distinction will clearly not do for Descartes, whose "things" must be clear and distinct. Therefore, his units of matter are not "atoms," but "parts" or "particles" (*parties*); his universe not infinite but indefinite.

This distinction leads to the final implication of the same short passage, which is that the "matter that God shall have created" in (and as) our "fixed space" extends "*to an indefinite distance.*" Now it seems here that Descartes wants conflicting things. On the one hand, he insists that the "place" we are imagining is "determinate"; on the other hand, he says that it is indefinite. So either Descartes is contradicting himself, or he is suggesting that the indefinite *is* determinate—that the indefinite, unlike the infinite, can be clarified, distinguished, and known. Hence the discrete "parts" rather than the holographic atoms. Moreover, the indefinite extension of matter prevents any hypothetical Atomist or even Stoic from suggesting that there might be a void beyond Descartes's fixed place. To say that the plenum extends indefinitely is to say that there *is* no outside, that the plenum is what there is. And, of course, this is the reason "there cannot be a plurality of worlds": because there is nowhere else for them to be.[75]

To return to our fable in *Le monde*, then, Descartes has taken us to his cosmic primordium: a determinate, indefinite, corporeal space. This is the Cartesian "chaos" (*cahos*), but our demiurge insists that it is a perfectly comprehensible chaos. In fact, "there is nothing simpler nor easier to know among inanimate creatures," for whereas our knowledge of all other bodies relies on the famously unreliable faculty of sense perception, our knowledge of "the matter from which I have composed the chaos" relies on intellect alone (55). All we need to do, then, is to subtract the qualities we know through sense perception—such as wetness and heat and color and light—and assume the chaos to be bare extension itself: "Let us [then] conceive of it as a real, perfectly solid body, which uniformly fills the entire length, breadth, and depth of the great space at the center of which we have halted our thought" (53). A body with no qualities might seem hard to imagine, but Descartes assures us that it is not. In fact, precisely *because* it possesses no qualities, he claims, this chaos is clearer and more distinct than anything we can see, smell, or taste. Or, as he puts it here, "chaos contains nothing that is not so perfectly known to you that you could even pretend not to know it" (55).

How, then, does a solid lump of chaos become a cosmos? Descartes backtracks again: it turns out that we need a bit more from God. In order to imagine that unqualified matter can form a world, we must also assume that when God creates the material, he divides it into parts (*parties*) of different size and shape *and* that he sets these parts in motion, making "some begin to move in one direction and others in another, some faster and some slower (or indeed, if you wish, none at all)[.]" "[T]hereafter," Descartes concludes, "he makes them continue their motion according to the ordinary laws of nature" (55). At this point, then, God can leave the scene, for once he "so wondrously establishes" these laws, they will be "sufficient to make the parts of that chaos untangle themselves [*se démélent d'elles mesmes*] . . . [into] a most perfect world" (55).

This is how it works: because there is no such thing as empty space, the chaos particles that God will have made must "touch one another on all sides, without there being any void in between" (60). This means that they cannot move in a straight line, as they would otherwise "naturally" do, and so must begin to move in circles instead, forming a vortex (*tourbillon*) (71, 79). But because God also made these particles distinct from one another, endowing them with different types of motion, "we should not imagine that they all came together to turn around a single center," forming one vortex alone. Rather, the particles begin to collect "around many different [centers], which we may imagine to be diversely situated with respect to one another" (79–81). The centers of Cartesian vortices, unlike those of Democritean vortices, are occupied by the *lightest* particles, which are most "agitated" and therefore move in the tightest circles. They are then held in place by the heavier particles, which move in slower circles around them and whose sheer bulk prevents the light particles from moving toward the peripheries (81–83, 109–11). And once the different kinds of particles have winnowed out in this manner, they have become the three basic elements that Descartes discerns in the cosmos. The lightest particles become the "luminous element" that composes the stars; the heaviest particles constitute the "opaque element" that composes the planets; and the "middling" particles form the "transparent element" that fills the heavens: "the element of air" (37–39). Each vortex, in other words, becomes a solar system, whose planets orbit a central star through a vast expanse of space.

Even in this state, however, space is never empty, and so the vortices are not exactly separate. Rather, they border one another by means of a "surface without thickness," which Descartes calls the "firmament" (87). As we can see in figure 4.1, the firmament (the line stretching from *H* to *H*, for example) is a permeable membrane, and so bodies can travel between vortices. Most of these traveling bodies are the phenomena we call "comets" (see the river-shaped path at the top of the image). Furthermore, the permeability between vortices means that the sun of one vortex can be seen from the planets of other vortices. This explains why the residents of the earth can see so many stars: each of the stars in "our" sky occupies the center of its own vortex. Descartes's hesitations and protestations notwithstanding, then, one can begin to see why his vortices became so popular among his pluralist contemporaries—and why they were quickly condemned in the French universities. Although Descartes might have maintained that *this* indefinite expanse of things is the only indefinite expanse of things, he also filled it with innumerable systems, each revolving around a center resembling our sun.[76] And for this reason, even as he insists that there is only one "world," Descartes imagines, as he tells us in *Le monde*, that "there are as many heavens [*cieux*] as there are stars, and, since the number of stars is indefinite, so too is the number of the heavens" (87). This brief gesture toward the vast plurality of the universe, along with the instability of the term *indefinite*, set Fontenelle, Huygens, and their enthralled audiences wondering: What sort of plants, animals, atmospheres, and astronomers might inhabit such an "indefinite" number of worlds? The possibilities seemed infinite.

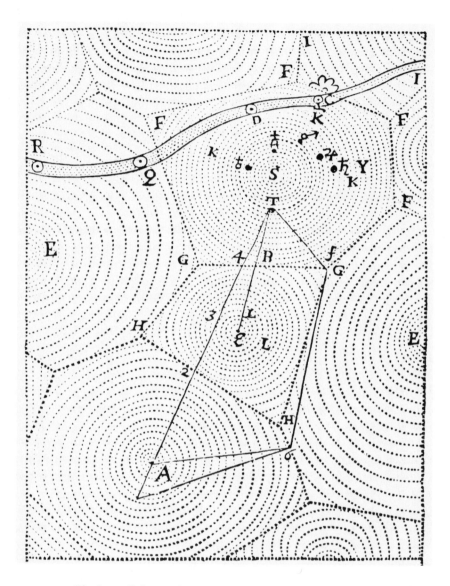

FIGURE 4.1 The "vortex" theory of René Descartes. (From René Descartes, *Le monde, ou Traité de la lumière* [1633], trans. Michael Sean Mahoney, Janus Library [New York: Abaris Books, 1979]. Reproduced by permission of the publisher)

Grace and Gravity in Isaac Newton

As history would have it, however, the craze over Cartesian cosmology was as short-lived as it was fervent. Just one year after the appearance of Fontenelle's Cartesian *Conversations*, Isaac Newton (1642–1727) published *Principia* (1687), whose synthesis of Galileo's physics and Kepler's orbital laws left Descartes's vortices in ruin.[77] In this work, Newton offers his Law of Universal Gravitation, which says, first, that the force keeping planets in orbit is the same as the force keeping our feet on the ground, and, second, that the magnitude of this force is determined by the mass of bodies and the distance between them ($F = gMm/r^2$).[78] Against Descartes, then, Newton showed that bodies need not touch one another to influence one another. In fact, he argued, a seamless fabric of particles such as those Descartes describes would prevent planetary motion altogether. In order for bodies to remain in orbit, gravity must pull them through *empty* space.[79] Newton therefore reinstated the distinction that Descartes had abolished between matter and space, arguing that "absolute space" is an inert, extended background through which objects may or may not move. Similarly, he insisted, "absolute time" flows universally and inexorably, "without regard to anything external."[80]

Gottfried Leibniz (1646–1716) rigorously contested Newton's "absolute" view of space and time, insisting that these terms were just descriptions of relations between objects. Space and time, Leibniz argued, have no independent existence; they are, as he wrote to the Newtonian Samuel Clarke in 1716, "purely relative" to each other and to the matter "within" them.[81] As we will see in chapter 5, Leibniz's more Cartesian view would be vindicated two hundred years later by Einstein's theory of general relativity, which, according to the *Times* article that broke the story, rendered "Newtonian ideas overthrown."[82] In the meantime, however, "Newtonian ideas" had won: Newton's absolute matrix was far more commonsensical than Leibniz's odd relations, and at the scales to which Newton was confined, it *worked*.[83] Although Newton infamously had no idea what gravity was (he steadfastly refused to formulate such a "hypothesis"),[84] he could predict its behavior with remarkable accuracy. By the end of the seventeenth century, then, Newton's principles had become entrenched as the new laws of cosmic motion.

Frustratingly for many would-be Newtonians, however, Newton never provided a cosmogony, primarily because he believed that the laws he had discovered were insufficient to such a task. Gravitation alone could neither create nor sustain the universe. This insufficiency comes across most clearly in the four letters that Newton wrote in the winter of 1692/1693 to Richard Bentley, a Cambridge theologian in the process of delivering a lecture series titled "A Confutation of Atheism." Bentley was convinced that Newtonian mechanics provided the most helpful combination of "natural laws" and "supernatural acts"; in other words, Newton's laws constituted *scientific proof* of the existence of an extracosmic God.[85] Newton opened his first letter with an enthusiastic endorsement of such a project: "[N]othing," he wrote, "can rejoice me more than to find it useful for that purpose."[86]

And, indeed, over the course of the letters to Bentley, Newton uncovered three significant gaps in his gravitational theory—gaps that, he believed, could be filled only by appealing to God. The first concerns the division of matter "into two sorts"—that is, "shining" and "opaque," or stars and planets. Are the two substances created differently from the outset? Do shining bodies begin as opaque but then light up through some mysterious process along the way? "I do not think [the two types of matter] explicable by meer [*sic*] natural causes," Newton professed, "but am forced to ascribe it to the counsel and contrivance of a voluntary agent. . . . Why there is one body in our solar system qualified to give light and heat to all the rest, I know no reason, but because the author of the system thought it convenient."[87]

The second Newtonian phenomenon that requires a divine supplement, we learn, is planetary motion itself. Under the influence of gravity, Newton explained to Bentley, matter is attractive. So gravity alone can account for the planets' movement *toward* the sun, but not for their movement *around* it; in fact, left to its own devices, gravity would cause all the opaque cosmic bodies to crash into the shining one. So "gravity may put the planets in motion," he reasoned, "but without the divine power it could never put them into such a circulating motion as they have about the sun."[88] Orbital motion, he concluded, requires an "intelligent agent."

Finally, assuming that the "fixt stars" are bodies akin to our sun and that they may or may not have their own systems of planets and moons miraculously processing around them, Newton found it remarkable that they all were sufficiently far apart from one another to avoid attracting one another. Much as unassisted gravity would cause the planets to crash into the sun, it would also cause the suns to crash into *one another*, forming one giant "mass" of shining material. Something, then, must have separated these systems from the outset and must continue to keep them separate. This "something," interestingly enough, is not only or even primarily God. Rather, the first principle Newton invoked to save the universe from gravitational collapse is *infinity*.

Cosmic infinity is, in fact, the matter with which Newton opened his first letter to Bentley. If the cosmos were finite, he began, then "the matter on the outside of this space would by its gravity tend towards all the matter on the inside, and by consequence fall down in the middle of the whole space, and there compose one great spherical mass."[89] What Newton was envisioning here is similar to the phenomenon that twentieth-century physicists would come to call a "big crunch"—except that for Newton, space itself would remain extended throughout such an event, whereas for the post-Einsteinian (and neo-Leibnizian) tradition, a "big crunch" would draw space-time *itself* into a gravitational collapse. At any rate, Newton had come to believe that if the cosmos were finite, gravity would cause all the matter in the universe to implode.[90] If, however, the universe were infinite and homogeneous—if space and matter were to extend forever—then "it would never convene into one mass, but some of it would convene into one mass and some into another, so as to make an infinite number of great masses, scattered at great distances from one another throughout all that infinite space."[91] In this brief passage, one can detect the sketchy beginnings of a cosmogonic narrative: once God has created, distinguished, and distributed matter throughout an infinite universe, gravity draws the opaque bodies toward the shining ones. But, again, Newton never pursued this line of thinking. His central concern was rather to establish cosmic infinity as a kind of outward "push" to counteract the gravitational "pull." Once the fixed stars are created, however they are created, they never collide because they are equally attracted to one another in all directions forever.

The infamous atheist Edmund Halley (1656–1742), Newton's friend and collaborator, attested that this explanation does, in fact, ensure the noncollapsing nature of the cosmos. As Halley explained it, "[I]f the whole be infinite, all the parts of it would be nearly *in equilibrio*, and consequently each fixt star, being drawn by contrary powers, would keep its place; or move, till such time, as . . . it found its resting place."[92] For Newton, however, Halley's "nearly" amounted to impending chaos; as he went on to explain to Bentley, an infinite universe would be highly unstable and would remain in equilibrium only if each component of the universe were not "nearly" but *absolutely* equally attractive, in all directions, forever. Absolute stability, however, is difficult to attain; as Newton confessed, "I reckon this as hard as to make not only one needle only, but an infinite number of them (so many as there are particles in infinite space) stand accurately poised upon their points." So it is here that he appealed, once more, to God. It is impossible for gravity alone to keep the cosmic bodies balanced on their needle points, "yet I grant it possible, at least by a divine power; and if they were once to be placed . . . they would continue in that posture without motion for ever, unless put into new motion by the same power."[93]

In sum, then, Newton appealed to God in order to account for the creation, distribution, and repulsion of matter—the same functions for which Descartes appealed to God. Unlike Descartes, however, Newton believed that God was required to *keep* the cosmos in balance. Whereas Descartes was happy to call God onto the stage at the beginning and then push him off for the rest of the cosmic show, Newton thought that the laws of nature could not function on their own; in particular, they depended on God to intervene periodically and correct irregularities. This was another point on which Leibniz would disagree vehemently with Newton: "[A]ccording to [his] doctrine," Leibniz chided Clarke, "God Almighty needs to wind his watch from time to time, otherwise it would cease to move. He did not, it seems, have sufficient foresight to make it a perpetual motion."[94] But, again, theology was not Newton's primary or even secondary concern (it is telling, for example, that he delegated to Clarke the responsibility of debating these points). And considering how reluctant he was to imagine the origins of *this* world, Newton certainly had no desire to imagine the birth of other worlds. He did briefly entertain the notion in his penultimate letter to Bentley that "there might be other systems of worlds before the present ones, and others before those and so on to all past eternity."[95] There might, in other words, be Stoic cycles built into the infinite, quasi-Epicurean universe. But just like the rest of Newton's cosmological principles, such cyclicality could be accomplished only by God: "[T]he growth of new systems out of old ones, without the mediation of a divine power, seems to me apparently absurd."[96]

Although Newton did not write a cosmogony, then, he opened the possibility of such a thing through his brief references to an infinite universe filled with infinite worlds—worlds, moreover, that *might* undergo periodic destruction and re-creation. And although a number of astronomers and theologians, including Bentley himself, would go on to explore these tenuous openings,[97] the most extravagant Newtonian cosmogony would emerge in the early writings of Immanuel Kant.

An Immeasurable Abyss: Immanuel Kant's
Cosmic Symposium

Failure to Launch

Published anonymously when Immanuel Kant (1724–1804) was thirty-one years old, the *Universal Natural History and Theory of the Heavens* (1755) announces itself as "an essay on the constitution and mechanical origin of the whole universe [*Weltgebäude*], treated according to Newtonian Principles."[98] In the words of the philosopher Milton Munitz, this book "ventures where Newton feared to tread," offering a full cosmogony based on the law of gravitation: a Newtonian replacement for Descartes's demolished vortices.[99] Unlike Descartes's cosmogony, however, Kant's was almost completely unknown in his own time. Just after printing the *Natural History*, his publisher went bankrupt and was forced to impound all the unsold copies. A year later, the book was reprinted under Kant's own name and dedicated to Frederick the Great, but neither the king nor anyone else seemed to notice. Johann Lambert (1728–1777) had not even heard of Kant's *Natural History* when he published many similar (and many different) ideas in *Cosmologische Briefe* (1761); and even after Lambert's book gained a degree of recognition, Kant's did not.[100]

As Munitz has written, the *Natural History* "had to wait for more than a century for its true greatness to be appreciated."[101] And appreciated it has been: it is now not uncommon to hear Kant's early theory hailed as "the first scientific cosmology"—a cosmology, moreover, that anticipates "the essence of modern models."[102] But, of course, not everybody shares this view. The book is almost never taught in philosophy or history of science curricula, and there is painfully little secondary material on it—especially compared with the vast body of work on Kant's three critiques, *Religion Within the Limits of Reason Alone*, and late essays.[103] Perhaps the book's most trenchant modern critic has been Stanley Jaki, who in a lengthy introduction to his own translation of the *Natural History* charges Kant with displaying a "tenacious amateurism" in physical science, indulging in "willful and confused speculations," and "constantly referring to Newton without ever studying him seriously." In short, Jaki concludes, "Kant's parading in Sir Isaac's cloak [is] a shabby performance," and whatever modern principles he anticipated are the result of a "lucky guess."[104]

But the most likely reason for this text's continued obscurity is neither that it was ahead of its time nor that it was behind in its science, but that it ventured beyond "the limits of reason alone." As numerous scholars have noted, the wildly imaginative *Natural History* of Kant's youth is not only overshadowed by his later work but invalidated by it.[105] After all, the critiques confine our epistemological reach to "phenomena," things as they appear, whereas the *Natural History* accounts not only for things as they are, but also for the process by which all things have come to be in the first place. As Kant himself admits in a preface to the work, "[S]uch ideas seem to surpass very far the forces of human reason."[106] And yet, he assures us, "I do not despair. I have ventured, on the basis of a slight conjecture, to undertake a dangerous journey and I already see the promontories of new lands" (81/39).

Atomism, Enlightened

Like any good cosmogonist, Kant starts the *Natural History* with chaos: "I set the first state of matter [*Zustand der Natur*] . . . in the universal dispersion of the basic stuff [*Urstoff*], or atoms [*Atome*]" (85/45). And so we are back with the Atomists, to whom Kant is inexorably drawn throughout this work, no matter how fiercely he assails them.[107] Their godlessness is loathsome; their principle of "accident," a joke. Yet "even in the most senseless opinions," Kant concedes, "one can at times notice something true" (85/45). This, it should be said, is a colossal understatement. The "something true" that Kant can "at times notice" in the Atomists is nothing less than Atomism itself, right up to its infinite universe and infinite worlds. The only significant difference is that the whole thing is set in motion by God and executed according to "necessary laws" (86/46). Kant calls these laws "Newtonian" and offers them as a correction to the "blind concourse" of the Epicureans. Nevertheless, the laws *are* Epicurean.

To return to our cosmogony, what we have so far is a chaos of uniformly distributed atoms. Kant continues: "Epicurus posited a gravity which drove these elemental particles to sink" (85/45). Here, we might note that Kant is calling to mind the "second" type of Lucretian chaos: rather than a primordial *tempestas*, we have the downward cascade, what Michel Serres calls the "laminar flow" that pulls atoms down "in a straight line through the void."[108] Kant calls this falling force "gravity" (*Schwere*) because it "seems not to be very different from the Newtonian attraction [*Anziehung*], which I assume" (85/45). This, then, is the way Kant will cosmogonize Newton: by mapping gravity onto Lucretius's laminar flow. As we will recall, however, laminar flow alone will never make a world. The atoms have to "swerve a little."[109] And so Kant borrows from the Epicureans a certain "deviation [*Abweichung*] from the rectilinear motion of the fall"— even though he calls their account of it "absurd" (85/45). Kant's swerve will happen not by chance, but by "the repulsive force of the particles [*der Zurückstoßungskraft der Teilchen*]," which acts in tension with the attractive force of gravity (85/46, translation altered slightly).

Unfortunately for Kant, however, there is no Newtonian equivalent of this "repulsive force"—as we have seen, in fact, this lack was part of what prevented Newton from thinking cosmogonically himself. Gravity accounts for matter's attraction, but not for its dispersion or rotation, and so as Newton wrote to Bentley, he could never explain how matter "became evenly spread throughout the heavens, without a supernatural power."[110] In this light, Kant's appeal to a naturally "repulsive force" constitutes a significant deviation from Newton in two senses: it introduces a non-Newtonian law, *and* that law has nothing to do with a "supernatural power." But it falls right in line with the Epicureans. In effect, Kant's repulsive force is none other than the Epicurean "swerve." And when atoms are subjected to the attraction that makes them fall and the repulsion that makes them swerve, Kant tells us, they form a vortex (*Wirbel*), and a vortex makes a world (85/46).

Being "Newtonian" (and Epicurean), Kant's vortices differ significantly from those of Descartes. First, they form not in a gapless plenum, but "in a non-resistant space," which Kant, himself swerving around the void debate, describes as "completely empty, or at least as good as empty."[111] Second, Kant's vortices are not the places of solar systems, but the processes that produce them. And third, they form not around the fastest particles, but around the most attractive ones.

Beginning again from the laminar flow, Kant imagines a "point" (*Punkt*) in the midst of a "very large space . . . where the attraction of elements . . . works stronger than anywhere else" (115/89). This most attractive point will draw innumerable atoms toward itself, forming "a body in that center of attraction which grows . . . from an infinitely small seed" into an increasingly massive core, whose weight draws even more particles toward itself (115/90). As the body grows, it begins to pull particles toward itself from greater and greater distances, at which point even the slightest repulsion between them can "bend" their velocity "sideways," producing "great vortices of particles [*große Wirbel von Teilchen*]" (116/90). At first, these vortices intersect one another, the flood of particles colliding in different circles through all imaginable planes. But this very conflict gradually brings the particles "into equilibrium" so that they all begin to move in the same direction in "horizontal, that is, parallel-running orbits" around this strongly attractive body at the center, which has become a sun (116/90). The bodies moving in orbits, of course, become the planets, moons, and comets.

Theological Detour

This, then, is the way a solar system comes into being for Kant, and it is striking that God does not seem to figure into the process at all, let alone the particular places that Newton had reserved for God. In his preface to *Natural History*, Kant explains that this godlessness had been a significant concern from the moment he began the project at hand. If a chaos of atoms can bring itself into perfect order, if "nature is sufficient to herself," then it seems to have no room for God: "[D]ivine government is unnecessary, Epicurus lives again in the midst of Christendom and an unholy philosophy [*Weltweisheit*] tramples underfoot the faith which . . . enlighten[s] philosophy itself" (82/40). This, we may recall, is precisely what Lucretius says Epicurus has done: freed himself from the "crushing weight of religion" to leave it "trampled underfoot." Such trampling allows Lucretius to travel "beyond the fiery ramparts" of the world to see an infinite number of others,[112] and indeed, this seems to be the passage Kant is channeling when he tells us his "dangerous journey" has allowed him to "see the promontories of new lands" (81/39). Finally, just as Epicurus's journey made him "a god,"[113] Kant promises that those "who have the courage to continue the investigation" will "designat[e] with their own names" the lands that they find there—mastering, naming, and in that sense possessing them all (81/39). All this notwithstanding, Kant insists that his account does not do away with the existence of a "highest being" but, to the contrary, *proves* it.

Kant proves the existence of God not by filling mechanical gaps with the deity, but by demonstrating God's evidence within the whole cosmogonic process. Impishly, Kant accomplishes this task by using the Atomists' two antitheistic principles, accident and infinity, against them. In the face of the Atomist doctrine of "accident," Kant sets forth a straightforward teleological proof (the very one that Lucretius undermines, David Hume ridicules, William Paley reasserts, and Kant himself will dismantle less than a decade after writing the *Natural History*).[114] The order and harmony among the elements of the universe, Kant argues, testify to its having been designed by a supremely intelligent, good, and powerful creator. To the objection that his own physics seems to indicate otherwise, Kant cleverly replies that the very appearance of a lack of design is a proof of design. That is, the sheer fact that a primal cacophony of particles can produce anything at all is "undeniable proof" of their having originated in a brilliant being with a master plan. In short, *"there is a God precisely because, even in chaos, nature can proceed in no other way than regularly and orderly"* (86/47, translation altered slightly, emphasis in original).[115] Kant then goes on to unravel the Atomists' atheism through their principle of "infinity." In an infinite universe, Lucretius tells us, everything that can happen will happen, so there is no need to appeal to the gods when asking why things are as they are; rather, everything is bound to happen somewhere, given enough time and space.[116] Again, Kant's response is that this boundless field of possibilities demonstrates precisely what Lucretius thinks it repudiates: the very infinity of creation is "a witness of that power which can be measured by no standard."[117] In other words, nothing but an infinite cause could produce an infinite effect, and so the infinite universe must be the product of divine contrivance.

It is a strong sign of the times that Kant does not hesitate at all to call the creation "infinite." Unlike Descartes, who a little more than a hundred years earlier called the universe "indefinite" to signal its inferiority to God, Kant insists in the *Natural History* that the revelation (*Offenbarung*) of God is just as infinite as God himself (151/145). One can even hear Brunian echoes as Kant says that anything short of an infinite creation would be an insult to an infinite God: "Now it would be senseless to set God-head in motion with an infinitely small part of his creative ability and to imagine his infinite force . . . locked up in an eternal absence of exercise" (151/144). It is "far more proper," Kant concludes—even "necessary"—to find the infinite glory of God reflected in an infinite creation (151/144). What, then, does Kant's infinite creation look like?

Cosmic Ascent

We embarked on the foregoing theological detour after having witnessed the formation of a sun and its planets. Together, Kant explains in *Natural History*, these bodies constitute a "system" because they move in a common plane around a common center (112/85). The resulting "solar world" (*Sonnenwelt*) is sustained by the continued operation of our two primeval forces: the attractive force that pulls bodies toward the center and the repulsive force that pushes them away (98/65). It is the steady tension between these forces that keeps the planets and moons in orbit, with the attractive force preventing them from flying outward into infinity and the repulsive force preventing them from crashing into the sun. Moreover, Kant explains, the repulsive force provides the "small deviation" from circularity that results in Kepler's ellipses (118/94).

From here, Kant throws off Newton's typical restraint on the matter, announcing with little fanfare that our solar world cannot possibly be the only one. Rather, "the fixed stars, as so many suns, are centers of similar systems in which all may be arranged just as greatly and orderly as ours" (101/70). Thus begins an ascending meditation on the staggering scope of creation, an "infinite cosmic space [that] swarms with worlds [*von Weltgebäuden wimmele*] whose number and excellence has a relation to the inexhaustibility of their creator" (101/70). Now, as far as Kant is concerned, the plurality of solar systems is simply not up for debate; everyone "since Huygens' time believes that [all] the fixed stars are . . . suns" (101/70). What is less clear is whether these suns bear any relation to one another; indeed, from the seemingly haphazard positions of the stars, most people are led to think that they exist in no order at all (101/70).[118] But the notion that systematic regularity might apply only on "the small scale" and not "among the members of the universe [*Weltall*] at large" is perplexing to Kant. He therefore appeals to Thomas Wright of Durham, whose *Original Theory or New Hypothesis of the Universe* (1750) proposes that the stars of the Milky Way compose a system analogous to our solar system. The book had been reviewed a year after its publication in a German periodical, and although Kant had not read the work itself, the review had made such an impression on him that he credits Wright with having "first" given him "the prompting to look upon the fixed stars not as a scattering swarm with no visible order, but as a system" (88/50–51). Yet Wright, Kant claims, was not clear about the shape of this system, nor had he articulated a resemblance between the Milky Way and other cosmic phenomena.[119]

In order to "improve" upon Wright, Kant therefore employs a mixture of observational tactics and strange geometry, the conclusion of which is that all the stars of the Milky Way occupy a common plane. For Kant, this means that they constitute a system; in effect, the Milky Way is a solar world writ huge, an ordered set of relations that Kant calls a "world-system" (*Weltsysteme*) (103/72–73). By the rule of analogy, he infers that this higher system must be sustained by the same forces that order the solar world. Mindful, however, of how far one star is from any other, Kant postulates that each of them cannot exert a gravitational pull on each of the others. Rather, "the attraction of the sun should reach to about the nearest fixed star," which in turn tugs on its neighbor in what William James might call a "strung-along" fashion, so that "the entire host of [stars] would be driven to approach one another" through the attractions each exerts on its nearest neighbors (103/73). But, Kant reminds us, the attractive power of gravity alone would cause the "world-system" of the Milky Way to "fall together sooner or later into one lump," just as it would do in our solar system, if there were not that other non-Newtonian force, the Epicurean repulsive force that Kant now also calls "centrifugal." This repulsive force works against gravity to produce "the eternal orbital motions whereby the edifice of creation is secured from collapse and made fit for an imperishable duration" (103–4/73–74).

If it is the case that the Milky Way is a "system" analogous to (but greater than) our solar system, then three further conclusions follow. The first is that the galaxy must have a center (Kant proposes the star Sirius as the most likely candidate) (104/74). The second is that the "fixed stars" must actually move around this center—not once a day, as they did around the Ptolemaic earth, but perhaps one degree every four thousand years (which would explain why twenty-five hundred years of observations had not recorded any change) (105/76).[120] And the third is that just as our solar system is not the only solar system, the Milky Way cannot be the only Milky Way (106–7/77–78).

With this last conclusion, Kant had come upon a truly new idea. Wright had not proposed it directly (although he had gestured toward it in his penultimate and more or less unexplained diagram [figure 4.2]), Johann Lambert stopped short of it, and no one much before them would have had the language to propose it.[121] In 1786, the observational astronomer William Heschel would present evidence that the nebulae were "no less than whole siderial systems, [which might] well outvie our Milky Way in grandeur."[122] Nevertheless, this multigalactic proposal would lose all credence during the nineteenth century. As we will see in chapter 5, Albert Einstein believed for half of his life that there was only one galaxy; it was not until Edwin Hubble's discoveries of the Andromeda and Triangulum galaxies in the early 1920s that a different view became plausible. Unlike Hubble, of course, Kant did not observe other galaxies directly; rather, he reasoned to them from analogy. If the Milky Way is structured like a solar system, he argued, and if there are countless other solar systems, then surely

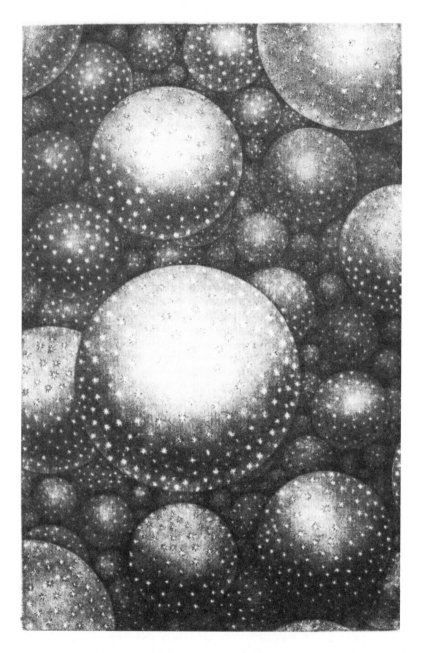

FIGURE 4.2 The "finite view of infinity" of Thomas Wright of Durham. (From Thomas Wright, *An Original Theory or New Hypothesis of the Universe [1750]: A Fascimile Reprint Together with the First Publication of A Theory of the Universe [1734]* [New York: American Elsevier, 1971], 174, pl. XXXI. Reproduced by permission of the publisher)

there are countless other Milky Ways as well. In an infinite universe, what would prevent that from being the case? And would countless Milky Ways not reflect the omnipotence of God far better than a lone galaxy would?

But Kant's conjecture is not strictly analogical (or, for that matter, theological). With help from Pierre-Louis Moreau de Maupertius's astronomical descriptions (published mostly in the 1740s),[123] he reasons in the *Natural History* that if there *were* another Milky Way, it would be so distant from us that it would appear "as a small space illuminated by a weak light, [a space] whose figure will be circular when its plane presents itself directly to the eye, and elliptically when it is seen from the side" (106/77–78). And, of course, such phenomena do exist. They are the cloudy figures astronomers call "nebulae," which Galileo thought were small clusters of stars and Maupertius thought were single stars that were "less luminous" than others and "somewhat flattened."[124] Kant's reasoning is that if the nebulae were single stars, and if they were as distant as the other stars, then they would have to be immense in order to appear so large to us. But if they were immense, then why would they look so pale? Kant therefore proposes that the nebulae are neither giant stars nor small clusters of stars, but *galaxies*—vast structures of "so many thousands" of stars separated by vast stretches of space (107/97).[125] Again, this idea would not be taken seriously until a century and a half later, when Hubble realized that all the nebulae *are* structures like the Milky Way, so that far from there being only one galaxy in existence, there are "hundreds of thousands" of them.[126] But Kant does not stop there.

Why, he wonders in the *Natural History*, should galaxies be considered the highest-order systems? The law of analogy (along with the findings of somewhat more tenuous observation) can lead one, Kant reckons, to produce the "conjecture that even these higher world-orders are not without relation to one another and . . . constitute a still more immeasurable system" (108/79–80).

At this juncture, Kant ventures momentarily beyond the bounds of reason, leading his reader on an upward journey through the successive ranks of cosmic systems—a journey that fills him with increasing wonder. "If the greatness of a planetary world-edifice [*Weltgebäude*], in which the earth as a grain of sand is hardly noticed, moves the intellect to admiration [*Verwunderung*]," he begins,

with what astonishment [*Erstaunen*] will one be enchanted if one considers the infinite [*unendlich*] amount of worlds and systems which fill the total of the Milky Way; but how this astonishment increases when one realize[s] that all these immeasurable star-orders again form the unit of a number whose end we do not know and which perhaps just as the former is inconceivably great and yet again is only the unit of a new number system. We see the first members of a progressive relation of worlds and systems, and the first part of this infinite [*unendlich*] progression makes already known what one must conjecture about the whole. There is no end here but an abyss [*Abgrund*] of a true immeasurability in which all ability of human concepts sinks. (108/80)

In a striking topographical reversal, the journey upward ultimately casts Kant down. The higher he is able to climb through the power of intellect, the more dramatically he finds himself sinking into the unintelligible. So "human concepts" will never be able to grasp the entirety of this "progressive relation of worlds and systems," plunging us instead into "an abyss of a true immeasurability." The only options, it seems, are to fall silent in astonishment or to try to dust oneself off from the journey and resolve to measure the immeasurable. Kant opts for the latter.

Climbing down from his mountain and up from his abyss, Kant tells us that if the space of creation is infinite, if the stuff of creation is infinite, and if cosmic bodies are arranged in greater and greater systems, then "the cosmic space will be enlivened with worlds without number and without end" (152/146). As we have just seen, however, our philosopher—much like those in the century that preceded him—is both amazed and terrified by these worlds without end. So one way to overcome this riot of emotion is to ask whether this endless progression of systems might *itself* constitute a system—whether there might be an übersystem within which all the supergalaxies, galaxies, and solar systems are somehow nestled. As Kant phrases the question, "[W]ill now that systematic connection extend also to the whole and hold together the entire universe [*Universum*], the all of nature [*das All der Natur*], in one such system through the connection of attraction with centrifugal force?" He answers immediately, "I say yes" (152/146). And so here Kant begins a complicated retreat from infinity.

Out of the Abyss

There must be a single System of systems, Kant reasons in the *Natural History*, because if there were none, then there would be no overarching force to regulate the attractions among different systems. Left to the whims of gravitational variation (we have heard this story before), the bodies of a hypothetically polysystematic universe would fall into a massive cosmic implosion. And even if these separate systems happened to have equal gravitational force, Kant imagines, then "the smallest displacement in the universe would . . . hand it over . . . to collapse" (152/146). This would make the continued existence of the universe a perpetual miracle (*Wunder*), and Kant, taking a sudden Leibnizian swipe at Newton, says that God does not work by means of intermittent wonder working or by holding up the stars on needle points. "One . . . hits the mark far more appropriately," he declares, "if one makes from the entire creation a single system" so that the harmonious order among cosmic bodies becomes a function of the laws of nature themselves (152/146). Kant therefore decides that the endless "chain of members" he has been contemplating cannot be "*truly infinite [wirklich unendlich]*"; rather, it finds its end in a single, universal System (152/146, emphasis added). In other words, the world is not "immeasurable" at all, and "human concepts" need not give way to helpless astonishment.

Back on solid ground, Kant continues: if God has indeed made the world a single System, then he has made "related to *a single center* all worlds and world-orders [*Welten und Weltordnungen*] that fill the entire *infinite space*" (152/146, emphasis added). This regrounded line of thinking immediately runs into a serious difficulty, though. In order to say that there is one universal System, Kant has decided that the progression of worlds cannot be "truly infinite." Now, just a few sentences later, he says that space *is* infinite, so presumably the progression of worlds *could* in principle go on forever, yet Kant insists that the whole thing ends in one all-embracing System. At what point, then, does the endless progression end? Where, in infinite space, might one find the border of the übersystem (and what happens if Lucretius throws a spear at it)? Kant diverts our attention from this problem by focusing not on the periphery of this System, but on its "single center," a *Mittelpunkt* around which "all worlds and world-orders" must be organized. But this middle point is equally perplexing: as any reader of Cusa or Bruno knows, an infinite universe has no center.

Kant does acknowledge that, geometrically speaking, there is no center in an infinite universe. But if there is no center, then there is no System, and there must be a System. So Kant resolves to designate as "center" the place where creation began: that "lump of . . . exceptionally large attraction" that first pulled the falling atoms toward itself (152/147). Because worlds first began to form around this lump, Kant tells us, they will be more densely clustered there, growing sparser as they get farther away. And because the "primeval stuff [is] piled up considerably thicker . . . and increases in its scattering with distances from that point," Kant concludes that "such a point can have the privilege to be called a center" (153/148). In effect, he just displaces the spatial problem temporally, "centering" the infinite around its purported place of origin. But *where*, exactly, can an infinite universe be said to have begun?

In a footnote to this perplexing sentence, translator Stanley Jaki writes that "this reasoning shows Kant a most uncritical thinker, though neither the first or [*sic*] the last of those who, being infatuated with infinity, lose their critical faculties" (284n.23). I would suggest, however, that the problem is not Kant's infatuation with infinity, but his fear of it. The problem is Kant's infatuation not with infinity, but with *oneness*, which requires him to have an all-encompassing System, which requires him to posit a center, which necessitates his *retreat* from the infinite. Moreover, I would submit that the error seems to stem not from Kant's having lost his critical faculties, but from his desperate attempt to regain them. He makes no such blunder as he is ascending in amazement through an infinite progression of worlds. It is when he comes back down and tries to assemble them into a single, nested System that he goes wrong and tries to center the centerless.[127]

Either way, however, Kant *thinks* that he has managed to organize the infinite universe around one gravitational center. He believes this center to be the starting point of creation, with worlds, galaxies, and supergalaxies fanning out from it in all directions, becoming sparser and sparser the farther they find themselves from the center. Of course, if worlds began to form at one place in one time, and if they thin out from that "center," then there may be an immense number of them, but there is not an *infinite* number of them. Space, as it turns out, is infinite for Kant, but the worlds within it are numbered.[128] Because these worlds are not genuinely innumerable, one might therefore conclude that creation is *not* as infinite as its creator. It is perhaps to prevent our drawing this conclusion that Kant, at this late stage, suddenly adds another major principle to his cosmogony. "Creation is never completed," he asserts. "Though it has once started[, it] will never cease. It is always busy in bringing forth more scenes of nature, new things, and new worlds" (155/150). This, then, is the sense in which there *are* an infinite number of worlds: worlds are *temporally* infinite. As Kant puts it, creation "needs nothing less than eternity . . . to enliven the entire limitless reaches of infinite space with worlds without number and without end" (155/150–51).

If one were able to see the Kantian universe as a whole, it would therefore look like a vast sphere expanding into infinite space, with fully formed worlds, galaxies, and supergalaxies at the center; half-formed world-systems around them; and chaos on the periphery—where an infinite number of atoms lie (or fall) in wait, ready to form a never-ending number of new worlds (155/151).

Just as we have begun to wrap our minds around this image, however, Kant tosses in yet another new principle. Having established that creation is never finished, he suddenly announces that "every world that has been brought to completion gradually shows [a tendency] toward its destruction."[129] So it seems that systems do not enjoy "imperishable duration," as he said earlier, in line with Bruno and Plato (103–4/73–74). Now he channels Lucretius instead to tell us that just as animals and plants perish, so will worlds themselves (157/154). The corollary, of course, is that just as nature brings forth new plants and new animals, so does it build "new formations in other celestial regions . . . repairing the loss with gain" (157/154). Therefore, the death of the old worlds is compensated by the birth of the young ones. And assuming that worlds live longer than the time it takes to make them, Kant figures there will always been a net gain of *kosmoi* (159/157).

So with this new insight, a revised bird's-eye view of the universe would show a vast sphere expanding into infinite space, with destroyed worlds, galaxies, and supergalaxies at the center; older world-systems in a ring around them; younger world-systems in a ring around *them*; half-formed world-systems in a ring around *them*; and an infinite sea of *Urstoff* (basic material) on the periphery. The upshot of all this is that "the developed world finds itself confined between the ruins of a collapsed nature and the chaos of an undeveloped one" (159/157, translation altered slightly).

Lest we think that this is the final vision, however, Kant adds one last big idea to the end of his increasingly extravagant cosmology. The idea, he says, is "just as probable" as the notion that worlds perish in one place to be created elsewhere, and it is just as "befitting the constitution of divine works" (159/157). It has to do with those ruins that the dead worlds leave behind. "Can one not believe," he asks, "that nature, which was capable of placing herself from chaos into regular order and a skillful system, will not be in the position to restore herself from the new chaos . . . and to renew the first combination?" (159/157). Can one not believe, in other words, in the eternal return of the same? So, with very little warning, Zeno is back on the scene, now mingling awkwardly with Newton, Leibniz, and Lucretius. In case we are uncertain that there really is a Stoic in our midst, Kant proceeds to tell us how this cosmic recycling will take place. When the systems of our world break down and come to a standstill, he explains, all the "planets and comets" (presumably still under the force of gravity) will crash into the sun. The sun, in turn, will "obtain an immeasurable increase through the mixture of so many and great lumps" (159/157). The resulting conflagration will "not only dissolve again everything into the smallest elements, but these will again expand and scatter . . . into these same distant spaces which they occupied before the first formation of nature" (159/157). Thus rebegins the cosmogony, with "a new world-edifice" emerging from the ashes of the old "through the combination of . . . attraction and repulsion" (160/158). And toward the end of this very strange paragraph, Kant calls the "astonishing" phenomenon he has just introduced "the phoenix of nature" (160/158).

He unfortunately says very little else about this phoenix, pausing just long enough to make it clear (pace Augustine) that human souls do not undergo such cycles before he ends the section (161/159–60). And so we do not learn, for example, whether this conflagration happens just at the level of solar systems, whether it extends out to galaxies, or whether it involves the whole universe. Nor do we learn how to reconcile this sudden cyclicality with the remaining sections of the *Natural History*, which go on to arrange the inhabitants of Kant's "worlds without end" into an Enlightenment-flavored Great Chain of Being, with the least intelligent beings in the old worlds toward the center and the most intelligent in the new worlds on the periphery (167–68/169–70). One is therefore left wondering: After the conflagration, do the old worlds become new? Does the center thereby become the periphery? Might a fuller reflection on the phoenix somehow save Kant from his strange hierarchy, his impossible center, his infatuation with oneness? Kant never lets us know.

What he offers us instead is a riot of incompatible elements, a swarm of cosmological contradictions that has been gathering throughout this treatise and that gets louder and more frenetic as it ends: "Newtonian" gravity with a Lucretian swerve; Lucretian cosmogony with an omnipotent God; Leibnizian theology in Newtonian space; Timaean eternity with an Atomist demise; Brunian infinity with a Ptolemaic center; endless worlds in a Thomistic hierarchy; and, at the end of it all, a Stoic conflagration that burns the world into Epicurean *atoms*, pulling material bodies toward itself only to fling them, in a burst of fire, out to the infinite. In this final contradiction and expulsion, one can moreover detect proleptic hints of the oscillating big bang model, but Kant's retention of Newtonian space prevents it; space, for Kant, stands still, whereas matter is drawn to the "center" of infinity and then somehow thrown outward again. The *Natural History*, in the end, is remarkably, ingeniously *crowded*: a late Enlightenment kitchen sink of past, present, and future cosmologies. You might even call it chaos.

5

BANGS, BUBBLES, AND BRANES

Atomists Versus Stoics, Take Two

There is chaos, there is circumstance, and suddenly there is the whole
foundation. There is the background noise, then a noise in the midst of that
background noise, and suddenly there's the whole song.

MICHEL SERRES, *GENESIS*

There is an embarrassing vagueness about the very beginning.

STEVEN WEINBERG, *THE FIRST THREE MINUTES*

Let There Be Light

The century and a half that followed Kant's uncelebrated *Natural History*
was a time of almost total multiversal dormancy. This dormancy was both
marked and reaffirmed by the publication in 1796 of Pierre Laplace's *Exposition du système du monde*, which ordered the Kantian chaos into what
became known as the "nebular hypothesis," according to which a hot,
rotating nebula produces a sun and its planets.[1] What the next generations
took this to mean, although Laplace (1749–1827) himself waffled on the
matter,[2] was that the far-off nebulae were at worst just clouds of gas and at
best "potential star-clusters" of "inchoate suns."[3] Laplace's cosmology became "the standard Victorian model of the universe," to such an extent
that one hundred years later the historian of astronomy Agnes Clerke
would write that "the question whether nebulae are external galaxies hardly
any longer needs discussion. . . . No competent thinker . . . can now, it is
safe to say, maintain any single nebula to be a star system of coordinate
rank with the Milky Way." Rather, she insisted, all the visible objects in
the night sky, "stellar and nebular . . . belong to one mighty aggregation,
and stand in ordered mutual relations within the limits of one all-embracing
scheme," which is to say the Milky Way itself.[4] With very few exceptions
(including, remarkably, Edgar Allan Poe),[5] this "galactocentric" viewpoint

held well into the 1920s, so that when Albert Einstein (1879–1955) published his theory of general relativity in 1916, he believed that the universe consisted of nothing but the Milky Way—a single "island" surrounded by an endless void. He also believed that the universe was eternal. On the largest cosmological scales, then, the twentieth century began more or less where Aristotle had left off: with a single, static world that had existed forever. Then the 1920s hit, and the whole model was suddenly blown apart.

The uniqueness of the Milky Way was dramatically dismantled in 1924, when the *New York Times* announced that Edwin Hubble (1889–1953) and his assistant Vesto Slipher had determined "the spiral nebulae, which appear in the heavens as whirling clouds," to be galaxies in their own right. Kant, we will recall, named these structures "world-systems"; Hubble tended to call them "island universes."[6] Hubble had confirmed Kant's guess by observing luminous stars in the Andromeda and Triangulum (or Messier 33) galaxies, but he also extended his conclusions by analogy to all the other nebulae in the cosmos, which seemed to him even then to number in the "hundreds of thousands."[7] Although it would take until the mid-1990s for the world to *see* these hundreds of thousands of galaxies—which turned out to be *billions* of galaxies (figure 5.1)—Hubble's discovery immediately increased the size of the known universe and the number of worlds within it by a factor of half a million, give or take a few hundred thousand.[8] This cosmos, to borrow an Irigarayan turn of phrase, suddenly seemed not to be *one*.[9]

As for cosmic eternity, it also began to look implausible in the light of Hubble's discovery. As he determined in the years that followed, these newfound galaxies were not remaining in place; rather, almost all of them were moving away from one another. More precisely, the space between them was growing and pushing them apart.[10] To be sure, this was a strange proposition to early-twentieth-century ears. As a depressed young Alvy Singer explains to a therapist in Woody Allen's *Annie Hall*, "The universe is *everything*."[11] So, by definition, there is nothing "out there" for the receding galaxies to recede *into*. What Alvy realizes, what prevents him from doing his homework ("What's the point?" he asks his mother), is that something is pushing the whole world "out" into nothing at all. In more familiar terms, and as Alvy intones miserably, "the universe is expanding."

Although Hubble observed this phenomenon directly, the possibility of an expanding universe had been opened a decade earlier by Einstein's theory of general relativity, which had shocked the world in 1916 by revolutionizing

FIGURE 5.1 The Hubble Deep Field, in which each spot of light is a galaxy. The first of these images was produced in 1995, when the Hubble Telescope was trained for ten days on a patch of "empty" sky. (Space Telescope Science Institute, NASA, ESA, G. Illingworth, D. Magee, P. Oesch, R. Bouwens, and the HUDF09 Team)

Newtonian gravity.[12] According to general relativity, space and time do not constitute the impassive, "absolute" background that Newton had assumed; rather, they compose a dynamic, neo-Leibnizian "space-time" that can grow, shrink, bend, and warp in relation to matter and energy.[13] Prior to Hubble's observations, then, Einstein's own theory suggested that space-time might be either expanding or contracting. Parts of the universe—even the whole universe itself—might be either racing outward or retreating inward thanks to the warping effects of gravity.

Einstein's own calculations notwithstanding, this possibility was notoriously difficult for him to abide. One Newtonian belief he retained was that the universe must be static. Something must be regulating the elasticity of space-time, otherwise gravity, which exerts an attractive force on matter, would draw the whole world into a fiery collapse. At this point, we have seen numerous thinkers struggle with this problem. It is clear that matter acts attractively under the influence of gravity, but not at all clear what acts repulsively to keep matter distributed through the cosmos. Both Descartes and Newton ended up relying on God, not only to create matter, but also to provide the outward push that scatters it through the universe. Kant, by contrast, found his repulsive force not in God, but in the Epicurean "swerve," or *clinamen*. Einstein, faced with a similar difficulty, appealed neither to God nor to the swerve, instead positing a negative pressure, which he called the "cosmological constant" and designated by the Greek letter lambda (Λ), to offer an equal and opposite push to gravity's pull.[14] With gravity's attraction and lambda's repulsion in perfect proportion, Einstein was able (theoretically at least) to keep the universe from stretching out or caving in.

Upon hearing the news of Hubble's discovery—that space-time was not standing still, but instead expanding "outward"—Einstein immediately retracted his cosmological constant, calling it his "biggest blunder."[15] In the meantime, the work of a young Belgian physicist and priest named Georges Lemaître (1894–1966) was vindicated. Beginning in 1927, Lemaître had posited an expanding universe based on Einstein's equations, provoking Einstein to tell him, "[Y]our math is correct, but your physics is abominable."[16] The Russian mathematician Alexander Friedmann (1888–1925) had reached a similar conclusion five years earlier in his groundbreaking solutions to Einstein's gravitational equations, but he died before Hubble's discovery could confirm his findings.[17] As for Lemaître, he took courage from Hubble's fortuitous observations, going on to suggest that if the universe is expanding now, perhaps it has always been expanding—from the beginning of things. Perhaps if we could play time backward, we would see galaxies getting closer and closer together, the space between them contracting, until we would finally reach a moment when the whole universe would have been crammed into *one point*. The scientist-priest, soon to be made monsignor, concluded: the whole universe, in all its multiplicity, seems to have burst forth from one tiny ball of nuclear fluid, which he called the "Primeval Atom."[18]

This story is so familiar to us now that it is hard to imagine that it was ever controversial. But the middle decades of the twentieth century witnessed impassioned debates not just over the particulars of this strange new cosmogony, but also over the possibility of a scientific cosmogony in the first place. From the dawn of the seventeenth century through the dawn of the twentieth, observational astronomers had refrained from speculating about the origins of the universe, leaving such meditations to philosophers and theologians. Clerke summarized this restraint in 1905, saying, "[W]ith the infinite possibilities beyond, science has no concern."[19] By Einstein's time, then, the very notion that the universe could have *begun* at all had become scientifically distasteful; as Steven Weinberg puts it, "[A]n aura of the disreputable always surrounded such research."[20] After all, if the world had a beginning, then what was there before the world? And how did the "before" give way to the "after"? The problem is that if we assume that the universe had a beginning, we are almost forced to appeal to some kind of supernatural force to get it going. Hence, the Victorian scientific consensus that the universe must have existed from eternity: if the world does not *have* a beginning, then we do not have to worry about what might have begun it.

The rise of the big bang hypothesis thus staged a return of the mythological at the heart of modern science. To the horror of many midcentury physicists, positing a beginning to the universe seemed to open their discipline constitutively to something beyond it. The most contested site of this opening was the big bang's beginning itself: $t=0$, when general relativity predicts that the whole universe would have been squeezed into a "point" of no size. Containing everything in nothing, the temperature and density of this point would be infinite, and, unfortunately, most laws of physics break down at infinity. Physicists call this "pathological circumstance," when the values hit infinity and the calculations jam, a "singularity."[21]

Although cosmologists spent the rest of the twentieth century debating the nature and necessity of such a singularity,[22] the floodgates to philosophy and, more troublingly, religion were most definitely open. In particular, many Christian theologians of the mid-twentieth century quickly saw in the big bang's infinite starting point a confirmation of orthodox doctrine.[23] As early as 1952, for example, Pope Pius XII declared to the Pontifical Academy of Science that "present day science, with one sweep back across the centuries, has succeeded in bearing witness to the august instance of the primordial *Fiat Lux* [let there be light], when along with matter, there burst forth *from nothing* a sea of light and radiation, and the elements split and churned and formed into millions of galaxies."[24]

And the big bang hypothesis, in its standard form at least, does seem to stage an uncanny recapitulation of Christian creation theology. Just as the church has taught, the universe is not eternal but had a temporal beginning. Just as the church has taught, it was born in a sudden flash of light. And just as the church has taught, the whole thing seems to have come out of nothing at all.[25]

Scholars of varying theological commitments agree that the doctrine of "creation out of nothing" is not exactly biblical; rather, it emerged during the second and third centuries C.E. as early church scholars vied for dominance over the so-called Gnostics, and the teaching was adopted centuries later into a number of strands of Jewish and Muslim theologies.[26] Nevertheless, the resemblance between modern science and what physicists considered a "biblical" mythos was so strong and so disturbing that it prompted the renegade British astronomer Fred Hoyle (1915–2001) to look for a different story to tell—preferably one that did not "aid and abet" religion.[27] As far as Hoyle was concerned, the chief danger of positing an absolute beginning to the universe was that "a 'something' outside physics can then be introduced at $t=0$," and an infinite "something" beyond the reach of physics is almost always called "God."[28] It was Hoyle, in fact, who called his opponent the "big bang" in the first place, intending to ridicule the theory he ended up baptizing. As an alternative and with the collaboration of the mathematicians Thomas Gold and Herman Bondi, Hoyle posited the "steady-state" model, according to which the universe has been cooled and extended from eternity, continually producing new matter to fill the expanding intergalactic void.[29] Insofar as this new matter would also be generated out of nothing, this model did not sidestep the Christian narrative altogether, but the key for Hoyle and his colleagues was that if we extrapolate backward from a steady-state universe, the cosmic density comes nowhere near infinity, and there is never a need for an imponderable, incalculable, *godlike* singularity at the beginning. There is, in fact, no beginning at all.

Unfortunately for these alternative cosmologists, their model was ruled out when radio astronomers Arno Penzias (b. 1933) and Robert Wilson (1927–2002) accidentally discovered the Cosmic Microwave Background (CMB) in 1965 (figure 5.2).[30] Known as "the surface of last scattering," the CMB is a sphere of radiation left over from the big bang—more precisely, from 380,000 years after the bang, when electrons and protons had cooled sufficiently to form hydrogen atoms, allowing light effectively to "decouple" from matter and roam the universe. As such, the CMB provides a thermal record of the aftermath of the primordial blast,[31] establishing that the world as we know it cannot have been eternal. Rather, all of it had a

FIGURE 5.2 The nine-year microwave sky. (NASA/WMAP Science Team)

beginning—and the same hot-as-hell beginning—in a burst of light, in time. As American astronomer Robert Jastrow sees it, the big bang's mid-century victory therefore seemed to confirm the "biblical view of the origin of the world." Jastrow infamously goes on to attest that "for the scientist who has lived by his faith in the power of reason, the story ends like a bad dream. He has scaled the mountains of ignorance, he is about the conquer the highest peak; as he pulls himself over the final rock, he is greeted by a band of theologians who have been there for centuries."[32] It's enough to drive a scientist mad.

As we have seen, the first point of contested connection between modern and "biblical" cosmology lies in the notion of an initial singularity: a timeless, immaterial, and infinite power that fires forth space and time out of nothing at all.[33] The persistent resemblance of the big bang's infinite creative force to the Abrahamic "God of power and might" prompted physicists over the decades to try to find an alternative model—preferably one that might rid itself of this "singularity" and its weighty theological baggage. With Hoyle's alternative off the table, some physicists therefore posited an "oscillating" universe as a new way around the singularity.[34] According to this model, the amount of matter in the universe is large enough to allow gravity to counteract the initial outward thrust, drawing the universe back after a few billion years into an increasingly smaller, denser state—until it collapses into what some have called a "big crunch" and others a "gnab gib" ("big bang" backward). At the moment of tightest compression, the cosmos "bounces" and explodes again, producing another rush of radiation and elementary particles, another batch of galaxies that race outward before they are drawn back in, and the cycle repeats eternally.

By making "The Big Bang" a mere moment in a succession of other bangs and crunches (a succession sometimes referred to as the "big brunch"), the oscillating model avoided the unsavory notion that anything—let alone the universe—might be created out of nothing. As the astrophysicist John Gribbin wrote in 1976, "The biggest problem with the Big Bang theory of the universe is philosophical—perhaps even theological—what was there before the bang? This problem alone was sufficient to give a great initial impetus to the Steady State theory; but with that theory now sadly in conflict with the observations, the best way round this initial difficulty is provided by a model in which the universe expands from a singularity, collapses back again, and repeats the cycle indefinitely."[35]

Interestingly enough, in this very gesture of circumventing Christian cosmology, the oscillating model collided head on with *Hindu* cosmology, which posits our age as just one in an infinite series of cycles. Friedmann himself, when considering the possibility of a bouncing universe in 1923, wrote that "one cannot help thinking of the tales from the Indian mythology with their periods of life."[36] This resonance, for many theorists, was far more welcome—perhaps even a tacit confirmation of the cyclical model's transhistorical, cross-cultural truth.[37]

As it turned out, however, the oscillating model was even shorter-lived than the steady-state model had been. Weinberg notes that although the cyclical universe "nicely avoids the problem of Genesis," it also runs headlong into the problem of entropy.[38] According to the Second Law of Thermodynamics, the entropy (or measure of disorder) of a closed system always increases. This means that each universe that bangs will begin with a higher measure of entropy than the one that crunched before it. More entropy amounts to more radiation, and more radiation means a longer period of expansion before the cosmos contracts again.[39] Each cosmos, in other words, would last longer than its predecessor. If the lives of universes get longer into the future, then one can only conclude that they were shorter in the past. Tracing these shortened universes back, we eventually come to a universe of no length at all. This means that the oscillating model leads just as inexorably as the big bang hypothesis to an absolute beginning, when the cycles must have started.[40] So much for avoiding the singularity. Moreover, if the total entropy increases with each universe, then it becomes impossible to begin each cycle in the kind of well-ordered (which is to say, low-entropic) state needed to produce a decent cosmos; the universe, it seems, *cannot* simply collapse and rise again without bringing along its "baggage" from the previous cycle. And so the physics community gradually gave way to the story from which it had spent decades recoiling: the universe does, indeed, seem to have begun with a bang.

Although the story has undergone countless revisions since it was first told, it goes more or less like this. One day—but words fail us here already because there were no days—13.8 billion years ago, a tiny nugget of immeasurable density issued forth a searing white light. The light radiated from an ultrahot, ultradense plasma, producing a great flood of energy as particles raced apart from one another.[41] This roiling chaos took 380,000 years to cool down sufficiently for atoms to emerge, leaving us with the thermal record of the CMB. Such cooling is attributable to a regime change that took place 75,000 years after the big bang, when the fires of radiation gave way to the gentler tug of gravity. From the crazy primordial plasma, gravity began to draw nuclei into atoms and atoms into molecules, eventually coaxing them into stars, galaxies, and eventually planets.

Then, at the turn of the third millennium, physicists began to realize that about 7.5 billion years after the big bang, the inward pull of gravity was overcome by the outward push of empty space, and "dark energy" began to push galaxies away from one another at a faster and faster rate (see introduction). So Einstein was not wrong after all. There *is* a countergravitational force (Λ), but it does not exist in a happy homeostasis with gravity to keep the cosmos steady. Rather, it is accelerating cosmic expansion with every passing moment. According to the now standard model, lambda currently constitutes about 73 percent of the universe.[42] This means that as time goes on and space continues to expand, the energy of "empty" space will occupy an increasingly greater percentage of the cosmic mass–energy, accelerating the acceleration at a dizzying clip. Galaxies will eventually race apart from one another faster than the speed of light until they disappear from one another's view.[43] As Robert Kirshner explains, "[I]nstead of seeing more of the contents of the universe, we will see less and less," our night sky darkening as galaxies slip from sight.[44] Even galaxies will eventually unravel, stars will burn out, and matter will become a "thin gruel of particles," dissolving into a void of dark energy that races eternally outward into nothing.[45] And so the world that began with a big bang seems destined to end in what physicists call a "big whimper."[46]

As we shall see, however, there are other options. If the value of dark energy is greater than −1, then dark energy is not the "cosmological constant," but one form of a hypothetical substance called "quintessence." Its density will decrease with time, eventually producing a cosmic contraction and drawing the universe into an apocalyptic "big crunch."[47] If the value is less than −1, then the force is a different kind of quintessence called "phantom energy," whose density will *increase* with time, eventually tearing the universe into a "big rip."[48] But these theories are more or less our three options for the end of the world: a continually expanding void, an infernal implosion, or a great cosmic shredding.

And the Darkness Has Overcome It

As we saw in the introduction, the rise of the multiverse as a scientifically viable possibility is often traced to the discovery of dark energy. The most common explanation of the multiverse's sudden respectability is that dark energy's baffling, quantum-field-theory-defying weakness transformed the old fine-tuning problem into an all-out crisis: if lambda were much stronger, it would have torn the universe apart before anything could form; if it were any weaker, gravity would have collapsed the early universe in on itself. The almost impossible just-rightness of lambda—the painstaking calibration needed to allow life to emerge—prompted theorists to take seriously the possibility that there might be a slew of other universes with different cosmological constants.[49] If every value of lambda is tried out somewhere, then it makes sense to say we just happen to live in a universe whose lambda lets us live. I would also suggest, however, that there are also aesthetic, even existential motivations for turning from dark energy to the multiverse. Faced with the equally awful possibilities of the whimper, the crunch, and the rip, it seems that some physicists are looking for a different story to tell.

Of all the universe's mass–energy, approximately 73 percent is dark energy, 23 percent is dark matter (a persistently unidentified substance that neither absorbs nor releases light),[50] and a meager 4 percent is visible matter.[51] That is to say, all we can see—tables, particles, puppies, and stars—everything that seems to *be*, is only 4 percent of what is.[52] Because the discovery of dark energy revealed the extent of our cosmic insignificance, theoretical physicist and cosmologist Lawrence Krauss has called it

"the ultimate Copernican Revolution." As we saw in chapter 3, Copernicus unsettled the Platono-Aristo-Christian fantasy that the earth occupied the center of the solar system. Subsequent centuries of astronomy
showed that our solar system does not occupy the center of its galaxy and
that our galaxy does not occupy the center of the universe. And now with
the discovery of dark energy, it seems that the vast majority of the universe is totally inaccessible to us. Far from being the center of anything,
Krauss shrugs, "we're just a bit of pollution" made of cosmic leftovers and
living nowhere in particular.[53]

As we have also seen, it has traditionally been the church that has resisted Copernican revolutions—insisting that there are no "antipodeans,"
for example, or subjecting Galileo to lifelong house arrest, or terrifying
Descartes into posthumous publication, or burning Bruno at the stake.
But when it comes to dark energy, it is not the churches that are in crisis—
for the most part, the churches either do not know or do not care about
it—but (some) physicists themselves.

Not only was dark energy, from all accounts, "unexpected," but once it
was found, it was also infamously miscalculated by 120 orders of magnitude, prompting Marcelo Gleiser to call it "ugly and unexpected."[54] On
top of that, it is still not clear what dark energy *is*. "The term doesn't mean
anything," quipped David Schlegel in 2007. "It might not be dark. It might
not be energy. The whole name is a placeholder for the description that
there's something funny that was discovered nine years ago now that we
don't understand."[55] So the discovery of this way-too-small, ugly something has initiated what one might call an epistemological crisis in the
physics community. As Krauss recently told a group of physicists and astronomers, "You are liable to spend the rest of your lives measuring stuff
that won't tell us what we want to know."[56] Strikingly, however, the way
physicists communicate this discovery also reveals the presence in their
midst of something akin to an *existential* crisis.

Assuming dark energy remains at a steady density forever, it will cause
galaxies to disappear from our view as time goes on, effectively extin-
guishing the faraway lights in the sky. Our "descendants" (whatever that
means on the time scale of hundreds of billions of years) will eventually
see nothing more than a supercluster of the Milky Way, Andromeda, and
a few dwarf galaxies, swimming in an endless, empty sea. In other words,
as Lawrence Krauss and Robert Scherrer imagine, "for these future astronomers, the observable universe will closely resemble the 'island universe' of 1908: a single enormous collection of stars, static and eternal,

surrounded by empty space."[57] What Krauss and Scherer find frightening, first, is that the observations by these "future astronomers" will produce an "incorrect" vision of the universe: with no galaxies to see receding, they will not even know that the cosmos is expanding.[58] Second, future astronomers will be profoundly *alone* in the universe. And third, after a long period of deception and isolation, even galaxies will start to unravel, and, as Michael Lemonick puts it, "all that will be left in the cosmos will be black holes, the burnt-out cinders of stars and the dead husks of planets. The universe will be cold and black."[59] Robert Kirshner, a member of one of the two teams that discovered dark energy in 1998, imagines the universe's final scene as "lonely, dull, cold, and dark."[60] Theoretical physicist Brian Greene sees it as "vast, empty, and lonely."[61] And Brian Schmidt, who led the Nobel Prize–winning High Z Supernova Team, of which Kirshner was a member, calls this end-time scenario "the coldest, most horrible end to the universe I can think of. I don't know—it's creepy."[62]

"So this, then, is the story of the universe," writes the Search for Extraterrestrial Intelligence Institute's senior astronomer Seth Shostak. "A Big Bang, a hundred billion years of light, life, and late-night television, and then an infinitude of nothingness. Am I getting through to you? Not a long time—not a really long time—but an *infinitude*. A flash of activity, followed by a never-ending darkness. Our universe is destined to spend eternity in hell, without the fire."[63]

Something about this "creepy," hellish, dark-energetic end of the world seems to have prompted physicists to rethink its beginnings. This is not to say that any specific cosmologist is directly motivated by existential panic; it is simply to mark the coincidence at the broadest level of new projections about the end of the universe, on the one hand, and new models of its beginning, on the other. In particular, as some physicists have projected the inexorable unraveling of *this* universe, others have raced to map a profusion—even an infinity—of others. Some theorists have even sought ways to get new cosmic beginnings out of the end itself: a burst of new worlds out of even the most lifeless, dark-energetic void.

But, again, the past decade's fascination with the multiverse did not emerge out of nothing. Rather, most of these theories had been posited— and more or less ignored—decades beforehand. So what was so suddenly attractive about the multiverse? There are a number of ways to answer this question, but one possibility emerges clearly in the work of Alexander Vilenkin, an early theorist of "eternal inflation." In the wake of the discovery of dark energy, "the long-term prospects for any civilization appear rather bleak," he writes. "Even if a civilization avoids natural catastrophes and self-destruction[,] it will, in the end, run out of energy. The stars will eventually die, and other sources of energy will also come to an end. But now eternal inflation appeared to offer some hope."[64] And it did so by offering an infinite number of other universes in which life might emerge.

The Rise of Inflation

A Remedy for Fine-tuning

The most widely accepted multiverse scenario comes out of what is called "inflationary cosmology." Alan Guth first posited the theory of inflation in the early 1980s, borrowing an all-too-familiar economic term to describe the exponential growth of the earliest universe.[65] He offered *inflation* as a way to resolve a number of problems with standard big bang cosmology. The first problem, as Brian Greene puts it, was that "the big bang leaves out the bang. It tells us nothing about what banged, why it banged, how it banged, or, frankly, whether it really banged at all."[66] In its standard form, the big bang hypothesis could explain in dazzling detail the processes of the early universe from about $1/100$ of a second onward. But there remained what Steven Weinberg admits was "an embarrassing vagueness about the very beginning"—an uncertainty as to what happened at "$t=0$" to get the universe hurtling outward to begin with.[67] As Paul Steinhardt and Neil Turok summarize this problem, "[T]he universe is simply assumed to have appeared out of nothing, filled with all kinds of exotic matter and energy, at nearly infinite temperature and density."[68] But where did all of this matter–energy come from, and how?

The second shortcoming of the standard big bang hypothesis was that it left unresolved the "flatness" and "horizon" problems, both of which concern the appearance of fine-tuning in the early universe. The "flatness" problem refers to the shape of the universe itself. As Alexander Friedmann showed in the early 1920s, space as governed by the laws of general relativity can take on three different topographies, depending on the amount of matter and energy in the universe. If the amount is greater than a "critical density" of "about six hydrogen atoms per cubic meter," then the universe will have a positive curvature, like the surface of a sphere. If the amount of matter and energy in the universe is less than the critical density, then space will have a negative curvature, like a saddle or (to use a remarkably common illustration) a Pringles® potato chip. If, however, the mass–energy of the universe is equal to the critical density, "the equivalent of a raindrop in every earth-sized volume,"[69] then space will be "flat" in three dimensions. Every indication over the past few decades has shown that the density of matter is almost exactly equal to the critical density, so space is nearly perfectly flat on large scales.[70] The problem is that if a split second after the big bang the density had been *at all* lower or higher than the critical density, then the universe would have taken on either severely negative or severely positive curvature, the former spreading matter out too far to form any major structures and the latter drawing all matter together into a collapse. So, Greene explains, "some mechanism . . . must have tuned the matter/energy density of the early universe *extraordinarily* close to the critical density," and as Martin Bucher and David Spergel charge, "the big bang theory offers no explanation apart from dumb luck."[71]

The "horizon" problem refers to the striking thermal homogeneity of the early universe. Since the discovery of the CMB in 1965, physicists have marveled that regions too far apart ever to have been in contact (that is, regions that lie beyond one another's cosmic horizons but within our own) nevertheless have almost identical temperatures. "So," as Andrei Linde relates the thought process, "one could only wonder what made these distant parts of the Universe so similar to one another."[72] The first solution was to appeal to the "cosmological principle," a neo-Atomist dictum that simply states "that the Universe *must* be uniform" in all directions.[73] But to say that the universe is uniform because of a principle that makes the universe uniform is tantamount to saying, with Molière's doctor, that opium puts people to sleep by virtue of its "dormitive property";[74] it is not really an explanation at all. Moreover, although the universe is uniform on the largest scales, it is not *perfectly* uniform; after all, there are galaxies in some places and leagues of dark space in others owing to slight thermal *in*homogeneities in the early universe.[75] How, then, are we to account for the extraordinary flatness, homogeneity, *and* inhomogeneity of the early universe, all of which allowed existence as we know it to exist?[76]

Once again, these problems are familiar, boiling down to the mystery of matter's creation and distribution throughout the universe. Now, of course, one can always repeat the Cartesian–Newtonian solution and say that a skillful, intelligent creator-God made and scattered matter and energy with the values that God's eventual creatures would need to emerge. *Or* one can posit an *immanent* cosmic principle to fulfill these functions, as Kant did with his "repulsive" or "centripetal" force and as Einstein did with his cosmological constant. But although the cosmological constant has been vindicated in recent years, it is not nearly powerful enough to perform these cosmogonic functions. So we need *another* repulsive force—a stronger, faster energy that Guth posited in the 1980s, calling it "inflation."[77]

Inflation in its most basic form is a brief burst of insanely rapid universal expansion that kicks in right after the big bang—a hyperactively repulsive gravity that, "according to even conservative estimates," blows the universe up by "a factor of 1,000,000,000,000,000,000,000,000,000,000 in [0].0000000000000000000000000000000001 second."[78] As Stephen Hawking and Leonard Mlodinow explain the process, "[I]t was as if a coin 1 centimeter in diameter suddenly blew up to ten million times the width of the Milky Way."[79] Because inflation happens so quickly, it irons out any positive or negative curvatures and leaves the universe flat. Moreover, because it expands space-time faster than the speed of light, inflation explains how regions that *appear* never to have been in contact could have "thermalized" (that is, reached the same temperature) before inflation flung them apart. And finally, as inflation stretches space-time outward, it stretches quantum fluctuations into slight temperature variations that then become the seeds of galaxies,[80] solving all of the big bang's fine-tuning problems at once. But this first formulation of inflation still did not explain the bang. In the years that followed the publication of Guth's landmark paper in 1981, a number of theorists therefore set out to answer a number of unanswered questions—most pressingly, what got inflation going and what turned it off. In the process, they unintentionally opened the door to a slew of multiple universes.

Inflation and Infinity

Theorists working on inflationary cosmology collided with what they now call the "multiverse" from two directions. First, they discovered that the universe that inflation produces appears to be spatially infinite. Second, they discovered that once inflation gets under way, it does not seem to stop. These two insights opened, in turn, onto what mathematician Max Tegmark calls the "Level I" and "Level II" multiverses, respectively.[81]

Our visible universe is composed of the expanse of space-time that light has been able to travel since the big bang banged 13.8 billion years ago. Because space has been expanding since then (and accelerating for the past 5 billion years), the farthest visible object–events are now about 40 billion light-years away, even though they took place fewer than 14 billion light-years ago. The sphere 40 billion light-years in radius, centered wherever a given satellite is located, is called our "Hubble volume," or "observable universe." Now according to even the most standard cosmological models, there are regions of space-time beyond our Hubble volume; light just has not had enough time to reveal them to us (and if dark energy stays constant, it never will). So there is *something* beyond what we can see, and one of the central predictions of inflation is that this "something" is infinite.

Although a number of physicists and cosmologists contest this prediction and are currently seeking to establish the finitude of the universe, many others take it almost as given.[82] They assume, moreover, that the infinite space beyond our Hubble volume is similarly filled with matter and energy: this contemporary articulation of the old "principle of plenitude" is now known as the *cosmological principle*.[83] And as we will recall from the Atomists, Giordano Bruno, and even Nicholas of Cusa, the moment the universe is said to be both spatially and materially infinite, a number of remarkable conclusions follow. For Epicurus and Lucretius, for example, an infinite number of atoms with a finite number of shapes ensured not only that there were an infinite number of worlds "beyond the fiery ramparts" of our own, but also that some of them would be exactly like ours (see chap. 2, sec. "Fire and the Phoenix"). After all, if there is a limitless amount of material but a limited number of ways to configure it, then all possibilities will have to repeat; in fact, they will have to repeat an infinite number of times. Greene invokes the same principle when he argues that the universe is most likely infinite and that there are an unfathomably high but nevertheless *limited* number of arrangements that particles can assume. Sounding like a latter-day Lucretius, Greene concludes that "an infinite number of occurrences combined with a finite number of possible configurations ensures that outcomes are infinitely repeated."[84]

The popular multiverse literature therefore almost always opens with a guided meditation along the following lines: imagine that somewhere out there, there is another "you" reading this book.[85] The two of you have made all the same decisions in your lives, from college to shoes, except that just now she (or he) has stopped reading to check Facebook, whereas you have dutifully decided to read on. She reads a disturbing news story about a natural disaster and decides to jump on the next plane to join the

FIGURE 5.3 The Level I, or quilted, multiverse, in which the bright circles represent the Hubble volume of any given universe. (Illustration by Kenan Rubenstein)

relief effort, eventually changing careers forever, whereas you hold on to your current job and apartment. In the meantime, somewhere else "out there" another "you" has just gotten up to make the fourth cup of coffee that you have just decided you really should not have, while another is doing exactly what you are doing and, in fact, is just now wondering about the other "you" who is just now wondering about her. In an infinite universe, all these possibilities play themselves out, and each of them does so an infinite number of times.

Where are these other yous? As in Atomist cosmology, they are spatially arrayed, far beyond any possible connection to our world. Tegmark estimates that your first doppelgänger is nearer than $10^{10^{28}}$ meters away: an unfathomable distance, but nonetheless finite. Even farther out, "about $10^{10^{118}}$ meters away should be an entire Hubble volume identical to ours"— that is, an observable universe with the same distribution of matter, the same Milky Way, the same rings of Saturn, the same *Brontosaurus* that turns out to have been an *Apatosaurus* all along, and the same disturbing Twitter feed that mashes up quotations from Søren Kierkegaard with the wisdom of Kim Kardashian.[86] And in between that world and ours, there are countless other Hubble volumes, some of them very similar to ours with a few minor differences, others completely different. This whole ensemble of universes constitutes what Tegmark calls the "Level I multiverse" and what Greene calls the "quilted multiverse" (figure 5.3). Like Cusa's perspectival multiverse, this multiverse is composed of an endless and overlapping ensemble of observable universes, each of them centered around any given cosmic body and all of them issued forth from the same big bang nearly 14 billion years ago.

Inflation and Eternity

But we still have not accounted for the bang itself—in particular, for where it came from or what it was. Such an account began to emerge in the mid-1980s with the "new" and "chaotic" inflationary scenarios that developed the work Guth had begun.[87] Although there are some significant mathematical differences between these two models,[88] they are often grouped together cosmologically because they translate into very similar stories. The first major insight that the models share is that inflation takes place "outside" our universe—before the bang, as it were. And the second is that inflation is eternal; once it starts, it never ends.

Depending on the author, this story takes one of two very different points of departure. Alexander Vilenkin tends boldly, even gleefully, to proclaim that his account begins "from nothing!"[89] As I have argued elsewhere, what he means by "nothing" is not nothing at all, but a vacuum, which according to quantum mechanics is not empty, but seething with "virtual particles" that flash into and out of existence.[90] This vacuum, then, is Vilenkin's primordial scene: a spaceless space that he himself calls "chaotic" and "foamlike"—neither nothing nor something, but a bustling fray of particles appearing and disappearing at random. One day, the story continues, one of these particles seems to have flashed into existence and then failed to flash back out, tunneling through the energy barrier that should have caused it to collapse and inflating instead to form space-time as we know it.[91] Edward Tryon initially proposed this idea in 1970, apparently blurting out during a seminar on quantum mechanics, "[M]aybe the universe is a vacuum fluctuation!" As Vilenkin relates the story, "[T]he room roared with laughter,"[92] yet a little more than a decade later Tryon's weird idea would provide the basis for Vilenkin's own weird idea: that the universe quantum-tunneled into being once upon a time.

In order to grow into a universe, this initial region of space-time would have been filled with inflationary energy, a repulsive gravity that expands the region exponentially at every moment. This state of renegade expansion is called the "false vacuum"—"false" because it is unstable and tends to decay into a lower-energy state, like the one in which we exist with our tiny cosmological constant. This false vacuum, incidentally, is where most other inflationary scenarios begin: with a rapidly stretching inflationary sea.[93] Again, this state is unstable; regions of space-time cannot sustain this level of energy for long and so will soon "roll" down to a lower-energy state, forming "bubbles" or "pockets" of "true vacuum," which is to say regions of space-time with a low-energy cosmological constant. Crucially, at the point of this roll-down, the excess energy is converted into matter in the form of radiation, which is flung into the newly stable pocket of space-time as "a hot fireball of elementary particles."[94] The end of inflation, in other words, *is the big bang*, providing the primordial plasma that gravity will eventually assemble into a universe.[95] Inflation therefore fulfills both of the functions for which Descartes and Newton had to appeal to God: the creation and the distribution of matter in the universe.

In the meantime, although inflation has "turned off" in one particular region, it continues unabated (and, for Andrei Linde, jumps to even higher energy levels) elsewhere, sending space-time "outward" faster than the speed of light. In other words, as Guth and a host of others have shown, once inflation turns on, it never turns off.[96] In fact, the false vacuum is always producing exponentially more of itself, relaxing into isolated bubbles of true vacuum here and there, but also sustaining an ever-growing sea of runaway space-time between them.[97] In this "fractal" cosmology, then, "one inflationary universe sprouts other inflationary bubbles, which in turn produce other inflationary bubbles."[98] And because the process is eternal, inflation "produces not just one universe" and not just a *lot* of universes, "but an infinite number of universes."[99]

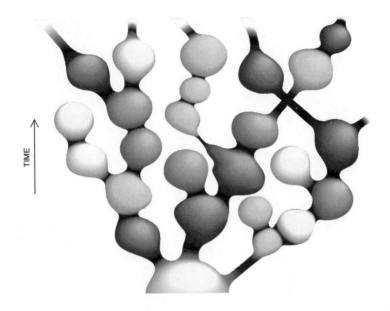

TIME

FIGURE 5.4 The inflationary multiverse of Andrei Linde. (From Andrei Linde, "The Self-Reproducing Inflationary Universe," *Scientific American*, November 1994, 48. Reproduced by permission of Andrei Linde)

This infinite collection of universes, which Tegmark calls "Level II" and Greene calls the "inflationary multiverse," looks more Brunian than Cusan—an infinite number of worlds are arrayed throughout infinite space, held apart by an ever-expanding "void" (figures 5.4 and 5.5). For nonspecialist audiences, the inflationary multiverse is often compared to "island universes" floating in a vast sea or to "gas pockets in a loaf of rising bread" or to holes in a "gigantic block of Swiss cheese."[100] Each island, pocket, or hole is a universe; the sea, bread, or cheese is the inflationary region that separates the worlds from one another. Unlike the Level I multiverse, each of these Level II universes arises out of different big bangs and, as such, might have very different "constants" from ours, including the strength of the electromagnetic force, the number of spatial dimensions, and the value of the cosmological constant (Λ). With a lambda value much smaller than ours, a new bubble would "expand for the shortest of time before collapsing again," whereas with a large lambda, a bubble would "expand so fast that matter is spread far too thin for stars and

FIGURE 5.5 The inflationary multiverse. (Illustration by Kenan Rubenstein)

planets ever to form. The overall result," as John Gribbin explains, "would be a chaotic mess of bubbles all expanding at different rates in different regions of spacetime."[101]

For the most part, these worlds will never come into contact with one another because the space-time between them is expanding faster than the speed of light. For the most part, worlds with a sustainable cosmological constant will live for a time and then, along Vilenkin's account at least, suffer a strikingly Epicurean death by decay.[102] But it is always also possible that a universe might suffer a more Democritean fate and collide with another universe before inflation has time to fling them apart, the "collision send[ing] destruction screaming through the walls of both bubbles."[103] Fascinatingly, though, such destruction might not be total; in fact, a cosmic collision might leave pockets of each universe intact, stamping a kind of thermal bruise on each of them. As we shall see in chapter 6 and "Unendings," some cosmologists are currently examining the CMB for signs of this sort of a collision, which would provide invaluable evidence in support of the inflationary multiverse.[104]

All told, the scene looks like this: from an "outsider's perspective" (assuming for a moment that it makes sense to speak of such a thing), each bubble universe appears as a bounded region within a larger multiversal expanse. Each universe, in other words, looks finite from the outside. But as noted in the preceding section, the view from *inside* one of these universes—ours, for example—seems to suggest that the universe is infinite. The explanation here is remarkable: just as Einstein's theory of special relativity says that "rest" from one perspective looks like "motion" from another and that neither perspective is wrong,[105] inflationary cosmology says that "each of the bubble universes appears to have *finite* spatial extent when examined from the outside, but *infinite* spatial extent when examined from the inside" and that neither perspective is wrong.[106] At this point, we might do well to bring Cusa back into this neo-Atomist scene. For Cusa, the term *contracted infinity* was a way of expressing the puzzling truth that what looks infinite from the perspective of creation looks finite from the perspective of an infinite God (see chap. 3, sec. "End Without End"). For inflationary cosmology, the explanation is that the infinite *time* outside each bubble is converted into infinite *space* inside each bubble, so that from the perspective of the sea each bubble is spatially bounded, whereas from our perspective the universe goes on forever. And although, observationally speaking, "our" perspective is the only one we will ever have, neither view is more "correct" than the other.

In conclusion, then, a Level I multiverse—that overlapping set of worlds with all those copies of "you" doing an infinite number of identical, similar, and different things—can be found *within* each of the bubbles of the Level II multiverse, an infinite number of which contain particles and creatures like us and an infinite number of which do not. From the "outside," each of these bubbles looks finite, but from the inside each looks infinite. It is as if a Cusan multiverse were contained within each of Bruno's infinite worlds, and then the whole mess were subject to Cusa's perspectivalism. Infinities here become nestled within infinities—or within finitude, depending on how you look at it.

Landscape Design

As both Vilenkin and Linde have testified, the eternal inflationary scenario was barely noticed when it was first proposed; in fact, it is said that
Guth fell asleep when Vilenkin first presented the idea to him.[107] Others
were more directly critical than Guth, accusing inflationary cosmology of
looking more like metaphysics than science. In Vilenkin's account, "[T]he
main objection against [eternal inflation] was that it was concerned with the
universe beyond the horizon[,] which is not accessible to observation."[108]
Stephen Hawking remains a vocal critic of any effort to determine what
happened "before" or "beyond" the bang; as he wrote in his mega–best
seller *A Brief History of Time* (1988) and has reaffirmed in his recently coauthored book *The Grand Design* (2010), nothing that might have taken
place before the big bang could have produced any "observational consequences" on the universe as we know it, so there is no use speculating about
such events.[109] Eternal inflation not only requires a robust pre–big bang
scenario, but also generates numberless regions of space-time forever
beyond our horizon. For these reasons, inflation was initially thought to
create more problems than it solved; an infinite number of universes seemed
a high price to pay for an explanation of the flatness and homogeneity of
this one.

Just around the turn of the millennium, however, the fortunes of
eternal inflation began to change. Vilenkin and his collaborator, Jaume
Garriga, along with a number of other theorists, attribute this change to
the discovery of dark energy in 1998.[110] The perplexing smallness of the cosmological constant, the extent to which it seems improbably calibrated
to enable life as we know it, seems far less mysterious if there are an infinite number of universes, with all possible values of lambda. Eternal inflation, as we saw in the introduction, suddenly looked like a promising
way to redeem the anthropic principle, which in its strong form verged
on theism and in its weak form could say little more than that "the world
we observe must be consistent with our existence, insofar as we are observing it." With a multiverse, the weak anthropic principle can become
far more robust and can now proclaim along with the Atomists that,
given an infinite amount of space, time, and material, every possible configuration is bound not only to emerge, but also to repeat. And so the
universe is the product of neither wild improbability nor divine action.
Thanks to the age-old pair of accident and infinity, "nature," as Lucretius
might say, is "rid of her proud masters."[111]

For his part, Linde attributes this millennial inflationary turn to developments in string theory that caused it, independently of inflationary theory, to collide with its own multiple-worlds scenario. But these two sources are not totally separate; as Linde explains it, string theory's multiverse emerged out of efforts "to solve the cosmological constant problem."[112] String theorists themselves would likely disagree, attributing the emergence of this scenario to the discovery of "fluxes" within string theory itself. But regardless of causation, the correlation is striking: the rise of multiverse theories at the dawn of the twenty-first century seems to have been the product of what political theorist William Connolly might call a "resonance machine" composed of inflationary cosmology, dark energy, and string theory.[113]

String theory is commonly said to be the most promising candidate for a "theory of everything"—that is, a theory that accounts for the universe simultaneously on the largest scales (general relativity) and on the smallest scales (quantum mechanics).[114] According to string theory, subatomic particles are not "points," but one-dimensional "strings," whose vibrations constitute their mass. String theory also posits higher-dimensional objects called "branes" that can assume all kinds of shapes, or "topologies," and that occupy up to nine spatial dimensions.

Until fairly recently, as Helge Kragh has explained, "string theorists . . . [had] hoped that the theory, when sufficiently developed and understood, would result in a unique compactification or in a 'vacuum selection principle,' from which the one and only vacuum state describing the universe would emerge."[115] They hoped, in other words, that string theory would eventually explain why the values of the cosmological constant, the gravitational constant, the nuclear and electromagnetic forces, and all the other seemingly fine-tuned parameters of fundamental physics have to be the way they are. What they discovered, however, was that string theory allows for a *staggering* number of possible "vacua" or "stable configurations of spacetime."[116] The most common estimates range from 10^{500} to $10^{1,000}$ possible vacua, which means that according to string theory there are between 10^{500} and $10^{1,000}$ different kinds of universe.[117] These numbers, although unassimilably huge, are still finite, and so (as Epicurus realized millennia ago) they ensure the repetition of any given universe, including our own.[118] All of this makes the fundamental parameters we observe seem a bit less impossible than they would otherwise appear. "In particular," explains Guth, "the cosmological constant (e.g. the vacuum energy density) would be expected to have different values for different vacua";[119] in

FIGURE 5.6 Universes on the string theory landscape. (Pam Jeffries and Mario Livio, Space Telescope Science Institute)

effect, the vacuum energy has had at least 10^{500} chances to obtain the wildly unlikely value it takes on in our universe.

In 2003, Leonard Susskind proclaimed that with the discovery of these vacua solutions, string theory had reopened and redeemed the anthropic principle, calling its set of manifold vacua the "landscape" of string theory.[120] Notoriously impossible to get one's head around, the landscape is a mathematical representation of string theory's $10^{500(ish)}$ different dimensional compactifications, which is to say its different types of universes. It is usually represented as a kind of mountainous terrain, with each "valley" corresponding to one of the stable vacua (figure 5.6).

Such representations, of course, are two or, at best, three dimensional, a limitation that makes visual depictions rather comically impaired from the outset. Susskind leads his nonspecialist readers in the following thwarted meditation: "Try to imagine a space of 500 dimensions with a topography that includes 10^{500} local minima, each with its own Laws of Physics and constants of nature. Never mind."[121] If the landscape is a schematic representation of all possible universes, then which cosmology describes their actual generation? As Steven Weinberg has explained, there are a number of candidates—including the most recent incarnation of Stephen Hawking and Thomas Hertog's "no-boundary" proposal, which we will encounter in chapter 6.[122] But the cosmological model that is most commonly paired with the landscape is eternal inflation. Linde remarkably made this connection in the mid-1980s,[123] but it seems that a broader acceptance of the idea had to wait for the "perfect storm" of eternal inflation, string theory, *and* the cosmological constant problem. If it is the case (as Guth, Raphael Bousso, and Joseph Polchinski seem to suspect and as Linde and Susskind wholeheartedly believe),[124] that these two theories describe the same multiverse—or "megaverse," as Susskind would have it, or "universe-of-mini-universes," as Linde prefers—then each of the types of bubbles of inflationary cosmology would correspond to one of the 10^{500} (give or take a gazillion) valleys in the string theoretical landscape. And each type of universe, as long as the number is finite,[125] would recur an infinite number of times.

Now a major problem posed by both the landscape and inflation multiverses is how to calculate the probability that any given type of universe will be generated—especially if any possibility can be said to be generated an infinite number of times throughout eternity. "Inflation plus the landscape" thus runs us up against the so-called measure problem: if the number of Type X universe is infinite and the number of all universes is infinite, then how do we calculate the probability that any given universe will be Type X? How is one to divide infinity by infinity?[126] This, then, is the challenge for landscape and inflationary theorists alike: "to determine how these potential [inflationary] universes are distributed across the corresponding Landscape Multiverse."[127] Greene stages this task as a kind of heroic ordeal, explaining that "detailed maps of this mountainous terrain have yet to be drawn. Like ancient seafarers, we have a rough sense of what's out there, but it will require extensive mathematical explorations to map the lay of the land."[128] But as far as Susskind is concerned, even before the details are fully worked out, the landscape has already done a great service: it has provided the ultimate solution to the fine-tuning problem *and* a vindication of the anthropic principle. Why is our universe the way it is? Because, the landscape tells us, "somewhere in the megaverse, the constant equals *this* number; somewhere else it is *that* number. We live in one tiny pocket where the value of the constant is consistent with our kind of life. That's it! That's all! There is no other answer to the question."[129]

Other Answers: The Return of the Phoenix

The Cost of Inflation

One of the earliest physicists to discover the eternal nature of inflation was Paul Steinhardt, a theoretical physicist and cosmologist who has gone on to become one of the theory's most noteworthy critics. In a recent popular article, Steinhardt runs through the case for and against inflation, admitting that the theory is "powerfully predictive."[130] It accounts beautifully for the flatness and homogeneity problems, predicts the slight thermal variations imprinted on the CMB, and tends to be "in exquisite accord" with the latest data streaming in from the Wilkinson Microwave Anistropy Probe (WMAP) and Planck satellites, which measure the CMB with increasing precision as the years go on.[131] For these reasons, inflation has become the most widely accepted theory of cosmic origins among physicists. But, Steinhardt reminds us, in the process of solving these problems and making these predictions, inflation comes with an "awkward corollary," which is that it never turns off. And as we have just seen, the upshot of eternal inflation combined with string theory is an infinite number of 10^{500} (or more) different types of universe, arranged in some unvisualizable landscape and eternally separated by a ballooning expanse of useless space.

One cannot avoid the sense that Steinhardt harbors an aesthetic revulsion to this multiversal excess; in one interview, he calls it "a dangerous idea that I am simply unwilling to contemplate," and he and Neil Turok warn that eternal inflation "seems likely . . . to drag a beautiful science toward the darkest depths of metaphysics."[132] As the latter quotation indicates, however, Steinhardt's aesthetic revulsion stems from a commitment to a particular kind of scientific rigor. Here, he explains, is the problem: "[I]n an eternally inflating universe, an infinite number of islands will have properties like the ones we observe, but an infinite number will not." And the minute this is said to be the case, he charges, the theory's predictive power is completely undermined: "What does it mean to say that inflation makes certain predictions—that, for example, the universe is uniform or has scale-invariant fluctuations—if anything that can happen will happen an infinite number of times? And if the theory does not make testable predictions, how can cosmologists claim that the theory agrees with observations, as they routinely do?"[133] The problem, as Steinhardt sees it, is that the theory designed to explain *this* universe has ended up explaining *any* universe, thereby undermining its most celebrated function. As for the additional claim that eternal inflation explains the fine-tuning of dark energy and the other constants, Steinhardt believes that the price of this explanation is simply too high. As he and Turok explain in their popular book *Endless Universe*, "[T]he two of us confess to being among those who are skeptical about the anthropic principle as a panacea for fundamental physics. The main problem is that it relies on a host of assumptions that cannot possibly be tested: the existence of the multiverse, the notion that different pockets have different physical laws . . . and so on."[134] Rather than relying on a host of extracosmic *kosmoi* to explain the one we are in, Steinhardt and Turok therefore set out to find a different story to tell.

The New Ekpyrotic Scenario

As Steinhardt and Turok recall, their new model was inspired by a lecture they invited Burt Ovrut to deliver to cosmologists on "M-theory," which encompasses all previously unreconciled branches of string theory.[135] The element that captured their imagination was not string theory's slew of vacuum solutions, they say, but its idea of membranes—those up to nine-dimensional objects of varying topologies. For each of these "branes," they learned, there is a corresponding "antibrane partner," and this realization got them thinking.[136] What if our universe is a three-dimensional membrane? What would its "partner" be? *Where* would its partner be? And what would happen if the two of them were to collide?

The resulting theory was first called the "ekpyrotic scenario," intentionally named after the signature aspect of Greek Stoic cosmology.[137] Since then, as details of the model have changed, it has been variously renamed the "cyclic model," "new cyclic model," and "new ekpyrotic model."[138] In the meantime, its reliance on string theory has weakened: one paper even considered abandoning the idea of branes before a follow-up paper reinstated them.[139] But the best-known model looks like this: two membranes of three spatial and one temporal dimension are suspended in a fourth spatial dimension. Our observable universe—everything we can see and not see between here and 40 billion light-years in any direction—"lies on one of the branes, often called our 'braneworld,'" explain Steinhardt and Turok. "It is separated by a tiny gap, perhaps 10^{-30} cm across, from a second 'hidden' braneworld. . . . All the particles we are familiar with, and even light itself, are confined to our braneworld. We are stuck like flies on flypaper, and can never reach across the gap to the 'hidden' world, which contains a second set of particles and forces with different properties from those in our braneworld."[140]

The only forces that *can* cross the gap are gravity and dark energy, which keep the two branes separated during one stage of the cosmic cycle and draw them together during the next. As the branes are drawn together, the energy between them increases, and they gain speed before finally colliding, at which point they separate again, their energy converted into a mass of radiation flung outward on each brane. The collision, in other words, is the big bang.

On each brane, the model temporarily follows the course of the better-known big bang hypothesis, except without the inflationary energy. First, there is the big bang, then radiation, then gravity, then dark energy, but at this point when the standard model predicts the gradual unraveling of the world, the ekpyrotic model changes course. Dark energy does not continue to speed the cosmos outward forever, either dissolving it into a void or ripping it apart. Rather, after a trillion years, the energy density of dark energy begins to decay. Its outward force starts to reverse, drawing the two branes back together, another big bang bangs, "and the cycle begins anew" (figure 5.7).[141]

Here, then, we have a temporal rather than a spatial multiverse. One universe (composed of two partner branes) undergoes continual destruction and rebirth throughout eternity. Steinhardt and Turok, like their mid-twentieth-century predecessors, note the resonance with Hindu cosmology; in fact, they even map the two worldviews onto each other mathematically, showing that one full ekpyrotic cycle corresponds almost exactly to a year in the life of Brahma (around 3.11 trillion years).[142] We may recall that the downfall of these midcentury models came at the hands of the Second Law of Thermodynamics: if entropy increases from cycle to cycle, then two problems ensue: first, a new universe will never begin in an ordered state comparable to the one before it, and, second, each cycle will last longer than the previous one, which means the cycles get shorter and shorter into the past, eventually reaching an initial singularity. The new ekpyrotic theorists avoid the first pitfall by pointing out that the problem for any new universe is not the *amount* of entropy, but the *density* of entropy. In the braneworld scenario, both branes increase in size from cycle to cycle, spreading the entropy across a larger surface area and decreasing the overall entropy density of each new universe.[143] This solution may or may not also address the second problem, that of an absolute beginning,

FIGURE 5.7 The cyclic universe of Paul Steinhardt and Neil Turok. (From *Astronomy,* April 2009. Courtesy of *Astronomy* magazine)

but Steinhardt and Turok seem untroubled by the outcome. The cycles may be past-eternal, or they may not be: "[O]ne could imagine the sudden creation from nothing of two infinitesimal spherical branes arranged like two concentric soap bubbles," which would collide and expand until they became large and flat.[144] Or at least Steinhardt and Turok could imagine such a thing.

Either way, what is crucial for these cyclical theorists is that their model addresses all the concerns inflation addresses, but without introducing unconnected regions of space-time governed by differing laws of physics. The reason the universe is so flat after the bang is that it was flat before the bang: both branes remain large and extended throughout the cycle. The reason the CMB is so uniform is that dark energy empties the branes of all matter before the contraction that gives way to the bang. And the inhomogeneities arise from small quantum fluctuations that wrinkle the branes slightly as they draw near to each other: the places where the branes collide first are the "hot spots" where galaxies will eventually form. All this with no (spatial) multiverse and no anthropic principle: "[A]ccording to this view, the universe is a single, coherent entity that exists in a stable cycling state whose properties can eventually be understood as a consequence of the basic laws of nature."[145] The physical constants, in other words, remain *constant* in this model.

What we have in the ekpyrotic scenario, then, is a neo-Stoic rival to the neo-Atomist theory of inflation. Like the Stoics, the new Ekpyrotics insist that there is only one world. Like the Stoics, they maintain that the one world is tensionally interrelated by means of a substance that suffuses and governs it—the role that *pneuma* played for the Stoics is here filled by dark energy, which "regulate[s] the cycling," acts both attractively and repulsively, and absorbs systematic shock.[146] Like the Stoics, the new Ekpyrotics describe this immanent force by means of pantheological metaphors of creation, governance, and providence, telling us that "stars, galaxies, and the larger-scale structures observed in the universe today *owe their existence* to the period of dark energy domination in the previous cycle. And the dark energy *dominating* the universe today is *preparing* similar conditions for the cycle to come."[147] And finally, like the Stoics, they teach that this one world will be destroyed and re-created throughout eternity.

The new Ekpyrotics are notably reluctant, however, to indulge in any fantasies of eternal return. Nowhere in the specialist *or* popular literature is there a meditation on the infinite number of times that you have read this book before or that Socrates has taught Plato or that the same demon has stolen into your loneliest loneliness to proclaim you the same speck of dust. Nevertheless, they do project that in the model's simplest form, "all the physical properties of the universe are the same, on average, from cycle to cycle."[148] A more complicated version of the theory entertains the idea that "some properties thought to be constants . . . could actually vary over long periods," but they would do so by means of a slow, evolutionary process—not a haphazard "decision" among wildly different vacua.[149] In short, although the new ekpyrotic model returns to the Stoics, it tempers this return with the "old" modernist efforts to avoid speculation about that to which we have no access and to explain physical constants as they are in this universe rather than as they might be in a multitude of other universes.

As it turns out, however, comparatively few theorists have been converted to the ekpyrotic cause.[150] Many admit that the scenario accounts for the appearance of the CMB as well as inflation does, but even so they raise a host of objections to it.[151] Some charge that the ekpyrotic model does not actually escape the problem of entropy at all. Others ask why there are only four large spatial dimensions or why there is only one braneworld and its partner. "Why stop at two branes?" asks John Gribbin. "The most exciting thing about M-theory, . . . the most compelling reason to take the idea of a multiverse seriously, is that it offers an infinite choice of possible worlds, not just one boring pair of cymbals clashing out the same old song."[152] And perhaps the most vocal critic of this cyclic model has been Andrei Linde, who a week after its first publication proclaimed *ekpyrosis* "a house of cards."[153] The problem, he charged, was the arbitrary starting point of two perfectly flat membranes. "If you start with perfection," he conceded, "you might be able to explain what you see . . . but you still haven't answered the question: Why must the universe start out perfect?"[154] In recent years, Steinhardt and his colleague Jean-Luc Lehners have suggested that the universe might not start out "perfect" at all, but might keep cycling until dark energy forms it into a "large, flat, habitable universe."[155] But even this possibility remains too speculative for Linde and the majority of his colleagues.

During the first few years of this inflation–*ekpyrosis* debate, it seemed as though the argument would be easily arbitrated once the data were clear. Unlike inflation, whose repulsive force would cause gravitational waves (B-modes) to become imprinted on the CMB, the cyclic theory predicted that its big bang would *not* produce such waves. If, then, the European Space Agency's Planck satellite were ever to detect these waves, the ekpyrotic theory would be ruled out; as Turok explained in 2007, their model had the scientific advantage of being falsifiable.[156] In recent years, however, Steinhardt and Lehners have complicated this issue—perhaps because Planck's increasingly refined results have still revealed no B-modes. In the new "new ekpyrotic" scenario, there is far more "instability" and destruction than in the original model (the early literature relies heavily on adjectives such as *gentle, regular, slight,* and *regulated*). In the new version, "an overwhelming fraction of the universe fails to make it through the ekpyrotic phase" and is simply incinerated at the collision. But as long as the preceding cycle undergoes a moderately long stage of expansion before collapsing, there will be "a sufficiently large patch of space" that can cycle successfully through the big bang. Everything else will turn to dust, but this one patch will survive the infernal bounce and go on to re-create the universe. "As with the mythical phoenix," Steinhardt and Lehners conclude, "a new habitable universe grows from the ashes of the old."[157]

In this revised scenario, which Lehners and Steinhardt claim has more right to the "phoenix" moniker than its predecessors, the less gentle, less regulated collision *will* produce gravitational waves, "but very weak and of a kind that *may* never be detected."[158] According to these new calculations, then, if B-modes were to be found, the discovery would neither disprove the ekpyrotic model nor confirm the inflationary model. So the neo-Stoic model is likely to remain an underdoggish rival to the neo-Atomist model as long as the former finds theorists with enough energy, enough willing-ness to be slightly overlooked, and enough distaste for the infinite bubbles of eternal inflation.

Alternative Cyclic Cosmologies

Such distaste has prompted the emergence of other cyclic models as well—lesser known than the ekpyrotic scenario (and, for the most part, far less respected), but stemming from equally fervent critiques of inflation. For example, Lauris Baum and Paul Frampton, admitting that their objections to inflation are "more aesthetic than motivated directly by observations," have set forth a cyclical multiverse that is at once temporal *and* spatial.[159] According to this model, dark energy is neither inflation's cosmological constant (which retains a constant density) nor *ekpyrosis*'s quintessence (which decays over time). Rather, lambda in the Baum–Frampton scenario is what Robert Caldwell has named "phantom dark energy," whose density increases with time.[160] Instead of "simply" dissolving all matter into an ever-expanding void (the "big whimper" ending), dark energy becomes so powerfully repulsive that it splinters the universe into an immense num-ber (10^{84} to 10^{103}) of "patches" of space-time, sending them all on a course toward a final "big rip."[161] The patches are flung farther than light can travel between them, at which point dark energy empties each of them of all its matter. Each patch is eventually so distended and emptied that it "contains no quarks or leptons and certainly no black holes."[162] And then, just before the patch is shredded in an apocalyptic cosmic rip, it contracts, collapses, and bounces outward again, "reinflating in a manner *similar to* the big bang."[163]

It is "similar" to the big bang, but it is not the big bang; in fact, as far as Baum and Frampton are concerned, there is no "big bang" at all insofar as the term connotes an absolute cosmic starting point. As one paper they coauthored with Shinya Matsuzaki puts it simply, "[T]ime never began."[164] Rather, the universe has been cycling from eternity, with the dark-energetic demise of one universe producing a slew of others—one for each causal patch that contracts and rebounds.[165] Of course, these theorists admit, some patches that are insufficiently emptied out will "fail to cycle," but the probability of at least one patch successfully cycling into a full-blown universe is 1.[166] So the demise of each universe produces anywhere between 1 and 10^{103} new universes. Without positing multiple vacua or extra dimensions, the Baum–Frampton model branches outward like a tree as the cycles go on, "leading to an infinite multiverse" in both space and time.[167]

But there are yet other models. Using the framework of loop quantum gravity (the leading alternative to string theory),[168] Martin Bojowald predicts "a universe collapsing as a result of overpowering gravitational attraction" and approaching what looks like a remarkably Empedoclean stage of "strife." As the universe approaches the quantum level, Bojowald explains, it enters a state of "utter turmoil, chaos, and confusion," or the so-called State of Hell that constitutes the loop quantum vacuum. Although "all forces struggle for dominance," the eventual victor is gravity, which acts repulsively at the quantum level, "calms the seas, and lets the universe expand to vent its anger."[169] Then there is radiation, there is matter, and Empedocles's stage of "love" returns for another round. As far as Bojowald is concerned, the advantages of this model over inflationary and string models are manifold: loop quantum gravity avoids initial and terminal singularities;[170] it requires neither six extra dimensions nor 10^{500} extra vacua; and, in the words of Helge Kragh, "the repulsive force . . . is not introduced ad hoc."[171] Rather than positing a new kind of energy to account for the gravitational "push" (whether it be Kant's centrifugal force, Einstein's cosmological constant, quintessence, phantom dark energy, or inflation), loop quantum gravity needs only one force, which acts attractively at the classical level and repulsively at the quantum level. As Bojowald explains it, "[Q]uantum gravity [itself] . . . provide[s the] forces counteracting its own classical attraction."[172]

Meanwhile, back in the string theory camp, Cristiano Germani and his colleagues have proposed a model they call the "Cosmological Slingshot" as an alternative solution to the flatness and horizon problems.[173] According to this scenario, our world is one of a number of three-branes traveling through a "Calabi–Yau manifold," an unstable ten-dimensional space with "spikes" and "throats." As the universe falls into one of these "warped throats," it collapses into a big crunch but then flies back up the throat and reexpands; "the kinetic energy of the brane is passed to matter fields," and the universe cycles through the stages of radiation, gravity, and dark energy before slipping down the throat again and being re-created.[174]

Or there is Roger Penrose's "conformal cyclic cosmology," framed in opposition to inflation and to cosmologies arising from what he disparagingly calls "string theory culture."[175] Inspired by Maurizio Gasperini and Gabriele Veneziano's "pre–big bang scenario" (but, Penrose is careful to reiterate, without their reliance on string theory),[176] Penrose's scheme is one in which the end of one universe becomes the beginning of the next so that "the universe as a whole is to be seen as an extended conformal manifold consisting of a possibly infinite succession of aeons."[177] Toward the end of one aeon, the accelerated expansion of the universe flings distant galaxies apart from one another. In the meantime, nearby galaxies circle one another and eventually collide. At this point, the black holes at their centers merge and produce "whacking big explosions," which produce ripples in the fabric of space and time.[178] These ripples from the end of one aeon appear as "families of concentric circles" imprinted on the new universe; in fact, Penrose and Vahe Gurzadyan claim that they have detected such circles on the CMB.[179] This claim, it should be said, has found very little support; although other researchers agree that the circles are indeed detectable, they attribute them to quantum fluctuations in this universe—not to the influence of a previous universe.[180] David Spergel summarizes a fairly common critique of Penrose: "[A] universe with dark matter, dark energy and inflation is bizarre enough—we don't, however, get to detect circles from alternative universes."[181]

But, of course, it all depends on how you define "bizarre." Even the multiverse scenarios that look most outrageous from some perspectives seem to be perfectly simple—even parsimonious—from others.

ASCENDING TO THE
ULTIMATE MULTIVERSE

We've got to think of infinities of infinities. Indeed, there's perhaps even a
higher hierarchy of infinities.

MARTIN REES, "IN THE MATRIX"

Worse still, there is no end to the hierarchy of levels . . . gods and worlds,
creators and creatures, in an infinite regress, embedded within each other.

PAUL DAVIES, "UNIVERSES GALORE: WHERE WILL IT ALL END?"

The Quantum Multiverse

"What If the Wave Function Never Collapses?"

The most widely known and popularly dramatized multiverse scenario—
the one that sends *Family Guy*'s Stewie and Brian through an endless series
of alternative Quahogs—stems from the Many-Worlds Interpretation of
quantum mechanics, often abbreviated MWI. Initially posited in 1957 by
an American graduate student named Hugh Everett (1930–1982),[1] the MWI
is an alternative to the Copenhagen Interpretation, associated with Niels
Bohr (1885–1962) and Werner Heisenberg (1901–1976), who were among the
first to account for the bizarre behavior of matter at subatomic scales.
One of the central laws of quantum mechanics is the principle of "comple-
mentarity," according to which a particle's position and momentum (to
name one "complementary" pair) cannot be determined at the same time.
The moment a physicist measures a particle's location with precision, it
becomes impossible to measure the particle's momentum, and the moment
she measures the momentum, the position becomes imprecise. Brian Greene
offers the "rough analogy" of trying to take a picture of a fly: "If your shutter speed is high, you'll get a sharp image that records the fly's location at
the moment you snapped the picture. But because the photo is crisp, the

fly appears motionless; the image gives no information about the fly's speed. If you set your shutter speed low, the resulting blurry image will convey something of the fly's motion, but because of that blurriness it also provides an imprecise measurement of the fly's location."[2] Perhaps the most central lesson of quantum mechanics, then, is that the kind of answer you get depends on the kind of question you ask.

According to the Copenhagen Interpretation, the reason it is impossible to determine position and momentum at the same time is that a particle does not *have* a determinate position or momentum until it is measured.[3] The state of a subatomic particle cannot be expressed as a simply located point or series of points on a Cartesian plane; rather, it is expressed in terms of a "wave function" that details the varying probabilities of its possible states. When the particle is not being observed, it exists in a "superposition," occupying each of its possible states at once. But then when an observation is made, the wave function "collapses," and the particle takes on a definitive place, speed, energy, or spin—whichever property is being measured. Quantum particles are therefore a bit like the main characters in *Toy Story* (or Jim Henson's earlier, underappreciated *The Christmas Toy*),[4] who race around the room when they are alone but then freeze when a human opens the door. This account, without the layperson's illustrations, was more or less the story that prompted Everett to wonder, "What if the wave function never collapses?"[5]

According to Everett's explanation, if the wave function never collapses, then every possible outcome *actually* happens—each in a different universe. So every time a particle "decides" on a specific position, the universe has "split" into multiple branches—one for every position the particle might possibly take.[6] Every time the wind blows one way or another in the vicinity of a sailboat, the universe has split into some worlds in which the boat gains speed, some in which the boat changes course, and some in which the crew is stuck in the doldrums for a day and a half. And every time you order a burger with fries, there is another universe in which you have just ordered a salad and another in which you have just seen a mouse run across the restaurant floor and fainted.

If these musings sound familiar, it is because such bizarre possibilities arose in the notion of a "quilted" multiverse, which operates under the sole assumption that the amount of matter and space-time in the universe is infinite. That was Max Tegmark's Level I. The compendium of the MWI's ever-branching universes constitutes what Tegmark calls the "Level III multiverse," which (metaphysically at least) looks strikingly like the infinite set of overlapping spheres in Level I. As Tegmark explains these strata, "[T]he only difference between Level I and Level III is where your doppelgängers reside. In Level I they live elsewhere in good old three-dimensional space. In Level III they live on another quantum branch in infinite-dimensional Hilbert Space."[7] The quantum multiverse, then, is neither spatially arrayed nor temporally sequential. Even in principle, one could never reach these other worlds by traveling far enough "out there" or long enough "back then."

Perhaps unsurprisingly, Everett's theory was not an overnight success. It took the work of numerous others—most energetically, Bryce DeWitt (1923–2004)—to develop the MWI further and then to *publicize* it to the broader scientific community and laity. In 1970, DeWitt enumerated the extraordinary implications of the MWI; in effect, he explained, "every quantum transition taking place on every star, in every galaxy, in every remote corner of the universe is splitting our local world on earth into myriads of copies of itself." He also conceded that the idea might be hard for most people to assimilate, acknowledging that "the idea of 10^{100+} slightly imperfect copies of oneself all constantly splitting into further copies, which ultimately become unrecognizable, is not easy to reconcile with common sense." "Here," DeWitt summarized his audience's fears, "is schizophrenia with a vengeance."[8] And yet, as David Deutsch would go on to argue fervently in the decades ahead, the early proponents of the MWI suggested that it was actually the *simplest* explanation of quantum phenomena.[9]

According to its proponents, the MWI is preferable to the so-called Copenhagen Interpretation for two major reasons. First, it maintains that objects are subject to the same laws, regardless of their size. The many-worlders accuse the Copenhagen adherents of dividing the world into large things that behave classically and small things that behave quantumly: pitch a thousand baseballs through two holes in a wall, and each ball will go through only one hole at a time. Do the same thing with photons and a light-sensitive screen, and all the photons will seem to go through both holes at once.[10] For the many-worlders, this strange behavior is not confined to the subatomic world; rather, it reveals what is going on at the macroscopic level as well (this case can be made for Copenhagen, too, but the many-worlders do not tend to pay the argument much attention).[11] According to the MWI, not just each particle but the whole *universe* exists as a wave function of all its possible states, and every decision—whether microscopic, human, or galactic in scope—sends worlds branching off in every direction.

The second advantage the MWI claims over Copenhagen is that its quantum phenomena do not rely on observational influence to become concrete. Taken to its fullest expression, the ontological implication of Copenhagen is that there is no "reality" independent of the acts of observation, measurement, and interpretation. As Karen Barad has argued with unparalleled clarity, Bohr's interpretation means that phenomena are not just given; rather, they are formed only by means of relationships between observers, objects, and the experimental apparatus that conjoins them.[12] To many-worlders such as Colin Bruce, however, the whole idea seems to be "solipsistic," anthropocentric, and ultimately absurd. What counts as an observer? Bruce asks. "A mouse, a frog, a slug? . . . Does the observer need a Ph.D.? . . . Was the entire universe waiting to collapse into a definite state until the first ape-man came along?"[13] In short, according to the MWI, *there is a world* that is independent of our (or of a slug's) interactions with it; as Bruce proclaims, "only in a[n] [MWI] quantum world does it become possible to measure something without affecting it at all."[14] So although this quantum world is composed of an unthinkable number of independent subworlds, all of them can be described in one fell swoop by what many-worlders refer to as the "wave function of the universe." Rather than "collapsing" in relation to differing circumstances, the wave function continues unimpeded, "smoothly and deterministically" forever, regardless of who is observing what when.[15]

In his recent blog post "Does This Ontological Commitment Make Me Look Fat?" Sean Carroll therefore maintains that "the ontological commitments of the . . . many-worlds interpretation are actually quite thin." Because the wave function is unified and totally imperturbable, he argues, the MWI is "*simpler* than versions of QM [quantum mechanics] that add a completely separate evolution law to account for 'collapse' of the wave function. That doesn't mean it's right or wrong, but it doesn't lose points because there are a lot of universes."[16]

When it comes to taking sides with Copenhagen or the MWI, the physics community seems (ironically) to be split. Each camp tends to claim that "everyone who is anyone" is on its side and that no one takes the other interpretation seriously anymore.[17] By anyone's account, however, the MWI works just as well as Copenhagen to explain and predict quantum phenomena in this world. So as science writer Martin Gardner points out, any given theorist's "preference" for the MWI might be more functional than it is ontological; she might consider the "other worlds" not to be "physically real" but "useful abstractions such as numbers and triangles."[18]

After all, a mathematician can easily say that she "believes in" the idea of a perfect triangle without affirming the *physical* existence of a perfect triangle. Maybe the other worlds of the MWI are more like mathematical abstractions than they are "parallel universes"?

No More Drama: Stephen Hawking's Model-Dependent Realism

Although some many-worlders such as Deutsch, Bruce, and Leonard Susskind would reject outright this "nonrealist" reading, it finds a complicated negotiation in the recent work of Stephen Hawking and Thomas Hertog. As far as Hawking is concerned, neither the inflationary scenario nor any of the cyclical models provides a "satisfactory" account of the origins of the universe.[19] He offers a number of reasons for this dissatisfaction, but most of them boil down to his contention that anything that might have happened "before" the big bang is both inaccessible and irrelevant to the universe we are in.[20] All these models, Hawking argues, work in the wrong direction, from the "bottom up." They begin with a hypothetical precosmic state and then show how that hypothetical state might have produced a cosmos like ours. But they leave unexplained how this hypothetical precosmic state came about in the first place and *why*. In sum, writes Hawking, "all the pre-big-bang scenarios can do is shift the problem of the initial state from 13.7 [billion] years ago to the infinite past."[21] He therefore suggests moving in the opposite direction, from the present backward, and in this sense following what he calls a "top-down approach" to cosmology.

Hawking's distinction between "bottom-up" and "top-down" cosmologies corresponds roughly to Michel Foucault's distinction between "history" and "genealogy."[22] Whereas history begins from a fixed point in the past and then follows a linear course to the present, genealogy works from the present backward through its multiple sources so that the story branches farther out the farther back it goes (think here of a family tree). Hawking and Hertog's cosmology similarly begins from our point in space-time and then branches into more and more possible states as it moves into the past, ultimately reaching a quantum "beginning" that comprises "every possible history" of the universe.[23]

Most of these possible histories correspond to a possible universe, each of which has "its own probability" of expanding or collapsing, of having a strong or weak electromagnetic force, of having three or ten dimensions, and so on. This scenario allows Hawking and Hertog to argue that each possible quantum universe corresponds to one of the vacua of the string theory landscape. But unlike in the inflationary scenario, each of these possible worlds lies "within" the same universe; that is, each of them emerged out of the same big bang. Moreover, according to the "no-boundary" proposal that Hawking set forth in the 1980s with James Hartle, the big bang is not the product of anything outside itself.[24] At the quantum level, they argue, the distinction between time and space disappears, so there is simply no such thing as the "beginning of time." Rather than being bounded by anything outside itself, the Hartle–Hawking universe is therefore "completely self-contained and not affected by anything outside itself."[25] It is this no-boundary proposal that Hawking and Hertog now offer as a cosmological candidate to "populate" string theory's landscape. Insofar as their scenario has no need to posit a primordial scene beyond our event horizon, they argue, it is far preferable to eternal inflation or any of the cyclical models.

Crucially for Hawking, this no-boundary proposal stems from epistemological rather than ontological commitments. To say that the universe is not affected by anything outside the universe is not to say "there is nothing outside the universe." Hawking and Hertog freely admit that there may well *be* an "original" metaversal sea before and beyond our universal bubble, but they insist that once the bubble forms, the original sea "is irrelevant for observers inside the new universe."[26] Because the eternal inflationary model spends so much time generating regions of space-time forever beyond us, Hawking and Hertog argue, it is not so much wrong as it is unhelpful; as far as they are concerned, "the mosaic structure of an eternally inflating universe is a redundant theoretical construction, which should be excised by Ockham's razor."[27] At this stage, one might feel compelled to ask how it is that if Hawking and Hertog affirm Ockham's razor, they can also affirm the many-worlds mantra that "the universe will not have a single history but every possible history, each with its own probability."[28] How is it that all possible worlds *within* our event horizon are less "redundant" than all possible worlds outside it?

Hawking's response to this query is relentlessly pragmatic. He acknowledges that "the no boundary proposal predicts a quantum amplitude for every number of large spatial dimensions from 0 to 10."[29] This means that "there will be an amplitude for the Universe to be 11-dimensional Minkowski space, i.e. with ten large spatial dimensions. However," Hawking continues, "the value of this amplitude is of no significance, because *we do not*

live in eleven dimensions." In other words, in the face of the swarm of possibilities predicted by both quantum mechanics and the string theory landscape, we can ignore all worlds apart from the ones that bear the attributes
of our universe. And we do not need to worry about how "rare" or "common," how improbable or probable, our universe turns out to be. "As long
as the amplitude for three large spatial dimensions is not exactly zero,"
Hawking argues, *"it does not matter* how small it is compared with that for
other numbers of dimensions." Again, the issue is not that universes with
different sorts of dimensionality do not exist; rather, it is that any consideration of those extradimensional worlds is *"irrelevant,* because we have
already measured that we are in four dimensions."[30]

It is probably important here to point out that Hawking's most recent
work with Hertog has not been widely accepted among physicists. Philosophically, however, it is a fascinating position to take—and it might have
implications that reach beyond this specific cosmological model. Hawking is making neither the metaphysical claim that all possible worlds actually exist (the "realist" position) nor the equally metaphysical claim that they
are simply mathematical abstractions (the "antirealist" position).[31] Rather,
he is saying that because the quantum no-boundary proposal matches
observations and predicts them, it is a good model. But there might be
other, equally good models. For comparison, as Hawking and Leonard
Mlodinow remind us, the sun is no more "at rest" in the heavens than the
earth is. "The real advantage of the Copernican system" is not that it tells
us how the universe really is, but "that the equations of motion and rest
are much simpler in the frame of reference in which the sun is at rest."[32]
But there is nothing *wrong* with using an earth-centered frame of reference as long as you are willing to slog through the unnecessarily difficult
calculations required to make it work. Hawking and Mlodinow therefore
call their position a "model-dependent realism," arguing that our access to
the world is always mediated by the model through which we access it. In
other words, *"there is no picture- or theory-independent concept of reality."*
And so it does not matter how "bizarre" the MWI of quantum mechanics
might sound. It "has passed every experimental test to which it has been
subjected," so there is no good reason to reject it.[33] But there is also no
good reason for ecstatic reflections on the numberless copies of slightly
different yous—or on the creatures that might live in eleven-dimensional
space. "Life may, after all, be possible in eleven dimensions," Hawking
acknowledges, but who cares? "We know how to live in four."[34]

Putting It Together: The Multiversal Bath

Hawking is not the only theorist to connect the MWI with the string theory landscape. Susskind has also claimed that the two map onto each other, although his cosmological bets are on inflation rather than the no-boundary proposal and on ordinary realism rather than model-dependent realism.[35] This possibility—that inflation, string theory, and the MWI might be describing the same multiverse—has found sustained exploration in the work of cosmologist Laura Mersini-Houghton. Although this work has gained a great deal of recognition since the publication in the early spring of 2013 of the most recent data from the Planck satellite, it has not yet generated widespread discussion. Nevertheless, I am bringing it under sustained consideration because, in addition to looking increasingly plausible in relation to the newest data, Mersini-Houghton's model integrates a number of different multiversal scenarios, and it more radically begins to think them through philosophically.

For Mersini-Houghton, the most puzzling "fine-tuning" question is how the universe began in such a highly ordered state. As we have learned in the context of cyclical cosmologies, the Second Law of Thermodynamics dictates that entropy, or disorder, always increases in a closed system. To offer some common examples, a glass of water with ice cubes is in a relatively ordered state; it has a relatively low measure of entropy. As the ice cubes melt, the entropy increases until all the water molecules reach the same temperature, attaining a relatively disordered, high-entropic, or "chaotic" state.[36] The same glass sitting on a table, in turn, has a low measure of entropy compared with the glass the cat has just knocked onto the floor, shattering it into a thousand pieces.

What is remarkable about these processes is their seeming inexorability: a glass will shatter when it is knocked to the ground, but those shards of glass, when jostled around, will be very unlikely to reconstitute themselves into a tumbler. Ice cubes melt but do not spontaneously freeze again; eggs break but do not gather themselves back together. To be more precise, the Second Law does not say that such things *never* happen; it just says such situations are exceedingly rare. As Sean Carroll explains this rarity, "[T]here are more ways to arrange a given number of atoms into a high-entropy configuration than a low-entropy one."[37] In other words, there are a staggering number of ways in which a glass can shatter into pieces, but only *one* configuration that will order all those pieces into a drinking glass, so the probability of this one configuration's occurring is extraordinarily low. For all intents and purposes, systems always move from order to chaos in a direction cosmogonists might call "de-creative."

This unidirectional movement from low entropy to high entropy constitutes what is known as the "arrow of time"—the inexorable passage of all things from the past through the present to the future. Insofar as our universe seems indeed to have an arrow of time—that is, insofar as eggs do not unbreak and lives do not play themselves backward—the level of cosmic entropy must be progressively increasing. This means, looking back, that the universe must have begun in a very highly ordered, low entropic state; the primordial plasma must have been something like a great, unbroken cosmic egg. The history of the universe would, in turn, be one of increasing disorder as it moves from radiation to gravity to dark energy. As Carroll explains, "[T]he universe starts out in a state of very low entropy; particles packed together smoothly. It evolves through a state of medium entropy: the lumpy distribution of stars and galaxies we see around us today. It ultimately reaches a state of high entropy: nearly empty space, featuring only the occasional stray low-energy particle." Carroll continues to explain that the question then becomes, "[W]hy was the entropy low to begin with? It seems very unnatural, given that low-entropy states are so rare."[38] In effect, starting with a highly ordered, highly energetic universe is as perplexing as shards of glass spontaneously assembling themselves into a tumbler; Mersini-Houghton cites Roger Penrose's estimate "that such an event has only one chance in $10^{10^{123}}$ possibilities."[39] And yet without such a low-entropic and high-energetic starting point, there would be no arrow of time—which is to say no universe as we know it.

According to Mersini-Houghton, the extraordinarily improbable event of our universe's *being* at all can be understood only "if we accept that other more probable events can be conceived"—that is, we can understand the birth of our universe only if we posit a multiverse.[40] But the inflationary multiverse alone is insufficient because, as Hawking has also argued, it *assumes* the high-energy inflation it needs rather than deriving it from physical processes. Unlike Hawking, however, Mersini-Houghton is not rejecting inflation; rather, she is simply suggesting that it starts the story too late. Inflation, she argues, must itself be explained, and this can be done only by getting back *behind* inflation to the quantum conditions that produce it.[41] Also unlike Hawking, Mersini-Houghton is therefore saying that understanding the processes "before" the bang is crucial to understanding the processes after it.

This, then, is Mersini-Houghton's starting point: a primordial "multiverse bath," a chaotic sea of virtual particles in which two forces vie for dominance, both trying to bring the system into equilibrium. "Matter degrees of freedom" operate attractively, trying to pull each "patch" of proto-space-time into a black-hole "crunch." The bath itself also exerts an attractive force, trying to keep each patch "entangled" with the multiverse. Meanwhile, "gravitational [degrees of freedom]" operate repulsively, trying to expand each patch outward into its own region of space-time. The scene, in short, is a precosmic "battlefield," an eternal tug-of-war between the forces of attraction and repulsion. If matter and the bath "win," then the patch collapses into a "terminal universe." But if lambda is strong enough to "survive the backreaction of matter and the bath," then the patch inflates into a "survivor universe."[42] Here, then, is the origin of inflation *and* the reason our universe started out with such low entropy (which is also to say high energy): the only universes that nucleate successfully are the ones "born" at energies high enough to survive the attractive pull of matter, which have a sufficiently strong repulsive push to inflate outward into "survivor" universes.[43]

For Mersini-Houghton, the initial bath, which she also calls the "underlying reality," corresponds to the landscape of string theory *and* the wave function of the MWI insofar as it comprises "the ensemble of all possible initial conditions and energies." Moreover, this scenario begins to provide a "super-selection rule" that explains how certain universes on the landscape come into being, but others do not. Thanks to the battle between matter and lambda, "only a fraction of them, initial states with high energies Λ, are selected as physically relevant 'survivor' universes."[44] As one of these survivors finally inflates, Mersini-Houghton explains, her metaphors understandably mixing and crossing, the universe becomes "disentangled" from the original bath that produced it and is "pinned down" into an independent "branch" of the quantum-string-inflationary multiverse. Because the resulting baby universe starts at such a low entropic state, it now has a unidirectional arrow of time and promptly "forgets" where it came from (figure 6.1). "An infant is a pucker of the earth's skin," writes the poet Annie Dillard, "so are we. We arise like budding yeasts and break off; we forget our beginnings."[45] Ratcheting this imagery out to the cosmos, Mersini-Houghton offers a similar, if more Latinate, explanation: "There is information loss about the underlying reality of the multi-

FIGURE 6.1 The birth of universes from the landscape multiverse, according to Laura Mersini-Houghton's model. (Illustration by Kenan Rubenstein)

verse," she writes. "As the bubble decoheres[,] such entanglement with the bath is deemed irrelevant, and these correlations ignored."[46]

And yet the information and correlations might not disappear completely. In fact, Mersini-Houghton predicted that the primordial entanglement of our universe with other universes as well as with the bath from which they all nucleated should have "left its imprints on the cosmic microwave background and large scale structures,"[47] and it seems that some of these imprints have been *found*. In 2006, for example, Mersini-Houghton's team claimed "that a giant void of a size of about 12 degrees in the sky, should be found at about 8 billion light years away. Amazingly," she continues, "this giant void was observed only a few months later in 2007" by the Very Large Array Telescope, and it has been confirmed by the 2013 Planck results.[48] The team predicted one other, comparably sized void and is searching for it. In the meantime, Mersini-Houghton and her colleagues also predicted a large set of inhomogeneities on the Cosmic Microwave Background (CMB)—an "unusual pattern in the motion of around 800 galaxy clusters" for which inflation cannot account.[49] According to Mersini-Houghton, these inhomogeneities would have been caused by "the backreaction of superhorizon matter modes"—that is, the attractive forces that tried to keep the universe from inflating beyond the bath in the first place.[50] A team led by astronomer Alexander Kashlinsky strikingly discovered such an "unusual pattern" in 2009, naming it the "dark flow"—another inhomogeneity confirmed in 2013.[51] As far as Mersini-Houghton and her colleague Richard Holman are concerned, these discoveries (along with two more technical predictions, both of which have also been discovered)[52] provide compelling "evidence for the birth of the universe from the landscape multiverse"; in fact, they offer a kind of photographic record of this birth, testifying to "the nonlocal entanglement of our universe with *all other patches* and massive modes beyond the horizon."[53]

To be sure, this claim is formidable in terms of its scientific and ontological implications alike. More precisely, scientific rigor and ontological complexity here become totally bound up with each other. The possibility that the multiverse might have observational effects on our universe certainly increases its standing as a scientific hypothesis: not only does the multiverse emerge from inflation, string theory, and quantum mechanics, not only is it "needed" theoretically to account for the unlikely parameters of our universe, but it can also (perhaps) be *seen*. But the condition under which it might be seen, given our inescapable situation *within* this world, is the "entanglement" of our universe "with all other patches" of the multiversal bath.[54] Put more simply, the only multiverse that can be a proper object of scientific study is what Mersini-Houghton calls a "connected multiverse."[55]

This "connected multiverse" occupies a complicated ontological terrain. It is not "all-one"; that is, all universes on the multiverse landscape are not immediately present or connected to one another. But neither is the multiverse a set of totally separate worlds, each with its own sets of laws, floating along in total indifference to one another. Rather, according to Mersini-Houghton, the "more plausible scenario" is that "correlations among different domains do exist,"[56] even though these domains are not simply accessible to one another. She derives the "plausibility" of these correlations from "the sole assumption that quantum mechanics is valid at all energy scales,"[57] specifically the "sacred principle" of quantum mechanics that says that information, even in a black hole, can never be totally lost.[58] If it is the case that even preinflationary information is retained in some form, then, Mersini-Houghton predicts, "the entanglement [of our universe with the multiverse bath] leaves its traces everywhere in the present observable sky." So although our universe has become its "own" branch of the multiverse, it also remains complexly connected to the entangled whole; in her words, "[I]f our universe started in a mixed state in the landscape multiverse, then it can never evolve into a pure state at late times." Of course, the effects of this impurity diminish as the universe develops; "at present the strength of nonlocal entanglements is very small."[59] But the strength grows the farther out and back one looks; the imprints of other universes and of the bath itself will be found at the "edges" of our own. So regardless of the age or size of the observable universe, its early history will always bear marks of its entangled beginnings.

As far as Mersini-Houghton is concerned, this complex connectivity does not seem to have a philosophical analogue; as she suggests in passing, we do not have an "ontology of the multiverse."[60] But the "connected multiverse" under consideration is not totally without philosophical precedent. Here we might think back to the Cusan cosmos, which we find is neither one nor many, nor is it both one and many,[61] but many from the perspective of any world and one from the perspective of God. "The perspective of God" would here translate into the perspective of Mersini-Houghton's multiversal "bath," the "underlying reality" in which "time has no direction, no beginning and no end."[62] From the point of view of what Mersini-Houghton calls "fundamental time" (a phenomenon that earlier philosophers would have called "eternity"), all worlds are part of a shared multiversal space-time.[63] From the perspective of any particular world, however, the arrow of time is irreversible, setting each universe on its own independent course forever, marked only by the scars of its early entanglements. So the "connected" multiverse is either one or many, depending on how you look at it.

In addition to establishing this link to Nicholas of Cusa, we might also consider plugging back into Giordano Bruno's "multimodal" ontology and to William James, who coined the term *multiverse* to name a specific kind of cosmic co-implication: neither all-one nor total fragmentation, but a "strung-along type" of shifting unity, never disbanded yet never "absolutely complete."[64] Of Mersini-Houghton's connected multiverse, which displays a "connectivity through the nonlocal entanglement of our domain with everything else on the multiverse,"[65] we might well be able to say, along with James, that "our 'multiverse' still makes a 'universe'; for every part, tho it may not be in actual or immediate connexion, is nevertheless in some possible or mediate connexion, with every other part however remote." More radically even than James, however, the theory of quantum entanglement suggests that remote regions can be "in inextricable interfusion" with one another *without* "hang[ing] together with [their] very next neighbors"—that is, without being materially connected, either directly or in a Jamesian cosmic chain.[66] To say, as Mersini-Houghton does, that the multiverse is "nonlocal" is to say that its connections transcend the logic of connectivity, that "here" is ontologically inseparable from the "there" without which here would not be here. The multiverse, then, is not one, but neither is it simply many; rather, it is many by virtue of its complex unity and united in its irreducible manyness.

Black Holes and Baby Universes

Numerous as the Stars

In the previous section, readers may have noticed the strikingly parental (specifically maternal) metaphors that surface in relation to Laura Mersini-Houghton's connected multiverse. The original state is a primordial sea, or "bath," from which universes are "born."[67] The birth process is one of competing pushes and pulls that, when successful, leave indelible marks on the neonate. This new universe becomes progressively independent as it ages and grows, but it never loses the physical marks of its original dependency—one might think of the discovered "void" as a great cosmic navel, the "dark flow" as a kind of port-wine birthmark left forever on the CMB. Such images are not limited to Mersini-Houghton's work, however; one finds a complementary set of metaphors in Lee Smolin's very different cosmological scenario, which suggests that universes might be born on the "other side" of black holes.

Following the publication of Einstein's theory of general relativity in 1916, a physicist named Karl Schwarzschild (1873–1916) calculated that if a large enough mass were crammed into a small enough space, it would attract the surrounding area so forcefully that it would rip through space-time itself, producing what John Wheeler later named a "black hole."[68]

Black holes are one of the possible products of a dying star: as a star ex-
hausts its energy, it may collapse into a singularity, drawing all the matter,
energy, and space-time around it into a vortex surrounded by an "event
horizon" from which nothing (at least nothing recognizable) can ever escape.[69] It is through this singularity that Smolin connects black holes to
cosmogony. As he explains in his book *The Life of the Cosmos*, "[A] collapsing star forms a black hole, within which it is compressed to a very dense
state. The universe began in a similarly very dense state from which it expands. Is it possible that these are one and the same dense state?" If it is
possible, then "what is beyond the horizon of a black hole" would be "the
beginning of another universe."[70]

Of course, we would not be able to *see* the formation of this other universe, says Smolin, because the star's collapse would give way to an explosion only "after the black hole horizon had formed around it" (88). But
from the inside of this black hole, the stellar explosion would look like a
big bang. If all of this is the case, then by extension our universe is sitting
on the inside of some other universe's black hole, connected to this "parent" universe by a vanishing "umbilical cord" that has torn through the
fabric of space and time.

Smolin places his black-hole cosmogony in the lineage of the cyclic
cosmologies of the 1930s and 1960s—those "phoenix" universes that posited a universal contraction, crunch, "bounce," and reexpansion.[71] "What
we are doing," he explains, "is applying this bounce hypothesis not to the
universe as a whole, but to every black hole in it" (88). The result is not a
single universe recycled throughout eternity, but a "continually growing
community of 'universes,' each of which is born from an explosion following the collapse of a star into a black hole" (88). Smolin more commonly
refers to this community as an ever-growing "family," with each universe
producing "progeny" that themselves "reproduce" by forming as many
stars (which is to say, potential black holes) as possible.

Smolin begins his reproductive cosmogony from what he admits is an
arbitrary starting point. It might be that universes have existed throughout eternity, or it might be that a vast number of them suddenly appeared
at the same time. As far as Smolin is concerned, we will never know what
was "there" at the origin of things; all we know is that at least one universe
has survived the cosmogonic process (101). So for simplicity's sake, he suggests, "let us imagine that the universes in our collection are all progeny
of a single, initial universe" (96). What would this single, initial universe
have looked like? Again, Smolin confesses not to know and so suggests

that we "make the best of our ignorance and assume that its parameters are chosen randomly" (96). The universe tries on various values of lambda, of matter, of the electroweak force, and the results are likely to be fairly disappointing for a long number of cycles. If its constants are "chosen" randomly, he writes, then "it is very unlikely that the parameters of this first universe are finely tuned to values that result in a big universe full of stars" (96); rather, the new world will most likely either collapse instantaneously or inflate too quickly for anything to form.

"For a long time," Smolin imagines, "the world is nothing but a series of tiny universes, each of which grows out of the one before it. . . . What is happening is that, one by one, different possible parameters are being picked randomly and the consequences of each tried out" (97). Now we have no idea, to channel David Hume's Philo, how many "worlds might have been botched and bungled . . . ere this system was struck out."[72] But we do know that at least one universe eventually worked, so each universe in our lineage must have had at least enough matter to collapse and bounce one time. One of these random universes will eventually obtain the parameters that allow it to form stars, which is to say black holes, which is to say new universes. If the reproduction of universes looks anything like the reproduction of plants or animals, Smolin speculates, then these new universes will have traits resembling those of their "parents," so most of them will go on to produce stars, black holes, and baby universes of their own (100–101).

For Smolin, the most attractive feature of this scenario is that it keeps a safe distance from the anthropic principle.[73] In fact, he argues, this scenario replaces anthropic reasoning with a better-established scientific process—that is, natural selection. Rather than suggesting that the universe takes on the parameters it needs to create life (or, more weakly, that life emerges in those universes that can support it), the black-hole scenario simply says that the only types of universe that survive are those that can produce new universes like themselves. So, as John Gribbin summarizes Smolin's anti-anthropic reasoning, our universe has the parameters it does "*not* because it is a good home for life, but *because it is good at making black holes.*"[74]

Crucially for Smolin, this scenario also steers clear of theism at all turns by configuring the cosmos as a self-contained system. Borrowing (wittingly or unwittingly) some early Derridean terminology, Smolin states elsewhere that the first principle of scientific cosmology is that "there is nothing outside the universe."[75] The black-hole scenario adheres to this principle by positing a strictly *immanent* principle of generation and selection. Unlike the inflationary model, which configures universes as the offspring of a sole "ancestor, which is the primordial vacuum,"[76] Smolin's scenario allows each baby universe to produce its *own* offspring, eliminating the need for any single, extracosmic origin at all. Viewed as a whole, then, Smolin's cosmos looks like an inversion of Hawking's, branching genealogically outward as it goes *forward* in time, not backward. And the genealogical metaphorics are not lost on Smolin; as he imagines "a multiverse formed by black hole bouncing," it "looks like a family tree. Each universe has an ancestor, which is another universe. Our universe has at least 10^{18} children; if they are like ours, they each have roughly the same number of children."[77] This scenario allows Smolin to avoid positing any sort of creator; worlds are eternally generated by means of an immanent reproductive process. Ironically, however, Smolin avoids the old extracosmic God by taking literally that same God's promise to make his chosen creature's descendants as "numerous as the stars of heaven."[78]

Cosmic Manufacturers

Smolin's proposal has not been widely adopted, but it is certainly well known and has produced a series of fascinating physicophilosophical speculations. One proposal—whose authors argue that it can apply to black-hole, inflationary, *and* quantum cosmologies—demonstrates that if one of the eventual "baby universes" has the same conditions as the original universe, then it might, in fact, *be* the original universe. Time eventually closes these worlds into an autocreative loop; in a catch-phrase, "the Universe can be its own mother."[79] Specifically, in being the mother of all universes, the universe can at one point give birth to itself, which is to say the universe that gives birth to all universes.

A better-known but no less speculative riff on Smolin's work involves conscious intervention into universe formation. The line of thinking goes like this: if it is the case that anything sufficiently compressed can form a black hole, and if, as Smolin suggests, black holes produce universes inside their horizons, then what is to stop people like us from *making* black holes and thereby creating universes? This possibility has been explored in greatest depth by cosmologist Edward Harrison, who combines Smolin's proposal with Guth's inflationary cosmology to suggest that "all" we would need to create a universe would be "a small spherical nugget of space filled with a high-value inflaton field"[80] and enough energy to compress it into a black hole. Then *we* could be the mother of a few universes ourselves.

Those who have been following the progress of the Large Hadron Collider in Geneva, Switzerland, may remember the news storm surrounding its launch in 2008; some scientists, the blogs and papers told us, were worried that the collider might create a black hole that would swallow our entire planet.[81] Two decades earlier, however, Steven Blau, Eduardo Guendelman, and Alan Guth had shown that even if humans *were* able to compress a patch of space-time into a black hole, the resulting bounce would not take place within our universe.[82] Instead, as Harrison appropriates and explains this process, the interior of a black hole "immediately inflates not in our universe, but in a re-entrant bubble-like spacetime that is connected to our universe via the umbilical cord of the black hole. The black hole rapidly evaporates by Hawking radiation and severs the connecting link with the new universe."[83]

As many physicists charge and as Harrison admits, scientists are excruciatingly far from being able to generate the amount of energy that would be required to make a universe in this fashion.[84] But Harrison suggests that our limitations in this regard are more quantitative than qualitative; even if, as Brian Greene estimates, "the compressive force we would need to apply is trillions and trillions of times beyond what we can now muster,"[85] Harrison insists that this problem remains one of degree rather than kind. *In principle*, Harrison argues, we know how to do it. As a thought experiment, then, he suggests we imagine "that these practical difficulties can indeed be overcome, and that civilizations vastly more advanced than ourselves are able to steer the course of eternal inflation in their vicinity by initiating the production of mini-universes in their laboratories." At first, they would most likely create a slew of failed universes, botching and bungling worlds until they finally created one that worked. Within this universe would eventually evolve "yet-more-advanced civilizations (perhaps inheriting information from their makers to accelerate them on their way)," whose inhabitants would, in turn, continue to improve the parameters of *their* baby universes.[86] The result would be a universe that would be impossibly well tuned to support life, much like our own universe. Once the manufacturers got the hang of their art, they would be able to create as many universes as they liked—one a day, perhaps, or even one an hour. In very little time, then, manufactured universes would outnumber naturally occurring universes by a large measure. If it is indeed possible to create universes at all, it is therefore likely that ours has been manufactured.

This, then, is Harrison's answer to the question of why our universe looks so fine-tuned: *it is*. Our universe is conducive to the emergence of life and is, in turn, comprehensible to the intelligent beings who emerge therein because the universe was created by intelligent beings like us (just of far *greater* intelligence) who carefully set the parameters so that planets and stars and beings like us might eventually come along and decode the secrets of the universe.[87]

If this argument sounds familiar, it is because it is uncannily akin to the "argument from design" so inimical to secular physicists. Harrison seems unaware of the extent of this resonance; in fact, he offers his "do-it-yourself universes" as a "third choice," an alternative to the equally unappealing options of "either a supreme being beyond rational inquiry or a multiworld wasteland of mostly dark and barren universes"—that is, the inflationary scenario.[88] John Gribbin, who is a great enthusiast of Harrison's proposal, amplifies this distinction, insisting that the "intelligent designers" in Harrison's laboratories look nothing like the "Intelligent Designer" of theism because these new "gods" are "not incomprehensible."[89] Rather, as Harrison argues, they are "comprehensible beings who had thought processes *basically similar* to our own."[90]

Although I appreciate the distinction, and although many "philosophical theists" do indeed proclaim the omnipotence and omniscience of the deity (characteristics that Harrison's intelligent designers do not share), it seems important to point out that there is nothing "incomprehensible" about the God of intelligent design. Far from being "beyond rational inquiry," this God is *proven* through rational inquiry, thanks to (1) the fine-tuning of the universe, what Hume's Cleanthes calls "the curious adapting of means to ends," and (2) the resemblance of this universe to what he calls "products of human contrivance, of human design, thought, wisdom, and intelligence." Far from being "incomprehensible," the old designer God is said to be "somewhat similar to . . . man; though possessed of much larger faculties."[91] Likewise, Harrison's new designer-gods are intelligent beings "like us" (specifically like "us physicists"), although possessed of superior brainpower, technology, and sources of energy.[92] It is not clear, then, that the designer universe really is a "third choice"; in fact, it sounds remarkably like the old first choice, just with some updated ancestor gods. And until we make a universe of our own, this "argument from manufacture" will make just as tenuous an analogical leap from the known to the unknown as does the argument from design.

Sim Cosmos

Brian Greene is among those physicists who believe it is highly unlikely that we are living in a manufactured universe. Even if it were physically possible to make such a black-hole universe (and he is fairly certain that it never will be), he cannot imagine why future scientists would *want* to. After all, as he explains in *The Hidden Reality: Parallel Universes and the Deep Laws of the Cosmos*, they would never even know if they had been successful. Because a new universe would exist on the other side of a black-hole horizon, its manufacturers would never be able to tell whether the new world had inflated at all, much less how long it would live, whether it looked like its parent, how many stars it had, and the like.[93] A far more likely possibility, Greene argues, is that future scientists might make a *simulated*, or virtual, universe.

The reason Greene considers it more likely that scientists of the future will create simulated worlds than baby black holes is that "we're already doing it" (288). From *Sim City* and *Second Life* to military-training software to *Sindome: A Cyberpunk Role-Playing Game Set 85 Years in the Future*, programmers have already created virtual worlds, laying out the infrastructure that ordinary humans (that is, not computer geniuses) can fill in with content of their own design.[94] So, Greene contends, "the question is how realistic the worlds will become" (288). Is it possible that the "Wii Mii" you created yesterday—designing her to have interests in soccer and astronomy, giving her a job as a nurse at a local hospital, and placing her in a two-bedroom house with a dog—might one day think that she has done all this for herself? Might technology one day reach the point of being able to simulate *consciousness*?

Some philosophers of science call the (allegedly impending) moment at which the capacity of computers will exceed that of a human mind the "singularity," borrowing the term from cosmology.[95] Such a singularity would issue in a "posthuman" era in which the distinction between organism and machine would break down completely. At this stage, it would presumably be relatively easy to make "realistic" simulations; an overcaffeinated afternoon's work on the part of an average posthuman would produce a virtual cosmos filled with planets, stars, and a few billion lonely sims, full of false memories of the past and wondering if there is anything else "out there."

With this scenario as even a far-off possibility, the "transhumanist" philosopher Nick Bostrom offers the following argument:

At least one of the following propositions is true:

1. The fraction of human-level civilizations that reach a posthuman stage is very close to zero.
2. The fraction of posthuman civilizations that are interested in running . . . simulations is very close to zero;
3. The fraction of all people with our kind of experiences who are living in a simulation is very close to one.[96]

Walking a bit more slowly through this argument, Bostrom is saying that it is possible that (1) civilizations approaching a posthuman phase will destroy themselves with their advanced weapons and technology before they ever reach the singularity. This would mean that conscious simulations will never be achieved, nor have they ever been achieved in the past. Or it is possible that (2) once civilizations attain the intellectual sophistication required to run a conscious simulation, their posthumans will not *want* to—either because they will judge it unethical to create beings who mistakenly believe they are free or because, to put it bluntly, they will have outgrown video games. If this is the case, then although there might be a *few* simulated universes, their number would be "very close to zero," so we can be sure that most worlds are real (that is, not simulated). If, however, intelligent beings *can* reach a posthumanist phase without obliterating themselves, and if they remain interested in creating simulated universes, then (3) they would presumably create universes *all the time.* Adolescent posthumans would be creating worlds on their wePhones between classes. This would mean that the ratio of "real" worlds to virtual worlds would be *minuscule*, and the probability that we ourselves are living in such a simulation would be "very close to one." In short, if we believe that it will *ever* be possible to simulate conscious beings, then we must also believe that we are probably among those simulated beings— our universe, as Philo imagines in a particularly manic section of Hume's *Dialogues*, "only the first rude essay of some infant deity."[97]

Bostrom is commonly described as the philosopher who proclaims our universe to be a virtual one, but his position is more complicated than this description indicates. As he explains in a follow-up article to his widely circulated article "Are We Living in a Computer Simulation?" he does not argue that we *are* living in such a simulation. "In fact," he writes, "I believe that we are probably *not* simulated."[98] But if this is the case, if condition (3) is not true, then either condition (1) or (2) would have to be true; that is, if we are not simulated, it must be either because humans tend to destroy themselves on the way to posthumanism or because they lose interest in simulations. The crux of the argument is simply that that if we affirm the *eventual* possibility of simulated consciousness, then our own consciousness is most likely simulated.[99]

The question, then, is whether it will ever be possible to simulate consciousness at all. For those who believe that it will, the major premise is that "life and complexity means [*sic*] information processing power"; that is, everything from algae to fir trees to the brains of elephants can be explained in terms of computing operations.[100] The minor premise is that the computers of the future will one day be sufficiently powerful to run all the operations in the universe at once. The conclusion, of course, is that at such a time simulated worlds will be possible, and from there Bostrom's third stipulation holds: we ourselves are probably living in a simulation.

Greene, who delightfully proclaims this possibility in various forms of popular media,[101] explains the calculations this way in *The Hidden Reality*. A human brain can perform 10^{17} operations per second. This number is baffling; Greene estimates that the processing power required to simulate a human brain would be "a hundred million laptops, or a hundred supercomputers" (283). Over a hundred-year lifetime, the same brain will perform a total of 10^{24} operations, and if we multiply that number by all the humans who have walked the earth—one is struck all over again by the power of compaction in exponential notation—the total number of human operations "since Lucy . . . is about 10^{35}" (286). According to Greene, a computer the size of the earth could perform between 10^{33} and 10^{42} operations per second; hence, "the collective computational capacity of the human species could be achieved with a run of less than two minutes on an earth-sized computer" (287). Such a scale may seem to put the proposal of a simulated humanity beyond reach, but if quantum computing becomes as sophisticated as some project that it will, then the earth-size machine it would now require could be shrunk to the size of a laptop (287). So, Greene concludes, there is nothing in principle to prevent the simulation, not only of humans, but of a whole universe containing them—again, given that the (often unacknowledged) major premise is true and that organic life can indeed be reduced to information processing.

In *Meditations on First Philosophy* (1641), René Descartes famously worries whether he can trust his senses or thoughts at all.[102] The people walking outside below his window might be automata, he imagines, and the mountains in the distance some sophisticated projection. For all he knows, it might even be that 2+2=7 and that the angles of a triangle *actually* add up to 193 degrees—but that the universe is being run by an "evil deceiver" whose psychotic will it is to ensure his creatures are miserably wrong about everything. In the meditations that follow, Descartes uses this fantasy as a means of doubting everything—from his body to his memory to his surroundings—seeking from this point of radical doubt to find one thing of which he can be certain. His first discovery is well known: "thought exists," he proclaims; even if everything he thinks is wrong, *thinking* is nevertheless going on (2.27). So although the evil deceiver might have tricked him into believing that he has legs or that objects fall downward or that 2 + 2 = 4 when the answer is really 7, "he will never bring it about that I am nothing so long as I shall think that I am something" (2.25). This, therefore, becomes Descartes's "Archimedean point," the one thing of which he can be certain. If thought exists, then he exists; and if he exists, then, as the rest of the *Meditations* demonstrate, anything Descartes "clearly and distinctly" perceives to be true "is necessarily true" (5.70). Again, this exercise is well known—from T-shirts to coffee mugs to puns about Descartes blipping out of existence when he declines to order dessert ("I think not")—most people have a sense that Descartes's epistemology, his theory of what we can know, hinges on the demonstration of his own existence as a thinking thing.

What is less commonly discussed is that Descartes's philosophy *also* hinges on a demonstration of the existence of God. Descartes devotes lengthy sections of Meditations 3 and 5 to proving not only that there *is* a God, but that this God is *good*. The reason for this undertaking is that if God does not exist, and if he (per Descartes) is not good, then there may well be an evil deceiver ("God" might even be that deceiver!), and Descartes may well be a brain in a vat in his diabolical laboratory. And who knows—maybe the demon *can* deceive Descartes into thinking that he is thinking, when he is actually doing no such thing; maybe the evil deceiver is just that deranged. On its own, then, the infamous "I think" (*cogito*) is insufficient. It is only with the knowledge "*that there is a God*, and . . . that everything else depends on him, and that he is not a deceiver" that Descartes can be *certain* that he thinks, and that he can think clearly and distinctly, and that he has a body, and that other beings exist outside his mind, and that apples fall downward, and that the angles of a triangle add up to 180 degrees (5.70, emphasis added). "And thus," Descartes concludes, driving semester after semester of philosophy students crazy, "I see plainly that the certainty and truth of every science depends [sic] exclusively upon the knowledge of the true God" (5.71).

The point of this sudden excursus through Descartes is simply to point out that with the simulation argument the evil deceiver has returned—and multiplied. Greene testifies to this outcome in *The Hidden Reality* (without mentioning Descartes) when he asks, "[O]nce we conclude that there's a high likelihood that we're living in a computer simulation, how do we trust anything, including the very reasoning that led to the conclusion? . . . Will the sun rise tomorrow?" (288–89). As Descartes feared, without the assurance that the universe was created by a loving God, it does become a real possibility (at least for some) that there might be "not a supremely good God, the source of truth, but rather an evil genius, supremely powerful and clever, who has directed his entire effort at deceiving me." In this event, says Descartes, "the heavens, the air, the earth, colors, shapes, sounds, and all external things" would be "nothing but the bedeviling hoaxes of my dreams, with which he lays snares for my credulity" (1.22). What, then, are we to do?

For Paul Davies, the simulation argument is the point at which the whole multiverse hypothesis becomes a "*reductio ad absurdum*," undermining itself completely. "If the perceived universe is a fake," he explains, "then so are its laws, and we have no justification in extrapolating the fake physics to the whole of reality. . . . And since we would have no idea at all what the laws of physics might be in the *real* universe—and no reason to expect them to resemble 'our' laws—then we cannot assume that the real laws will permit a multiverse."[103] So, he concludes, "while it may be true that our universe *is* a fake, it seems to me that drawing that conclusion would spell the end of scientific inquiry."[104] We will return to this and other critiques of the multiverse in the final chapter, but for the moment it will suffice to mention that other thinkers approach the simulation possibility a bit more lightheartedly. Along with Davies, they concede that the moment we *presume* we are being simulated, we must abandon the Enlightenment dream to know the world as it "really is"; all we can hope to know is confined to the world as it appears to us at this moment right now. So, as economist Robin Hanson suggests, the question might have to shift from an epistemological one to an ethical one. Rather than wondering how to gain knowledge in a simulation, he argues, we ought to be asking "how to *live* in a simulation."[105] Most pressingly, we ought to be asking: How are we to live in such a way that our simulators do not decide to pull the plug on our universe or replace us with a cuter, more interesting set of sims?

Hanson frames his argument much the way Blaise Pascal frames his famous "wager" on the existence of God.[106] We may or may not be living in a simulation, Hanson begins. It is likely that we will never know for sure. Therefore, "all else equal," you should live as if you *are* being simulated and do your best to ensure your continued (virtual) existence. "All else equal," Hanson suggests, "you should . . . expect to and try to participate in pivotal events, be entertaining and praiseworthy, and keep the famous people around you happy and interested in you." Although it is hard to know what "all else equal" means here, the main assumption behind Hanson's behavioral prescriptions is that our simulators are likely to keep running the simulations that entertain them. Our efforts should therefore be directed toward keeping them happy with us; as Hanson points out, "your motivation to save for retirement, or to help the poor in Ethiopia, might be muted by realizing that in your simulation, you will never retire and there is no Ethiopia." So, he concludes, "all else equal you should care less about the future of humanity and live more for today."[107]

Again, it is difficult to know how, exactly, "all else" might be "equal" under these circumstances. How does one weigh the possibility that people might be dying of starvation against the possibility that they might be "the bedeviling hoaxes" of some narcissistic simulators? Moreover, ought we to assume such narcissism on the part of our simulators in the first place? What if they are running a moral experiment to pull the plug on anyone who fails to respond to the suffering of other simulated creatures? Hanson is aware that he is making an unprovable assumption about the programmers' character, but he thinks that it is nevertheless a safe bet. The traits the simulators would encourage in us would likely be the same ones they themselves possess, and "it would seem inconsistent of them to greatly emphasize humility, for example." After all, they themselves are "willing to play God." So, Hanson exclaims, "be funny, outrageous, violent, sexy, strange, pathetic, heroic . . . in a word, 'dramatic.'"[108] Otherwise, the simulators might not let you *be* at all.

It seems important at this juncture to state the obvious: we have somehow again collided with theology. The moment we engage in speculations and wagers about the existence of Simulators,[109] the moment we say (as we did in relation to the black-hole producers) that "they" are just like "us," only smarter and more powerful, the moment we try to enumerate their attributes, divine their will for humanity, and attempt to conduct ourselves in accordance with it, there is no conceivable way to distinguish our technoprophetic ruminations from constructive theology in its most classic form. To make matters worse, not only do the Simulators look like the old gods in new lab coats, but they also prevent our knowing anything "true" about the world at all. Most physicists are therefore eager to point out that the simulation argument stems from *philosophy*, not physics, and even that "simulated realities are not welcomed into the scientific world-view."[110] That having been said, simulation arguments have not been shut out of the "scientific world-view" entirely; there are at least a few (particularly prolific and media-savvy) physicist-cosmologists who find the idea at least worthy of serious thought as they work their way through different theories of the multiverse.[111] What is ironic, however, is not only that the simulated multiverse dismantles any claim to know the world as it is (including whether there is a multiverse or not), but also that it undermines much of the philosophical impetus behind the recent multiversal turn in scientific cosmology. As John Barrow reminds us,

> [T]he multiverse scenario is favoured by many cosmologists as a way to avoid the conclusion that the Universe was specially designed for life by a Grand Designer. Others see it as a way of avoiding having anything to say about the problem of fine tuning at all. But now we see, once conscious observers are allowed to intervene in the Universe . . . there is a new problem. We end up with a scenario in which *the gods reappear in unlimited numbers* in the guise of the simulators who have power of life and death over the simulated realities that they bring into being.[112]

In a similar vein, Davies charges that "the simulated beings . . . stand in the same relation to the simulating system as human beings stand in relation to the God (or gods) of traditional religion."[113] The only considerable difference between these new creator-gods and the old creator-gods/God (whether demiurgic or omnipotent) is that the black-hole manufacturers and universe simulators make their worlds neither out of some primordial chaos of materials nor out of nothing, but out of other worlds—that is, the ones that they themselves are in. In other words, both the lab-designer and the simulation arguments rely on earlier cosmic processes to get "natural" worlds going before (post)humans gain the power and intelligence necessary to make their own.

Edward Harrison navigates this difficulty by relying on a combination of Lee Smolin's and Alan Guth's cosmologies. In the beginning, he ventures, collapsing regions of space-time inflate on the "other" side of black holes, evolve according to natural selection, and then eventually produce "at least one universe . . . with intelligence at about our level."[114] When the universe's humanoids become capable of creating universes, the number of manufactured universes begins to exceed the number of galaxy-rich "natural" universes—until the probability of any given creature's living in a simulation becomes "almost one." Martin Rees takes a similar tactic, using different resources. Integrating the simulation argument not with Smolin, but with inflationary cosmology and string theory, Rees argues that of the infinite number of 10^{500} universes, surely *some* will be able to produce the quantum computers that can simulate worlds. As he puts it, "[O]nce you accept the idea of the multiverse, and that some universes will have immense potentiality for complexity, it's a logical consequence that in some of those universes there will be the potential to simulate parts of themselves." But the fairly dizzying result of all this is that "you may get [a] sort of infinite regress, so that *we don't know where reality stops and where the minds and ideas take over*, and we don't know what our place is in this grand ensemble of universes and simulated universes."[115]

Multiverse of Multiverses (Forever and Ever)

Martin Rees's concern about the confusion between "reality" and "ideas" places this snowballing multiverse scenario into a philosophical lineage at least as old as Plato. The terminology, however, is a bit misleading: for Plato, "reality" *was* the "Ideas." The realm of the Forms was the realm of Being, Truth, Eternity, whereas the material world was in one or another way derivative: either a deceptive shadow of the Forms (as in the Allegory of the Cave) or a "mixed" incarnation and a "moving image" of the Forms themselves (as in the cosmology of the *Timaeus*).[116] But either way, the ideal is what is really *real* for Plato because it (presumably) never changes, (presumably) transcends spatial and temporal location, and (presumably) holds for all beings everywhere. The most common illustrations of this principle tend to be mathematical: an ideal triangle is more *real* than any material triangle in the universe; it is the Triangle itself, whereas all earthly triangles do their best only to approximate it. It is in this sense that theoretical physicist and multiverse theorist Max Tegmark proclaims himself a "Platonist"; as far as he is concerned, the reason mathematics "describes the universe so well" is *not* that math is a useful abstraction of physics (which would be the "Aristotelian" position), but that the physical world is an incarnation of mathematics itself: "[T]he universe," he declares, "is inherently mathematical."[117]

Declaring that "most theoretical physicists" share this view, Tegmark ascribes their incessant propensity to ask "why" to their thoroughgoing Platonism.[118] As he understands it, "the Platonic paradigm raises the question of why the universe is the way it is. To an Aristotelian, this is a meaningless question: the universe just is. But a Platonist cannot help but wonder why it could not have been different. If the universe is inherently mathematical, then why was only one of the many mathematical structures singled out to describe a universe?"[119] At this point, it becomes hard to know what Tegmark means by a "Platonist" and therefore hard to answer the question. For Plato, at least, the reason the universe has the specific mathematical configuration it does is that this configuration allows it most clearly to resemble the "perfect living creature"—that is, the Form that contains all other Forms within it.[120] This resemblance crucially hinges on the eternity and singularity of the cosmos; regardless of how seriously one takes the demiurge, Plato's explanation for the fine-tuning of the cosmos is that the universe has precisely the mathematical configuration it needs to ensure, first, that it lives forever and, second, that it has no other worlds outside itself.[121]

Perhaps unsurprisingly, this is not the way that Tegmark answers his own "Platonic" question about the properties of our universe. Rather, offering an overly concrete interpretation of the "reality" of the mathematical, he suggests "that all mathematical structures exist physically as well. Every mathematical structure corresponds to a parallel universe." And just as one might begin to respond, with some perplexity, that the "reality" of the Platonic Forms consists precisely in their *not* existing physically, Tegmark adds the further "Platonic principle" that "the elements of this multiverse do not reside in the same space but exist outside space and time."[122] This principle is what Tegmark calls the "Mathematical Universe Hypothesis": every possible universe actually exists—*physically*—in some kind of Ideascape "outside space and time." It is as if Timaeus's "Perfect Living Creature" were not just the form of creation, but the *sum* of creation.

I do not want to belabor the point of Tegmark's exceedingly strange interpretation of Platonism—merely to mark how important it seems to him to speak under that mantle (in one piece, he calls his stance a "radical Platonism" insofar as it asserts that the ideas "exist 'out there' in a physical sense").[123] Tegmark's clearer philosophical ancestors are the American philosophers David Lewis, whose theory of "modal realism" states that all metaphysically (not just mathematically) possible histories must be actual,[124] and Robert Nozick, who argues that the characteristics of our universe make sense only if every other set of characteristics is actual somewhere else. Without referring to the anthropic principle, Nozick operates according to what he calls the "fecundity assumption," a descendant of Arthur Lovejoy's "principle of plenitude," which of course took its cue from Lucretius.[125] From this genealogical perspective, then, Tegmark is far more of an Atomist than he is a Platonist.

Rather than signaling much of a metaphysical lineage, Tegmark's claim to Plato might instead have something to do with the foundational place that Plato occupies in the history of Western mathematics, philosophy, and cosmology, for, as far as Tegmark is concerned, his own Mathematical Universe is the consummation of all these fields of knowledge. His claim might *also* have something to do with the Neoplatonic notion of a progressive cosmic "hierarchy," a ranked Chain of Being that extends from rocks and worms to animals, humans, angels, archangels, and finally to the Absolute itself.[126] After all, Tegmark's multiverses are arranged in what he himself calls "the multiverse hierarchy": Level I is nestled within Level II; Level III brings Level I into infinite dimensional space; and the landscape describes them all.[127] But the great chain of *kosmoi* does not end here. Having walked his readers up this steep mountain of multiverses, Tegmark announces that Level IV, which is to say his own mathematical multiverse, is the highest level imaginable—even higher than the landscape, for although, in the words of Helge Kragh, "10^{500} universes are a lot, the number is infinitesimal compared to the number of possible universes."[128] A multiverse that comprises all mathematically possible universes "brings closure to the hierarchy of multiverses," Tegmark claims, "because any self-consistent fundamental physical theory can be phrased as some kind of mathematical structure."[129]

This is the reason that Brian Greene calls Tegmark's Mathematical Universe Hypothesis the "Ultimate Multiverse," and Davies calls it a "multiverse with a vengeance."[130] To wit, there is room in this multiverse for everyone: room for quilted multiverses and inflationary multiverses, room for Atomists and Stoics, room for Thales's water and Anaximenes's air, room for "one boring pair of cymbals clashing out the same old song."[131] In the compendium of all possible universes, some worlds will be stacked like slices of bread, and some will fly down the throat of a Calabi–Yau manifold: "A universe governed by Newton's equations and populated solely by solid billiard balls . . . is a real universe; an empty universe with 666 spatial dimensions . . . is a universe too."[132] "How about a universe that obeys the laws of classical physics, with no quantum effects?" asks Tegmark. "How about time that comes in discrete steps, as for computers, instead of being continuous? How about a universe that is simply an empty dodeca-hedron? In the Level IV multiverse, all these alternative realities actually exist."[133] So some worlds will be linear, and some will be cyclical; some will be singular, and some will be plural; some will be infinite, and some will be finite; some will branch forward, and some will branch back. Some worlds will be manufactured, and some will be simulated; some designers will be kind, and some will be cruel, some capable and some all but incompetent.

And, presumably, some of the set of all possible worlds will have a creator-god who breathes over primordial waters, who separates the seas from dry land.

How on earth did we get back here?

UNENDINGS

On the Entanglement of Science and Religion

What a comedy of errors! When the debate between science and religion is staged, adjectives are almost exactly reversed: it is of science that one should say that it reaches the invisible world of beyond, that she is spiritual, miraculous, soul-fulfilling, uplifting. And it is religion that should be qualified as being local, objective, visible, mundane, unmiraculous, repetitive, obstinate, sturdy.

BRUNO LATOUR, "THOU SHALT NOT FREEZE-FRAME; OR, HOW NOT TO
MISUNDERSTAND THE SCIENCE AND RELIGION DEBATE"

Which Side Are You On?

The Atheist and the Archbishop

At first blush, it looks as though the boundary between "science" and "religion" is clear when it comes to the multiverse. Just as the Atomist philosophers did 2,500 years ago, modern multiverse theorists proclaim an infinite number of worlds in part to avoid the conclusion that this world was somehow "designed" for us. If an endless number of all sorts of universes *actually* exist, the scientists reason, then it does not matter how improbable or razor's-edgy our fundamental parameters might be. The random generation of universes throughout infinite time and space ensures that even a cosmological constant as absurdly small as ours was bound to arise at some point. In short, the multiverse does away with the need for a creator-god—at least as an explanatory principle. In this light, some theologians have disparaged the multiverse as "the last resort of the desperate atheist."[1]

Among these theologians is Cardinal Christoph Schönborn, the current Roman Catholic archbishop of Vienna, who accuses "the multiverse hypothesis" and "[n]eo-Darwinism" of having been "invented to avoid the overwhelming evidence for purpose and design found in modern science."

In response to this invention, he resolves, "the Catholic Church will again defend human reason by proclaiming that the immanent design evident in nature is real."[2] At first glance, Schönborn's use of the adverb *again* might call to mind a history of torrid conflicts between religion and natural science, placing the multiverse in an unsavory lineage that stretches from the slandered Epicureans through the executed Giordano Bruno and the prosecuted John Thomas Scopes. On the surface, then, the question of infinite worlds seems to stage a simple, familiar drama between the forces of dogma and innovation, sacred doctrine and secular reason, God and the multiverse.

And yet these lines become increasingly crossed, twisted, and even knotted the more closely we examine the controversy. It is important to note, for example, that Schönborn does not accuse the multiverse of violating doctrine, but of violating "modern science." In response, he commits the church not to defending *God* against the multiverse, but to defending *reason* against it: "Throughout history," he writes, "the church has defended the truths of faith given by Jesus Christ. But in the modern era, the Catholic Church is in the odd position of standing in firm defense of reason, as well." Standing firm in this "odd position," Schönborn marshals neither the Bible nor the fathers or doctors of the church to assert the singularity of the cosmos. He never calls the multiverse hypothesis heretical, incompatible with scripture, or an insult to divine infinity. Rather, he calls it an "abdication of human intelligence." In particular, he charges, appealing to an infinite number of invisible universes amounts to "giving up the search for an explanation of the world as it appears to us."[3] It is a claim that we may remember from Walter Charleton, who argued against the neo-Atomists in the mid-seventeenth century that many-worlds cosmologies not only violate common sense but also, and more troublingly, prevent scientists from gaining objective knowledge about the "real" world (see chap. 4, sec. "From Infinity to Pluralism"). What are we doing dreaming about other universes when we have not managed to address global warming? or malaria? or the common cold?

In Schönborn's unwitting resurrection of Charleton, we can therefore detect a resurgence and subtle realignment of those camps that formed in the early modern period over the question of a "plurality of worlds." As we might recall, astronomers such as Kepler and Galileo began to exercise a kind of cosmic restraint, narrowing their focus to the bodies they could see through telescopes (see chap. 4, sec. "From Infinity to Pluralism"). In the meantime, their theological contemporaries began to generate a flood of pluralist cosmologies, which, they believed, served to magnify the glory of God. Considering that Bruno had been executed less than a century earlier, this early modern turn to pluralism represented a dramatic departure from previous theologies (with the lone exception of Nicholas of Cusa). So now, at the dawn of the twenty-first century, we can detect another positional shift: in Schönborn's op-ed piece, we find a theologian calling neither for the monistic orthodoxy of Scholasticism nor for the pluralist magnificence of the seventeenth-century divines, but for *scientific restraint*—specifically, the scientific restraint of the early modern period.

Insofar as the multiverse hypothesis is said to be neither provable nor "falsifiable," to use Karl Popper's term,[4] the archbishop and like-minded critics accuse multiverse cosmologies of falling outside the proper purview of science.[5] Of course, one can always accuse these critics of hiding fundamentally theological motivations behind a scientific smoke screen—of saying their concern is for "human intelligence" when it is really for intelligent design. But whether this is true or not, such an argument is of limited value, for a number of reasons. First, many schools of Christian thought (including the Catholic, Orthodox, and Anglican traditions, not to mention their deist, process, and Platonic offshoots) have never held the sacred and the secular to be in conflict in the first place.[6] According to these traditions, human reason participates in divine reason, and the mysterious workings of God are partially revealed in the order of the universe that we measure and observe. So it makes very little sense to split theological motivations from "rational" motivations; if reason is functioning properly within these traditions, there is no rift at all between science and religion.

Second, to reduce Schönborn's and others' critiques of the multiverse to "secret theology" is to overestimate the metaphysical power of the position they criticize. After all, while the principles of accident and infinity may do away with the *need* for a creator-god, they hardly disprove the existence of such a god, nor are they even incompatible with one. As Thomas Aquinas argued immediately after positing the "teleological argument" (which eventually produced the "intelligent design hypothesis"), it does not matter *what* sort of material process an atheist posits as the creative force of the universe; the theist can always retort, "God made that material process."[7] For this reason, there are plenty of self-professed Christian physicists and philosophers who affirm the existence of the multiverse with no theological difficulty at all.[8] As astrophysicist and "progressive creationist" Jeffrey Zweernick has argued, multiverse theories still leave us marveling at the infinitely generative force of inflation, or the perfect unitarity of the wave function, or the well-oiled mechanisms that keep the ekpyrotic cycles cycling. So, he concludes, "as research into multiverse scenarios advances, it appears that they may simply move the design 'up one level.' In other words," he explains, "instead of just one universe requiring fine-tuning to support life, it appears that any multiverse-generating mechanism also requires a high degree of fine-tuning to reproduce the observable universe in which we live."[9] The process of inflation, for example, must have been designed in such a way that quantum fluctuations *can* form pockets of true vacuum from time to time rather than just churning out an eternal sea of nothingness.[10] Simply put, a Christian can easily affirm that God created the multiverse that created the universe.

Another Theological Detour

That having been said, such an argument, however updated its evidence, would in principle be no more forceful than the traditional argument from design and would be subject to the same logical and ethical critiques that Philo levies against it in David Hume's *Dialogues*.[11] Depending on the thinker, it might even be subject to a trenchant *theological* critique because it is not clear that the god of the design argument bears much resemblance to the God who breathes into the nostrils of an earth creature, or who delivers an oppressed people from slavery, or whose prophets implore the nation to "do justice, love kindness, and walk humbly with your God" (Micah 6:8 [NRSV]). In short, there is a chasm between what some theologians, following a fragment in Blaise Pascal's *Pensées*, call "the God of the philosophers," on the one hand, and "the God of Abraham, Isaac, and Jacob," on the other.[12] The argument from design, along with the other traditional proofs of the existence of God, are concerned strictly with the former— with a philosophical "God" understood in a fairly bloodless fashion as a first mover, a self-caused cause, a lighter of the cosmic fuse.

One of the most powerful critics of this "God of the philosophers" was the German theologian Dietrich Bonhoeffer (1906–1945). Writing from his cell in a Gestapo prison (he had been part of a failed plot to assassinate Hitler and was executed just days before the end of the war), Bonhoeffer assailed the kind of thinking that makes God into a metaphysical principle at the edge of the world rather than an ethical force in the midst of it. "Religious people speak of God at a point where human knowledge is at an end," he explained in a letter to Eberhardt Bethge; "actually, [this God is] a *deus ex machina* that they're always bringing on the scene, either to appear to solve insoluble problems or to provide strength when human powers fail."[13] The problem with appealing to God at the limits of human knowledge is that as the scope of human knowledge expands, the space for God progressively shrinks, until God is edged out of the world completely. Now that humanity does not need God to explain the fall of rain or the rotation of the planets or even the creation and distribution of matter, God becomes confined, at best, to that one little ocular mechanism that biologists cannot explain, or to the bacterial flagellum, or to the 10^{-34} seconds before the big bang hypothesis kicks in to explain creation. God becomes nothing more than a first nudge, a "stopgap" invoked to plug the remaining few holes in human understanding. "Inevitably that lasts only until human beings push the boundaries a bit further and God is no longer needed as a *deus ex machina*," Bonhoeffer predicted.[14]

So the "God of the gaps" recedes with each passing age, disappearing from view like some far-off galaxy. And in the meantime, we have learned nothing about how we are to *live* in relation to this stopgap god: how to respond to genocide, for example, or what to do in the face of escalating global poverty, or whether the existence of an infinite number of worlds means we can let this one warm and melt with impunity. For Bonhoeffer, then, the whole endeavor to prove the existence of a God outside the world is flawed from the outset. "I'd like to speak of God not at the boundaries, but at the center," he wrote, "not in weakness but in strength."[15] Far from existing at the edge of the universe (or of the multiverse), Bonhoeffer's God "is the beyond in the midst of our lives," an indwelling force that, precisely in refusing to swoop in and solve our unanswered problems, calls people to "responsible action" in the world—even to life in a Gestapo prison—so that "the coming generation" might "go on living."[16] Perhaps needless to say, the god of the design hypothesis does none of this in his sporadic plugging of epistemological holes. It might be, then, that the argument from design constitutes not just unsatisfying science, but unsatisfying *theology*.[17]

The Atheist as the Archbishop

Finally, to dismiss those who criticize the multiverse as secret theologians would be to miss the extent to which multiverse theories *themselves* function theologically, colliding with invisible divinities and realms in the very gesture of trying to avoid them. We have seen this collision rather dramatically enacted in the manufactured and simulated multiverses, each of which updates the old god or gods into teams of brilliant physicists—just like us but smarter and more powerful and with faster supercomputers. But even those multiverse scenarios that stop short of such anthropomorphic designers still display a remarkable faith in what St. Paul would call "things hoped for" and "things not seen" (Hebrews 11:1).[18]

INFINITE IDENTITY

Taking this allegation up through Max Tegmark's cosmic hierarchy, we might recall, for example, that the Level I, or quilted, multiverse—the endless set of overlapping observable universes centered on any given cosmic body—relies on two unprovable claims. The first is that the universe is infinite. Although satellite observations have confirmed that the universe is either flat or very close to flat in three dimensions, they have not confirmed that this flatness goes on forever. It might be that the universe is shaped like a gargantuan doughnut (torus), so that if you were to run to the "edge" and throw Lucretius's spear, it would come back around the other side.[19] If this were the case, then light would eventually "lap" the universe, so that we would see the same galaxies repeated at different stages of their lives. Teams of researchers are currently seeking such patterns in the sky, but so far they have "found no repeating images within one billion light-years of the earth."[20] If such images were ever to be found, they would demonstrate (as Aristotle knew)[21] the finitude of the cosmos. But *not* finding them would never amount to demonstrating the infinity of the cosmos—just its enormity. It will always be possible that an infinite-looking universe is really finite but so large that we will never live long enough to see light looping back on itself. So it is impossible not just in practice but *in principle* to prove the infinity of the cosmos. As John Barrow puts it, "[Y]ou can discover whether the Universe is infinite, but the learning will take an infinite time."[22]

The second article of faith subtending the quilted multiverse is the "cosmological principle," which states that the universe is uniform on the largest scales. Although observations of the Cosmic Microwave Background (CMB) have confirmed that this principle holds for our observable universe, they can never confirm that it holds for regions *outside* our universe. Of course, such an assumption might be a good guess, especially at short distances, but as George Ellis has argued, it becomes a worse and worse guess the farther "out" one ventures from the observable universe. "The proponents are telling us that we can state in broad terms what happens 1,000 times as far as our cosmic horizon," he writes,

> $10^{1,000}$ times, $10^{100,000}$ times, an infinity—all from data we obtain within the horizon. It is an extrapolation of an extraordinary kind. Maybe the universe closes up on a very large scale, and there is no infinity out there. Maybe all the matter in the universe ends somewhere, and there is empty space forever after. Maybe space and time come to an end at a singularity that bounds the universe. We just do not know what actually happens, for we have no information about these regions and never will.[23]

These, then, are at least two of the "things hoped for" and "things not seen" subtending the spatially configured multiverse: cosmic infinity and cosmic homogeneity. It is not at all clear that regions far beyond our own look more or less like ours, forever.

EXTRACOSMIC *KOSMOI*

In response to this charge, multiverse proponents might argue that insofar as cosmic infinity and homogeneity are predicted by the theory of inflation, inflation delivers these two principles from the realm of the speculative and places them on solid ground. In predicting these principles, however, inflation also generates an infinite number of *other* universes, the ensemble of which constitutes Tegmark's Level II. The problem with this "inflationary multiverse" is not so much that inflation has not yet been proved or refuted,[24] but that the scenario assumes that the laws we have derived from our observable universe—quantum field theory, for example— hold not only throughout our (purportedly) infinite universe, but also throughout wildly different and forever separated domains of a haphazardly generated multiverse.[25] To be sure, this notion gains support from the 10^{500+} vacua solutions of string theory, but as many critics are quick to point out, string theory *itself* remains what one might call "hoped for and not seen." As beautiful and internally consistent as it may be, string theory has not been confirmed or even supported by any experiment or observation.[26] The reason it has seemed so promising for so long is that it integrates gravity with the electromagnetic and strong and weak nuclear forces. But in the very gesture of tying up this one universe with a "theory of everything," the critique goes, string theory ends up becoming a "theory of anything," positing a mind-bending number of alternative universes that we can neither measure nor see. And, according to Paul Davies, "appealing to everything in general to explain something in particular is really no explanation at all. To a scientists, it is just as unsatisfying as simply declaring, 'God made it that way!' "[27]

In a recent *Harper's* article, Alan Lightman distills the theological costs of the inflationary-plus-string scenario thus: "[N]ot only must we accept that the basic properties of our universe are accidental and uncalculable. In addition, we must believe in the existence of many other universes. But we have no conceivable way of observing these universes and cannot prove their existence. Thus, to explain what we see in the world . . . we must believe in what we cannot prove."[28] What is ironic is that inflationary cosmology initially seemed like the *remedy* for believing in what we cannot prove; after all, it gave us a mechanism for flattening and homogenizing the universe without the interference of a god. String theory similarly seemed like the ultimate answer to the problem of the fine-tuning of the universe, initially promising to find a single vacuum that explained why the parameters must be what they are. But then, as we have seen Paul Steinhardt demonstrate, both inflation and string theory have ended up postulating a host of invisible *kosmoi* in the very process of delivering the world in which we live from an invisible god.[29] The situation is almost like a cosmological version of the old lady who swallowed the fly: the moment physicists ask, "*Why* is the universe this way and not another?" which is arguably a metaphysical question to begin with, they can produce answers that only open more metaphysical problems. So we can swallow string theory to catch eternal inflation to catch the cosmological constant, but as we try to digest 10^{500+} invisible universes, we would do well to ask why we swallowed the fly in the first place.[30]

Responding to the pseudo-theological postulates of inflationary and string cosmologies, Martin Gardner argues, "[S]urely the conjecture that there is just one universe and its Creator is infinitely simpler and easier to believe than that there are countless billions upon billions of worlds, constantly increasing in number and created by nobody."[31] A number of theologians have issued similar arguments,[32] which Richard Dawkins rejected several years ago by insisting, "[Y]ou can't get much more complex than an Almighty God!"[33] I confess to having no idea which hypothesis is "simpler" or "easier to believe" than the other—or even how one would go

UNENDINGS ●●●

about measuring such values, assuming that they are values. But many physicists find the anthropic multiverse to be at least as speculative and outlandish as the god hypothesis that it tries to avoid. In response, Paul Steinhardt, Neil Turok, and their colleagues posit a new cyclical model that preserves the oneness of the universe. As we have seen, however, it does so, first, by relying (to varying degrees) on the still-speculative string hypothesis of braneworlds and, second, by fashioning dark energy into what is essentially a nonanthropomorphic, pantheist god. According to the ekpyrotic model, dark energy is the sole cosmic regulator—an indwelling force that creates, animates, destroys, and re-creates the "endless universe" it suffuses. So the scenario is *not* not theological; to the contrary, it is (fascinatingly) pantheological.[34]

Lee Smolin, for his part, calls the anthropic principle "unscientific" and builds his alternative on the allegedly sturdier principle of natural selection.[35] But even his black-hole scenario relies on another perpetually unobservable assumption—that the big bang singularity is the same thing as a black-hole singularity, so we can assume that there are universes inside *black holes*, which, in the words of William Dembski, "of all the objects in space . . . divulge the least information about themselves."[36] Moreover, this reproductive multiverse reinscribes the traditionally biblical claim that the ultimate purpose of the cosmos lies in procreation—specifically, in producing descendants that eventually become as "numerous as the stars."

THE GHOST IN THE WAVE FUNCTION

As for cosmologies stemming from quantum mechanics, they rely on a very different series of "things hoped for" and "things not seen" than the lower-level scenarios in Tegmark's cosmic hierarchy. Unlike string theory, quantum mechanics has passed every experimental test to which it has been subjected, and this holds for the Many-Worlds Interpretation (MWI) as well as for the Copenhagen Interpretation. The MWI crosses into speculative metaphysical terrain only when it asserts that all its possible branches are *actual*—that is, that there are physical universes "out there" in which each quantum "decision" plays itself out. On the one hand, it is this conviction that allows theorists such as Sean Carroll and Colin Bruce to argue that the MWI is "simpler" and therefore more scientifically desirable than the Copenhagen Interpretation.[37] For the many-worlders, if every outcome actually happens somewhere, then the wave function never "collapses" in relation to experimental peculiarities. Rather, "it evolves smoothly and deterministically over time without any kind of splitting or parallelism."[38] What this means is that particles, far from existing in a superpositional haze until they are observed, *do have* determinate properties independently of the experimental apparatus that measures them; it is just that these properties take on different values in different worlds. As Brian Greene summarizes the matter, "[T]he mathematics of Many Worlds, unlike that of Copenhagen, is pure, simple and constant."[39]

On the other hand, this very assertion can be said to reveal what physicist Evelyn Fox Keller diagnoses as a longing for a nonquantum universe among adherents of the WMI. By insisting that every possible outcome is actual, the many-worlders display what Keller calls "an unwillingness to let go of the basic tenets of classical physics: the objectivity and knowability of nature."[40] According to her, proponents of the MWI are engaged in an act of "cognitive repression,"[41] specifically a repression of Copenhagen's radical proposal that insofar as experimental phenomena are *produced* by the "intra-actions" between the observer, the observed, and the mechanism of observation, there is no such thing as an "individual," self-constituted entity—whether it be a particle or a physicist. As Karen Barad has made clear, this does not mean that there is no such thing as "truth" or even "objectivity" along the Copenhagen Interpretation; to the contrary, this reading of quantum mechanics demonstrates that all that "is" is ontologically dependent on a series of interrelated events and components; as she puts it, "*we are a part of that nature that we seek to understand.*"[42]

Along this view, the MWI can be seen as an effort to deny the quantum entanglement of knower and known—to replace the Copenhagen Interpretation's relational and contingent universe with a nonrelational, deterministic one (well, many). And, indeed, Bruce affirms that "only in [the MWI's] quantum world does it become possible to measure something without affecting it at all."[43] This, then, is the major article of faith subtending the MWI: that entities must be self-constituted; that is, they must possess determinate properties independently of one another or anything else. This micro-individualistic metaphysic would mean, in turn, that all decisions are determined ahead of time: that everything that can happen will happen in one world or another. More precisely, it would mean that decisions are not really decisions at all, because if all possibilities happen somewhere, then none of them is ever unpursued. Nothing, in other words, is ever *lost* in the MWI. Rather, all that might be is eternally held within the wave function itself: the timeless, unchanging governor of all things that actualizes all possibilities according to its sovereign will, unaffected by those innerworldly forces it controls and transcends.

A very different scenario emerges when one extends the *Copenhagen* Interpretation out to the level of cosmic formation. Neither Keller nor Barad mentions this scenario, and, indeed, it gets very little attention from anyone, remaining what even its proposer calls a "frail reed" and an "idea for an idea."[44] The theorist behind the idea is the legendary John Wheeler (1911–2008), who imagined a revision to the double-slit experiment. For a century now, this experiment has shown that particles behave differently under different experimental conditions. Fire a beam of photons toward a screen with two slits cut out of it, and the beam will produce a wavelike interference pattern on the photographic film behind it. Fire photons one at a time through the two slits, and, astonishingly, each photon will appear to pass through both slits at once, interfering with itself before landing somewhere in the wave pattern. But—and this is where things get very strange—if you install a "which-path" detector above the slits so that you can see how on earth a single photon can pass through both slits at once, the photon will straighten up and behave like a particle (or a baseball), heading through just one of them and lining up neatly on the film behind it.

Wheeler's revision to this setup is called the "delayed-choice experiment." In this scenario, the observer is able to decide *after* the photon has passed through the slit(s) how she will configure the final measuring apparatus. If she configures it in a way that produces a wavelike pattern on the screen, then the photon will have passed through both slits. If she configures it to produce a particle-like pattern, then the photon will have passed through just one slit. Wheeler's prediction—remarkably confirmed by a team at the University of Maryland in the mid-1980s—was that the experimenter's decision would condition the photon's decision even though the photon's decision would have preceded the experimenter's decision. In other words, the observer *retroactively determines* the photon's behavior: her decision obliges it *already* to have acted like a particle or a wave. Wheeler dubbed this mind-bending phenomenon "backward causation."[45]

Extending his thought experiment out to the universe itself, Wheeler then imagined "a quasar—a very luminous and very remote young galaxy" a few billion light-years away, which earth-bound scientists are trying to observe. The quasar in this scenario is analogous to the light source in the traditional double-slit experiment, while mirrors and telescopes on the earth serve as the "screen" on which the light lands. Between the quasar and the observers, moreover, are two large galaxies. Because these galaxies' gravitational fields will bend the path that light travels between the quasar and the earth, the galaxies become analogous to the two slits in the classic experiment: light from the quasar can travel through one galaxy or the other on its way to the earthlings' instruments. Now, as science writer Tim Folger explains it, "if the astronomers point a telescope in the direction of one of the two intervening galaxies, they will see photons from the quasar that were deflected by that galaxy; they would get the same result by looking at the other galaxy." In other words, light from the quasar would be passing through the "slit" of either intervening galaxy and landing in a particle-like position on the photographic film on earth. But the astronomers could also compel this light to act like a wave. If they outfitted their telescopes with a series of mirrors, they could force each photon to pass through both galaxies, interfere with itself, and produce "alternating light and dark bands" on the film, "identical to the pattern found when photons passed through the two slits." Just as in the classic setup, then, the photons from this young quasar would be behaving as particles or as waves, depending on the particularities of the experimental apparatus. The presence or absence of the mirrors would determine the path that the photons took through the intermediate galaxies. But, of course, the photons would already have made the journey through those intermediate galaxies; in fact, they would have done so billions of years ago. This, then, is backward causation at the cosmic level: "the measurements made *now* . . . determine the photon's past."[46]

The punch line, then, is that the universe might well *not* "exist if we're not looking at it"—that observers might co-create the very cosmic phenomena they observe. As Wheeler puts it, "[T]he observer is as essential to the creation of the universe as the universe is to the creation of the observer."[47] In this case, the answer to the question of why the universe seems so finely tuned to our existence would be that our existence retroactively tunes the universe, creating the cosmic conditions that, in turn, create our existence.[48] Wheeler calls this cosmological model the "participatory universe."

To be sure, this participatory scenario runs the risk of a rather flagrant anthropocentrism. Along Wheeler's account, at least, the scenario seems to suggest that humans can somehow *make* the universe they want to have—in the manner, say, of American politicians who proclaim, "When we act, we create our own reality."[49] It is at this point that the model, which everyone admits needs a bit of work, would benefit considerably from Barad's "posthumanist" critique of the Copenhagen Interpretation. Humans, she insists, are just as much the product of specific material–discursive configurations (including, but not limited to, experimental arrangements) as the photons or electrons (or in this case, universes) they purport to measure. In fact, Barad explains, "what we usually call a 'measurement' is a correlation or entanglement between the component parts of a phenomenon," which include the measured object, the measuring device, and the measurers themselves. Far from simply determining the shape and nature of the world, she counters, "humans are themselves specific *parts* of the world's ongoing reconfiguring."[50]

Unlike the MWI, then, the ontological assumption of Copenhagen cosmology is that particles (and all other entities co-constituted by and as the universe) do *not* possess properties independently of one another; rather, phenomena emerge only through the relations among cosmic components. Grafting Barad's language onto the theory of the participatory universe, one might say that the observer, the observed, and the instrument of observing "intra-act" to produce the universe "itself" *as* a set of provisional and shifting intra-actions.[51]

ECSTATIC MATHEMATICS

Finally, we make our way to Tegmark's Mathematical Universe Hypothesis, or Ultimate Multiverse, which operates on the absolutely undemonstrable assumption that all mathematical possibilities must exist physically. As Tegmark explains his scenario, this collection of these universes exists "outside space and time," much like the Platonic Forms or, for that matter, the divine ideas.[52] Again, it should be noted that although multiverse proponents *and* critics tend to express varying levels of tolerance for the unobservable, almost all of them stop short of Tegmark's proposal, calling it "pure speculation"—an "ecstatic . . . mathematical fantasy land" that lies "beyond any scientific support."[53] Even Greene, who confesses to having a "taste . . . for the expansive" when it comes to cosmology, says that he "draw[s] the line at . . . the full-blown version of the Ultimate Multiverse" because it, unlike all the other scenarios, does not emerge organically from any physical theory and therefore offers "no possibility of being confronted meaningfully by experiment or observation."[54] Moreover, unlike all the other scenarios, the Ultimate Multiverse offers no common generating mechanism (such as, according to Greene, "a fluctuating inflaton field, collisions between braneworlds, quantum tunneling through the string theory land-scape, a wave evolving via the Schrödinger equation").[55] In the absence of such a cosmogonic hypothesis, there is no way *even theoretically* to prove that these worlds exist. There is, in principle, no way to gain any knowl-edge about the Ultimate Multiverse at all . . . *unless*, of course, it becomes possible to simulate universes. In this case, Greene imagines, "an army of future computer users . . . could spawn this [mathematical] multiverse through their insatiable fascination with running simulations based on ever-different equations."[56] So if we ever do become (or have ever before become) "posthuman," then all mathematically possible, which is to say all "computable," universes *could*, in fact, exist. But then we would be right back where we started this ascending cosmic meditation: with the new anthropomorphic gods of the multiverse.

The point of this journey has simply been to show that every multiverse hypothesis opens in one way or another onto uncannily metaphysical—even theological—terrain. Each scenario requires us to assent to worlds, gods, or generative principles that remain, in the words of an old English hymn, "in light inaccessible hid from our eyes."[57] This is not to say that the theories are somehow scientifically invalid; as Helge Kragh reminds us, "speculations have always been an integrated part of the physical sci-ences, sometimes hidden under the more palatable term 'hypotheses.'"[58] It is simply to say that the distinctions between the purportedly inimical terms of "science" and "religion" are highly unstable when it comes to the multiverse, which, to borrow Bruno Latour's string of adjectives, is per-haps the most "spiritual, miraculous, soul-fulfilling, uplifting" hypothesis that modern science has ever produced.[59] What the debate comes down to, therefore, is not an argument between theism and atheism or between sacred and secular reason, but between vastly different understandings of what counts as "science."

The Bullet and the Blunderbuss

As we have begun to see, critics of the multiverse accuse these scenarios of venturing too far beyond the bounds of observation and experiment, of causing more problems than they solve, and of demanding just as much "faith" in the invisible as does the design hypothesis. In response, multi-verse theorists tend to remind their detractors that modern science is *teeming* with elements—from black holes to dark matter to neutrinos to superpositions—that lie beyond our ability to access them directly, but explain phenomena that no other theory can explain. As Max Tegmark argues, "[F]or a theory to be falsifiable, we do not need to be able to observe and test *all* its predictions, merely at least one of them . . . consider Einstein's theory of General Relativity [*sic*]. Because this has successfully predicted many things that we *can* observe, we can also take seriously its predictions for things we *cannot* observe."[60] If, then, the multiverse solves observational and experimental problems that nothing else solves (and its proponents argue that it does), then no matter how inaccessible the "multiverse itself" might be, it should be taken seriously as a scientific hypothesis. Andrei Linde is perhaps the most energetic of these defendants, saying in an interview that

we don't have any other alternative explanation for the dark energy, we don't have any alternative explanation for the smallness of the mass of the electron; we don't have any alternative explanation for many properties of particles. What I am saying is to look at it with open eyes. These are experimental facts, and these facts fit one theory: the multiverse theory. They do not fit any other theory so far. I'm not saying these properties necessarily imply the multiverse theory is right. But you asked me if there is any experimental evidence, and the answer is yes. It was Conan Doyle who said, "When you have eliminated the impossible, whatever remains, however improbable, must be the truth."[61]

From the perspective of the most fervent multiverse proponents, it is this improbability, rather than a coherent scientific commitment, that motivates multiverse critics. As Tegmark charges, "[T]he principle arguments against [multiverse cosmologies] are that they are wasteful and that they are weird."[62] And, indeed, some physicists and philosophers do display an almost visceral disdain for the multiverse. When, for example, Paul Davies calls it a "fantasy-verse" and an "infinitely complex charade," or when Nobel laureate David Gross says, "I hate it," or when Paul Steinhardt calls it "a dangerous idea that I am simply unwilling to contemplate,"[63] we might hear echoes of St. Augustine's anxious "God forbid" or even of the character in Giordano Bruno's *On the Infinite Universe and Worlds* who exclaims, "[E]ven if this can be true I do not wish to believe it, for this infinite can neither be understood by my head nor brooked by my stomach."[64] As far as some of the modern heirs to Bruno are concerned, however, the multiverse is no "weirder" or less stomachable than Copernicanism was in Galileo's day or than multiple galaxies were in Kant's. This multiverse is just the next step in what Bernard Carr sees as a progressive expansion of our understanding of the universe. "Every time this expansion has occurred," he reminds us, "the more conservative scientists have said, 'This isn't science.' This is just the same process repeating itself."[65]

Other multiverse theorists disagree that "this is just the same process" as the convulsions over Copernicanism and galactic pluralism, but claim right alongside their detractors that "the very nature of the scientific enterprise is at stake in the multiverse debate."[66] The seismic shift that these rivals identify concerns the effort to understand the nature of the universe "from fundamental principles"—to determine why our universe *had* to be the way it is. As Davies explains the problem, "[W]e should like to understand the bio-friendliness of *this* universe. To postulate that all possible universes exist does not advance our understanding at all. A good scientific theory is analogous to a well targeted bullet that selects and explains the object of interest. The multiverse is like a blunderbuss—hitting everything in sight."[67] It is this "bullet" model of the scientific enterprise that multiverse critics ranging from Walter Charleton and Cardinal Christoph Schönborn to Steinhardt and Gross are seeking to defend. And it is the same model that Linde suggests we are going to have to give up so that we can learn to work with the blunderbuss. "For a long time," Linde explains, "physicists have believed that there is only one world and that a successful description of this world should eventually predict all of its parameters, such as the coupling constants and the masses of elementary particles. The fundamental theory was supposed to be beautiful and natural. This was a noble, but perhaps excessively optimistic, hope. One could call this period 'the age of innocence.' I believe we are now entering 'the age of anthropic reasoning.'"[68]

Linde seems happily—even impishly—reconciled to this new age. Leonard Susskind is even more insistent a harbinger, saying that it would be "the height of stupidity to dismiss" the anthropic principle "just because it breaks some philosopher's dictum about falsifiability."[69] But other theorists display far more ambivalence toward Linde's "age of anthropic reasoning." Both Alan Guth and Steven Weinberg regard the anthropic principle and the multiverse it now entails as a "last resort," "plausible only when we cannot find any other explanation" for the apparent fine-tuning of the cosmos.[70] Both of them confess that they are not sure whether even the appalling smallness of the cosmological constant warrants our appealing to the anthropic principle and its infinity of other worlds.[71] But both of them

fear that it might, and they confess to feeling let down by the new anthropic age it has ushered in. "It would be a disappointment if [an ensemble of universes] were the solution of the cosmological constant problems," Weinberg says, "because we would like to be able to calculate all the constants of nature from first principles, but it may be a disappointment that we will have to live with."[72] As for Guth, he assures his interlocutors (not least of all himself) that "there will still be a lot for us to understand, but we will miss out on the fun of figuring everything out from first principles."[73] And although it might *seem* like an unprecedented shift in the scientific project, both Weinberg and Martin Rees say that this multiversal letdown will simply be a cosmic extension of the disappointment that Newton experienced when he realized that planetary orbits could not be deduced from first principles.[74] Rees explains the comparison thus:

> People used to wonder: why is the earth in this rather special orbit around this rather special star, which allows water to exist or allows life to evolve? It looks somehow fine-tuned. We now perceive nothing remarkable in this, because we know that there are millions of stars with retinues of planets around them: among that huge number there are bound to be some that have the conditions right for life. We just happen to live on one of that small subset. So there's no mystery about the fine-tuned nature of the earth's orbit; it's just that life evolved on one of millions of planets where things were right.[75]

In effect, Rees is arguing, all the multiverse does is to move this arbitrary positioning up from the level of planets to the level of the universe itself. "There is no mystery" about the cosmological constant, the mass of the electron, or the strength of the nuclear forces; it is just that life has evolved in one of an infinite number of universes where things are right. Even a blunderbuss loaded with wood chips and rocks is bound to hit something at some point.

Bidden or Unbidden

In one of his numerous critiques of multiverse cosmologies, George Ellis issues a word of warning to physicists. "There are many other theories waiting in the wings," he cautions, "hoping for a weakening of what is meant by 'science.' "[76] What Ellis has in mind, Helge Kragh explains, is a host of "pseudosciences such as astrology, intelligent design, and crystal healing." Is an infinite number of all possible worlds not *at least* as speculative as the twelve signs of the zodiac (not to mention one lonely creator)? "That is," Kragh asks, "if multiverse cosmology is admitted as a science, how can scientists reject pseudosciences . . . on methodological grounds?"[77]

In response to this line of questioning, multiverse centrist Brian Greene concedes that at the moment there is no (widely accepted) observational or experimental evidence for any multiverse scenario. He reaffirms the importance of withholding scientific assent to any model "not supported by hard data." And yet he is fascinated by the idea that our universe might just be one of a host of others—not because of the metaphysical or narrative implications of such an idea (at least this is what he claims), but because although the multiverse has not yet been confirmed by any scientific discipline, it has been *predicted* by a number of them. From quantum mechanics to inflationary cosmology to string theory, "numerous developments in physics, if followed sufficiently far, bump into some variation on the parallel-universe theme." For Greene, this "bump" is what distinguishes contemporary multiple-worlds cosmologies from astrology, intelligent design, and crystal healing. "It's not that physicists are standing ready, multiverse nets in their hands, seeking to snare any passing theory that might be slotted . . . into a parallel-universe paradigm," he maintains. "Rather, all of [these] parallel-universe proposals . . . *emerge unbidden* from the mathematics of theories developed to explain conventional data and observations."[78] In other words, no one was looking for the multiverse, and yet it has sprung up nearly everywhere.

On the one hand, this account risks mystifying the multiverse even further—it is hard not to hear in Greene's language an echo of Karl Jung's "bidden or not bidden, God will be present."[79] On the other hand, it also has the potential to render the multiverse debate a bit less dramatic than it often seems. If these scenarios *do* emerge from theories that emerge from observations, then they should ultimately be accountable to observations. In other words, it might be that multiverse cosmologies represent neither a total transformation nor an obliteration of the scientific project, but an organic development that just happens to open physics onto metaphysics at more or less every turn. But in that case, the multiverse is going to have to pass (or fail) a few tests.

The central theoretical task is to confirm or rule out the various frameworks that predict differing models of the multiverse. For example, inflationary cosmology would be substantially supported by the discovery of gravitational waves (B-modes) on the CMB. It would be invalidated (at least in its "eternal" formulations) if the universe were determined to be finite. The various braneworld scenarios would gain experimental traction if the supersymmetric particles they predict were to be found. The MWI would be confirmed as an ontological (not just a mathematical) reality if quantum computing were to become sufficiently advanced. At that point, it might also become possible to compute a whole world into being, a feat that would in turn lend considerable credence to the simulated multiverse. And the landscape of 10^{500} or $10^{1,000}$ or an infinity of different types of universe would become far more plausible if it were possible to find "statistical rules governing different string vacua"—that is, the probability that a particular cosmic configuration will emerge, given infinite time and space.[80] In Steven Weinberg's (characteristically understated) words, "[I]t would not hurt in this work if we knew what string theory is."[81]

It is in this spirit that Nobel Prize–winning particle physicist and multiverse skeptic Burton Richter issues a challenge to the colleagues he calls "landscape gardeners." "Calculate the probabilities of alternative universes," he suggests, knowing full well what a monstrous task this is, "and if ours does not come out with a large probability while all others with content far from ours come out with negligible probability, you have made no useful contribution to physics."[82] What Richter is looking for, in effect, is a justification for the metaphysical extravagance of the landscape. If, of all these possible universes, ours turns out to be a fairly common one—if it is highly likely that any given universe will have the kind of cosmological constant, electrons, nuclear forces, and so on that our universe does— then the gazillions of universes on the landscape are a justifiable expenditure; we genuinely need them in order to understand our universe as it is. At the moment, however, even the most fervent proponents of the landscape admit that life looks possible only in a small subset of these hypothetical vacua.[83] And if this remains the case, then as far as Richter is concerned, there is no reason to keep cultivating the landscape model. "It is not that the landscape model is necessarily wrong," he explains, "but rather that if a huge number of universes with different properties are possible and equally probable, the landscape can make no real contribution other than a philosophic one."[84] In other words, it might be fun to think about all the other ways the world might have been and the extreme improbability of the universe as it is, but such a scenario would leave us just as befuddled as ever by the miraculous fine-tuning of the universe we happen to be in.

That having been said, the physics community does not have to await the statistical "population" of the landscape for evidence of some sort of multiverse. To the contrary, as we began to see in chapters 5 and 6, observational astronomers and cosmologists are already claiming to have found such evidence on the CMB. To be sure, the patterns in question differ from team to team, as do the models they allegedly support, but the very existence of such work means that the multiverse does not lie totally beyond the bounds of observation, as its critics often allege. In 2010, for example, one group of researchers reported having found four "disc-shaped" temperature variations that they attribute to collisions between our universe and others during the inflationary period.[85] If such collisions do not destroy the universes in question, they would leave "bruises" in the form of "inhomogeneities in the inner-bubble cosmology, which could appear on the CMB."[86] Other researchers, seeking to confirm Roger Penrose's cyclical cosmology, claim to have found different sorts of inhomogeneities on the CMB—"concentric circles" that purportedly record the black-hole collisions of the previous aeon, which is to say the universe whose end produced our beginning.[87] And four of Laura Mersini-Houghton's major predictions have been found, including one void and the set of inhomogeneities now called the "dark flow."[88] According to Mersini-Houghton, both of these cosmic anomalies provide compelling evidence of "the birth of the universe from the multiverse"; they effectively are scars—not of the collisions between universes, but of their "entanglement," of the primordial boundupedness of our universe "with all other patches . . . beyond the horizon," which is to say with the multiversal "bath" itself.[89]

In one of his numerous defenses of multiple-worlds cosmologies, Max Tegmark reminds his queasy critics that "the borderline between physics and metaphysics is defined by whether a theory is experimentally testable, not by whether it is weird or involves unobservable entities."[90] And, indeed, if the examples given here count as such "tests," then it does not matter how extravagant their underlying multiverse scenarios may be; they ought to be taken seriously as scientific hypotheses. But I would respectfully argue against Tegmark that this testability does not deliver such scenarios from the treacherous realms of philosophy or religion. To the contrary, the very observations and experiments that promise to establish the multiverse as "physics" *also* establish it as metaphysics.

Here I am using the term *metaphysics* both in the broad sense of "beyond the physical" and in the narrower sense that Martin Heidegger uses it to refer to the Western philosophical tradition founded on the question

"What is?"[91] To the extent that multiverse theories can be called physics, they can also be called metaphysics—first, because they posit realms that, however imprinted on or entangled with our own, remain inexorably *beyond* it and, second, because they operate by means of a specific conception of *what is*. Viewed in one light (there are others), the history of Western metaphysics can be seen as an ongoing effort to understand whether "what is" is fundamentally *one* or fundamentally *many*, identical or different. Philosophers often trace this debate to the distinction between Parmenides and Heraclitus or Plato and Aristotle or rationalism and empiricism. In the introduction, we saw William James refer to it as a conflict between monism and pluralism. And throughout the book, we have seen this question play itself out cosmologically: Is "all that is" part of the same reality or different realities? As Timaeus asks, "[A]re we right . . . to speak of one universe, or would it be more correct to speak of a plurality?"[92]

The only possible answer for Timaeus is "one," and a great deal of his "likely story" about the origin of the world is dedicated to demonstrating such oneness. His motivation in this regard stems from his insistence that the universe must be eternal; if there were another world beyond, before, or after this one, then it would render our own world vulnerable to destruction—whether by collision or by periodic annihilation and rebirth. In order to assure his audience of the permanence of the universe, Timaeus therefore assures us of its singularity. As we have seen, however, our Platonic astronomer can assert the unity of the cosmos only by establishing it as a "mix"—even, as Michel Serres phrases it, a "mix of mixes"[93]—of the divisible and the indivisible, the same and the different, the many and the one. A strikingly similar thing happens in the *Metaphysics* when in the very process of proving the oneness of the world (again, to establish its eternity), Aristotle suddenly leaves us with the conclusion that there are either forty-seven or fifty-five worlds, but that the math is too hard for him.[94] Then there are Thomas Aquinas, who hinges the oneness of the world on the oneness of a God who also happens to be three, and René Descartes, who insists that "there cannot be a plurality of worlds" even as he provides a cosmology of a plurality of worlds.[95]

But just as all these defenders of cosmic singularity end up colliding with plurality, their pluralist counterparts end up colliding with unity. From Epicurus and Lucretius to the Stoics, Nicholas of Cusa, Giordano Bruno, and Kant, even the most outlandish proponents of a plurality of worlds assert the oneness of that plurality in light of their shared cosmogonic principle (whether it be atoms and void, the *clinamen*, *pneuma* and *hyle*, or a triune God). So it is that Bruno, having danced his way through an infinite number of purportedly disconnected worlds, proclaims them all to be one: "[I]t is unity," says the pluralist, "that enchants me."[96] What we have witnessed, then, is a series of diverse negotiations of the singular and the plural—each negotiation demonstrating with some mixture of intention and accident that the world is neither one nor many, but many in its oneness or one in its manyness or many in a certain light and one in another. "We must give it a new name, definitely," writes Serres of such cosmic multiplicity, "it is a mixture, tiger-striped, motley, mottled, zebra-streaked, variegated, and I don't know what-all, it is a mix or a crasis, it is a mixed aggregate, it is an intermittence."[97]

The reason, then, that the multiverse is only physics insofar as it is metaphysics is that the only "other worlds" that might compose a scientific hypothesis would be worlds that are in *some way*, however mediated and strung along, ontologically bound up with ours. After all, if they were *wholly* "other," sharing neither space nor time nor any generating principle and bearing no trace of one another, then they would be wholly inaccessible to observation and experiment (this is the reason that most physicists draw the line at Tegmark's Mathematical Multiverse Hypothesis, however "expansive" their sensibilities may be). At the same time, however, the "other worlds" of the multiverse remain to an extreme degree *other*: they are forever separated by an unrecoverable past, an unreachable future, an uncrossable distance, or an irremediable split. So it may be that, from a god's-eye view, there is only one world. It may be that there are many. It may be that every mathematically possible universe does in fact "exist outside space and time," whatever such existence might mean. But the only way to make multiple worlds an *object for science* would be to configure them as neither one nor many—neither undifferentiated from one another nor indifferent to one another. And so the metaphysical claim that the multiverse necessarily makes is that "all that is" exists in "some possible or mediate connection" with all that is, rendering being itself many-one, pluri-singular: multiple.[98] As such, neither our universe nor anything else is self-constituted, nor is it invulnerable to destruction or radical transformation. And although this cosmic interdetermination of physics and metaphysics might scandalize the modern secularist, it would come as no surprise to the thinker whose demon stole into our loneliest loneliness to proclaim that "all things are entangled, ensnared, enamored."[99]

On the Genealogy of Cosmology

"Hectics in Some Sense or Other"

In the third essay of *On the Genealogy of Morals*, Friedrich Nietzsche marvels at the persistence of what he calls "ascetic ideals."[100] These ideals include humility, restraint, poverty, chastity, meekness—that "whole train" of what David Hume calls the "monkish virtues"[101]—and are enforced through various practices of self-denial, such as fasting, flagellation, and sleep deprivation. For Nietzsche, the ascetic sees the physical world as some kind of mistake. Setting herself against everything that is life giving, which is to say *everything that is*, she is a literal nihilist, reducing the whole world to nothing. As Nietzsche explains it, these ascetic ideals are Europe's Christian inheritance. It was Christianity, he says, that globalized the allegedly "Jewish" values of "poverty," "impotence," and "wretchedness," spreading these values through the parts of the world that it conquered. By now, what Nietzsche calls "slave morality" has become the foundation of Western political and cultural systems to such an extent that "we no longer see it because it—has been victorious" (1.7).[102] ("The same evolutionary course," he says in a long parenthetical paragraph, has been followed "in India," where "five centuries before the beginning of the European calendar," the Buddha taught the ascetic ideal that would spread itself throughout the "Eastern" world as effectively as the Christians spread their asceticism throughout the "West" [3.27].) The point is that wherever and whatever we might call ourselves, "we" are not finished with asceticism just because we claim to be free from religion.

Almost no one escapes Nietzsche's charge of asceticism, but the third essay of the *Genealogy* focuses on three particularly guilty classes of people. The first class, unsurprisingly, comprises priests, who "heal" their flocks (or "herds") by making them sick in the first place, prescribing them a set of self-destructive practices to keep them docile and numb (3.15–18).[103] The next class of ascetics is a bit less obvious: philosophers, Nietzsche claims, are subject to the same ascetic ideal. Although he initially offers the less than compelling evidence that real philosophers never marry (Socrates, he says, is the only exception, and he "married *ironically*, just to demonstrate *this* proposition" [3.7, emphasis in original]), his lasting charge is that philosophers are ascetics because, like priests, they believe in a world beyond this one: an eternal realm of Forms or Ideas or things-in-themselves, of which this world is at best a pale reflection. Insofar as philosophers declare that "*there is* a realm of truth and being, but reason is *excluded* from it!" (3.12, emphasis in original), they are no different qualitatively from the priests who proclaim the eternal, mysterious providence of a God before whom we must abase ourselves. And then finally, Nietzsche intones, there are "scientists," a broad category denoting scholars of what we might consider the human, social, and natural sciences—any discipline one might study at a secular university. Of course, Nietzsche says, "modern science" *thinks* itself "a genuine philosophy of reality," having "up to now survived well enough without God, the beyond, and the virtues of the eternal" (3.23). In other words, modern science *believes* that it has liberated itself completely from religious delusions. And yet, he insists, "such noisy agitators' chatter . . . does not impress me: these trumpeters of reality are bad musicians. . . . [S]cience today . . . is not the opposite of the ascetic ideal but rather *the latest and noblest form of it*. Does that sound strange to you?" (3.23, emphasis in original).

Assuming the answer is yes, Nietzsche goes on to explain that "science today" remains ascetic (and even excels at asceticism) for two interrelated reasons. First, scientists believe that there is a truth outside themselves, and, second, they devote themselves to the unconditional pursuit of that unconditioned truth. Here we might begin to think of all those hours at the computer, in the archives, in the lab; the ritualistic precision over methods, materials, data, results; the caffeine that stands in for both sleep and food: What is all this if not asceticism? And so, as Nietzsche inimitably explains the asceticism of science,

> these hard, severe, abstinent, heroic spirits who constitute the honor of our age; all these pale atheists, anti-Christians, immoralists, nihilists; these skeptics, ephectics, *hectics* of the spirit (they are all hectics in some sense or other), they certainly believe they are as completely liberated from the ascetic ideal as possible, these "free, *very* free spirits"; and yet, to disclose to them what they themselves cannot see—for they are too close to themselves: this ideal is precisely *their* ideal, too . . . they themselves are its most spiritualized product, its most advanced front-line troops and scouts. . . . [I]f I have guessed any riddles, I wish that *this* proposition might show it!—They are far from being *free* spirits: *for they still have faith in truth*. (3.24, emphasis in original)

From this perspective—and, for Nietzsche, it is perspective all the way down—science becomes indistinguishable from religion precisely at the point that it thinks itself most free: in its pursuit of a purportedly objective, singular "truth." In a similar vein, physicist Marcelo Gleiser calls modern science "monotheistic." In its quest for a grand unified theory, a single explanation for every last physical phenomenon, science, Gleiser argues, remains "under the mythic spell of the One."[104]

Asceticosmologies

Part of what makes the field of cosmology such a fascinating case study for this Nietzschean hypothesis is that until very recently, cosmology was not considered a "real science."[105] The reason was not only that cosmology has traditionally been the purview of philosophy and mythology (after all, *every* field has traditionally been the purview of philosophy and mythology), but also that cosmology, unlike the other disciplines that branched off in the early modern period, was not considered objective. The problem is that whereas every other secular discipline studies objects *within* the universe, cosmology studies the universe *itself.* This means that cosmology, unlike every other discipline, remains hopelessly internal to that which it studies. Unlike the biologist, the anthropologist, the economist, and even the astronomer, the cosmologist cannot even pretend to stand as a knowing subject over against her known object. Rather, she is inescapably caught within *and irreducibly constituted by* the very thing she is trying to measure and observe.

There is also the problem of repeatability. For a hypothesis to hold water, it has to be tested again and again on a vast number of specimens. But cosmology's specimen is the *uni-verse*, which is to say that cosmology's specimen is *all there is.* Where might cosmologists find another "all there is"—much less hundreds of them—to make sure they get the same results each time? Here we might recall Philo's battery of questions to Cleanthes: "Have worlds ever been formed under your eye? And have you had leisure to observe the whole progress of the phenomenon, from the first appearance of order to its final consummation? If you have, then cite your appearance and deliver your theory."[106] So from the seventeenth century onward, cosmology was accused of subjectivism, imprecision, speculation—in short, of sounding more like philosophy or religion than science.

But then, as the legend usually goes, this scrappy little discipline finally came into its own with the accidental discovery of the CMB in 1965. The story is remarkable. Two radio astronomers at Bell Labs in Holmdel, New

Jersey, kept hearing interference hissing through their high-powered antennas. When they climbed up to the roof to see what the problem was, they found piles of pigeon droppings congealed around the equipment. So they power-washed the roof and climbed back down to the lab, only to keep getting the interference. Then, in consultation with some colleagues, they eventually realized that the hissing was not a result of pigeon droppings at all; it was the remnant of the big bang.[107] The Cosmic Microwave Background, as it has come to be called, is a snapshot of the temperature and density variations of the universe when it was just a few hundred thousand years old. Subsequent developments in telescopic and satellite technology led to the release in early 2003 of the Wilkinson Microwave Anistropy Probe's ovoid image of the CMB (see figure 5.1; and a brand-new image was released from the Planck satellite in March 2013),[108] which suddenly became the Object that cosmology had needed. It is, in fact, an object that cosmologists can measure and observe as often as they like. Granted, this object is rather grainily compiled by inexorably *situated* satellites; like everything else, the CMB is produced by means of specific and perspectival material configurations. Nevertheless, it seems to have done the work of disaggregating the subject of cosmology from something said subject can regard as an "object," because the field has come into wide acceptance as a "proper, quantitative science."[109]

Staying tuned with our Nietzschean antennas, however, we might notice that the moment cosmology entered the domain of "objective science," it also collided head on with Christian creation theology. After all, it was the CMB that confirmed that the universe had a beginning, that it began in a burst of light, and that it came out of something like "nothing"—all ideas that Jews, Christians, and Muslims had been teaching for centuries (see chap. 5, sec. "Let There Be Light"). Although this "big bang" hypothesis scandalized decades of physicists (Georges Lemaître says that upon hearing the idea, Einstein shot back, "*No, not that*, that sounds too much like creation"),[110] the resemblance is not all that surprising if we, alongside Nietzsche, think of "science today" as the "latest and noblest form" of the ascetic ideal. In fact, it is a remarkable sign of the entanglement of Western science and religion that when science finally had a creation story to tell, it told such a familiar one.

What, then, of modern multiverse cosmologies? Have these extraordinary revisions to the big bang hypothesis finally delivered the scientific enterprise from its ascetic past? Based on the foregoing discussion of physics as metaphysics, it is probably clear that my ultimate answer to the question is no. But one can certainly make the contrary argument, citing three major pieces of evidence. First, multiverse scenarios promise a genuinely "objective" view of our universe. The moment a model claims, for example, that "from the outside" our universe looks finite, or that from the outside it looks like a membrane flying down the throat of a Calabi–Yau manifold, or that from the outside the wave function progresses "smoothly" and never collapses, such a model is claiming an extraworldly perspective through which the subject can finally transcend the very universe that embodies him and see "all that is" as an *object*. Second, the multiverse finally does away with the necessity of a designer-god. Although, as we have seen, it will always be possible to tag such a god onto any cosmology, the multiverse is said—often with elation on the author's part—to render that god a useless appendage. And third, the multiverse finally gives cosmology all those "other specimens" it needed in order to understand this one. Once the cyclic model is confirmed or inflation is better understood or the landscape is populated, we will be able to understand this one universe in relation to all the other universes—botched, bungled, and otherwise— that emerged before or alongside it, and those that are still to come.

In promising a view from nowhere that gets rid of God *and* accounts for every possible everything, the multiverse seems to promise the ultimate scientific vision of reality. The early modern scientist attempted to transcend his senses, his socioeconomic positioning, his historical location, his dressing gown. But the twenty-first-century multiverse theorist goes further: past the planet, the solar system, the galaxy, the supercluster; past the plasmic CMB and the cosmic horizon that even light cannot reach; up through the hierarchical ranks of increasingly unfamiliar *kosmoi* until he gains a god's-eye view of all worlds bubbling out of the sea or bursting out of black holes or simulated by posthuman tweens or populating the 10^{500} types of universe on the landscape—keeping his eye on the sky until his gaze finally opens onto all possible worlds, actually existing, outside time and space. At this point, it once again becomes very hard to argue that any of these visions of reality genuinely frees modern science from philosophy and religion—not least because they all seek the ultimate, objective *truth* of creation.

From the perspective of these visions' critics, this inexorable collision with the metaphysical and even mystical has the effect of invalidating multiverse cosmologies. Or at least, they claim, it *should* have this effect because such theories represent a total violation of the principles of falsifiability, testability, and, above all, economy that undergird the scientific project. But here again, I find myself tuning back in to Nietzsche: "What, in all strictness, has really *conquered* the Christian God?" he asks in *On the Genealogy of Morals* (3.27, emphasis in original). Citing another of his own books (*The Gay Science*), Nietzsche tells us that the answer is not science and its attendant "atheism." Rather, the Christian God has been conquered by "Christian morality itself, the concept of truthfulness taken more and more strictly, the confessional subtlety of the Christian conscience translated and *sublimated into the scientific conscience*, into intellectual cleanliness at any price" (3.27, emphasis added). Christianity told its adherents to "view nature as if it were a proof of the goodness and providence of a God" (3.27) and therefore to study it with reverential attentiveness. Christianity told the world to go out and seek the truth, the objective truth, and when the world found the truth, it finally realized that "belief in God" was a lie (3.27). But even after the death of God, the devotion to some purportedly eternal, extraworldly truth has not disappeared; to the contrary, the search for it has only intensified in the hands of these extravagant new ascetics. From this (particular) perspective, modern science can therefore be seen as what Nietzsche calls "the self-overcoming of Christianity." Put more simply, Christianity *produces* modern science, in a staggering gesture of self-sabotage, as its consummation and its destruction.

Nietzsche concludes the *Genealogy* by expanding this vision, promising that "*all* great things bring about their own destruction through an act of self-overcoming" (3.27, emphasis added). This promise, then, has me wondering. If science can be regarded as the self-overcoming of a particular form of religion, might multiverse cosmologies be something like the self-overcoming of *science*? Might they mark the end of the fantasy that "science" has wrested itself free from "religion," "objectivity" free from subjectivity, and matter free from meaning?[111] After all, we have seen each of these multiverse cosmologies open onto metaphysics and mythology not in moments of lapse or weakness, but precisely where they are scientifically most compelling. I would like to be clear here: by pointing out this confluence, I do not intend to say that philosophers have already solved all these scientific riddles or that any particular theology can account ahead of time for the tiger-striped/zebra-streaked being of the multiverse. To the contrary, by revealing the persistent entanglement of all these disciplines, multiple-worlds cosmologies condition and even necessitate a renewed engagement among them.

Unscientific Postscribble

"Okay . . . ," you might be wondering, "but does the multiverse *exist*? And if so, which model is the right one?"

In the face of such questions, I find myself wanting to hide behind someone like Johannes Climacus, a character that Søren Kierkegaard dreamed up to write something that he could not quite write. In *Philosophical Fragments, or a Fragment of Philosophy*, Climacus presents himself as a trifler, a "loafer out of indolence," totally unqualified to contribute to serious philosophy. The tone of the preface gets increasingly cranky until finally Climacus responds to an imagined interlocutor, "But what is my opinion? Do not ask me about that. Next to the question of whether or not I have an opinion, nothing can be of less interest to someone else than what my opinion is. To have an opinion is both too much and too little; it presupposes a security and well-being in existence akin to having a wife and children."[112]

A security and well-being in existence, not to mention a degree in physics, a relationship with the angels, and a telescope that travels faster than the speed of light—I imagine that one would need all these things in order to construct an opinion about the existence of the multiverse. So, no, on this matter I have no opinion.

But I do have a hunch.

To be sure, a hunch is hardly the basis on which serious scholarship ought to be conducted,[113] and so I offer mine as an afterthought, a possibility opened by the foregoing analysis—not at all as its foundation. My hunch is that "everything" probably works the same way as anything does. Just as light will behave as a wave or a particle, depending on the question you ask it, and just as chemical and biological and psychological experiments help produce the phenomena they measure, so will the universe appear to be one or many, or linear or cyclical, or infinite or finite, depending on the theoretical and experimental configuration that examines it. In other words, the shape, number, and character of the cosmos might well depend on the question we ask it. Of course, this is not to say that every theory is *right*; some will be more internally coherent, mathematically reliable, and observationally demonstrable than others—and those coherent, reliable, and demonstrable models, I imagine, will be the ones that survive the decades ahead. But I doubt very much that we will or should emerge with only one of these theories. Would it even make *sense* to have a single account of cosmic multiplicity? To arrive at the one truth of the multiple ways worlds can be multiple?

In the meantime, what this cosmic loafer finds promising is not so much the answers, but the processes that produce and undo them: those endless cosmogonic efforts to derive *all this* from *that*, efforts whose very multiplicity signal a persistence of chaos amid anything that looks like order. And such persistence, I think, is the real promise of the multiverse. Tuned in to the background noise of many-worlds cosmologies—of their failure to disentangle physics from metaphysics from religion from science—one can pick up the faint but unmistakable signals of an ontology that entangles the one and the many; of an "order" constituted, dismantled, and renewed by an ever-roiling chaos; of a "truth" that remains provisional, multiple, and perspectival; and, perhaps, of a theology that asks more interesting and more pressing questions than whether the universe has been "designed" by an anthropomorphic, extracosmic deity.

So let us begin again . . .

NOTES

Introduction

1. Alex Vilenkin, "Creation of Universes from Nothing," *Physics Letters B* 117 (1982): 25–28, and "The Birth of Inflationary Universes," *Physical Review D* 27 (1983): 2848–55; Andrei Linde, "Eternally Existing Self-Reproducing Chaotic Inflationary Universe," *Physics Letters B* 175 (1986): 395–400, and "The Self-Reproducing Inflationary Universe," *Scientific American*, November 1994, 48–55; David Deutsch, "The Structure of the Multiverse," *Proceedings of the Royal Society of London A* 458 (2002): 2911–23; Martin Rees, "Exploring Our Universe and Others," in "The Once and Future Cosmos," special issue, *Scientific American*, December 1999, 78–83; Paul J. Steinhardt and Neil Turok, "Cosmic Evolution in a Cyclic Universe," *Physical Review D* 65 (2002) : 1–20; Max Tegmark, "Parallel Universes," *Scientific American*, May 2003, 41–51; Lauris Baum and Paul Frampton, "Turnaround in Cyclic Cosmology," *Physical Review Letters* 98 (2007): 1–4; Mark Buchanan, "Many Worlds: See Me Here, See Me There," *Nature*, July 5, 2007, 15–17; Bernard Carr, ed., *Universe or Multiverse?* (Cambridge: Cambridge University Press, 2007); Laura Mersini-Houghton, "Birth of the Universe from the Multiverse," September 22, 2008, available only through arXiv/0809.3623.

2. Brian Greene, *The Hidden Reality: Parallel Universes and the Deep Laws of the Cosmos* (New York: Knopf, 2011). See also Michio Kaku, *Parallel Worlds: A Journey Through Creation, Higher Dimensions, and the Future of the Cosmos* (New York: Doubleday, 2004); Alex Vilenkin, *Many Worlds in One: The Search for Other Universes* (New York: Hill and Wang, 2006); Leonard Susskind, *The Cosmic Landscape: String Theory and the Illusion of Intelligent Design* (Boston: Back Bay Books, 2006); Paul J. Steinhardt and Neil Turok, *Endless Universe: Beyond the Big Bang—Rewriting Cosmic History* (New York: Broadway, Books, 2008); Louise Lockwood, "Parallel Worlds, Parallel Lives" (manuscript, 2008); Jad

Abumrad and Robert Krulwich, "DIY Universe," *Radiolab*, WNYC, March 25, 2009, and "The (Multi)Universe(s)," *Radiolab*, WNYC, August 12, 2008; Nathan Schneider, "The Multiverse Problem," *Seed Magazine*, April 14, 2009; Heather Catchpole, "Weird Data Suggests Something Big Beyond the Edge of the Universe," *Cosmos*, November 24, 2009; John Gribbin, *In Search of the Multiverse* (London: Allen Lane, 2009); Sean M. Carroll, *From Eternity to Here: The Quest for the Ultimate Theory of Time* (New York: Dutton, 2010); Stephen Hawking and Leonard Mlodinow, *The Grand Design* (New York: Bantam, 2010); "Multiverse," ed. Terry Gross, *Fresh Air*, NPR, January 24, 2011; George Ellis, "Does the Multiverse Really Exist?" *Scientific American*, August 2011, 38–43; Steven Manly, *Visions of the Multiverse* (Pompton Plains, N.J.: Career Press, 2011); "The Fabric of the Cosmos: Universe or Multiverse?" *NOVA*, PBS, November 23, 2011; John D. Barrow, *The Book of Universes: Exploring the Limits of the Cosmos* (New York: Norton, 2011); and Helge Kragh, *Higher Speculations: Grand Theories and Failed Revolutions in Physics and Cosmology* (Oxford: Oxford University Press, 2011).

3. Seth MacFarlane, creator, Greg Colton, dir., and Wellesley Wild, writer, "Road to the Multiverse," *Family Guy*, season 8, episode 1, FOX, September 27, 2009. In addition to the *Family Guy* video game, see, for example, Michael P. Kube-McDowell, *Alternities* (n.p.: iBooks, 2005); Paul Melko, *The Broken Universe* (New York: Tor Books, 2012), and *The Walls of the Universe* (New York: Tor Books, 2009); *Super Mario Galaxy* (Nintendo, 2007); and *Super Mario Galaxy 2* (Nintendo, 2010). See also the newly released *Multiverse* magazine, a roundup of news from the world of superhero comics (volume 0 available at http://issuu.com/richjohnston/docs/multiverse__0 [accessed May 29, 2013]).

4. Kragh, *Higher Speculations*, 255.

5. Lorraine Daston, "The Coming into Being of Scientific Objects," in *Biographies of Scientific Objects*, ed. Lorraine Daston (Chicago: University of Chicago Press, 2000), 1–14.

6. The idea of multiple worlds finds slightly later but far more extensive elaboration in the Hindu and Buddhist traditions than it does in the West, and these cosmologies certainly suffer far less controversy at the hands of their host cultures. Although this book makes occasional connections to these schools of thought, its focus is on the Western tradition for two reasons. First, this tradition tends to provide the metaphoric registers on which contemporary Western-based scientists (many of whom are not Western *born*) tend, for better or worse, to draw. Second, I have neither the historical nor the linguistic training to offer careful readings of Hindu or Buddhist texts. I hope that this work might contribute in a small way to other scholars' projects on these resonances. In the meantime, a compelling historical connection between Atomist, Stoic, and pre-Mahayana cosmologies can be found in Akira Sadakata, *Buddhist Cosmology: Philosophy and Origins* (Tokyo: Kosei, 2004), 9–25. See also Jamgon Kongtrul Lodro Taye, *Myriad Worlds: Buddhist Cosmology in Abhidharma, Kalachakra, & Dzog-chen* (Boston: Snow Lion, 1995).

7. "Multiverse," *Oxford English Dictionary*, online version (2012), http://www.oed.com/view/Entry/123653 (accessed June 25, 2012).

8. William James, "Is Life Worth Living?" *International Journal of Ethics* 6, no. 1 (1895): 10 (emphasis added).

9. Wayne Proudfoot, "Pragmatism and 'an Unseen Order,'" in *William James and a Science of Religions: Reexperiencing the Varieties of Religious Experience*, ed. Wayne Proudfoot (New York: Columbia University Press, 2004), 32–33.

10. James, "Is Life Worth Living?" 10, 16, 23.

11. Ibid., 23.

12. Ibid., 24 (emphasis in original). On the possibility of making oneself believe in the moral order of the universe (or in anything at all), see William James, "The Will to Believe," in *Pragmatism and Other Writings*, ed. Giles Gunn (New York: Penguin, 2000), 198–218.

13. On this transformation, see William James, *A Pluralistic Universe* (Lincoln: University of Nebraska Press, 1996), 196–98, 206–7.

14. Ibid., 30.

15. Ibid., 321.

16. Ibid., 325, 327.

17. Ibid., 325.

18. Martin Rees, "Cosmology and the Multiverse," in *Universe or Multiverse?* ed. Carr, 59.

19. Andrei Linde, "Inflationary Theory Versus Ekpyrotic/Cyclic Scenario: A Talk at Stephen Hawking's 60th Birthday Conference, Cambridge University, Jan. 2002," in *The Future of Theoretical Physics and Cosmology*, ed. G. W. Gibbons, E. P. S. Shellard, and S. J. Rankin (Cambridge: Cambridge University Press, 2003), 801–38, arXiv:hep-th/0205259; Paul J. Steinhardt, "The Inflation Debate," *Scientific American*, April 2011, 36–43.

20. Tegmark, "Parallel Universes."

21. This is especially the case when it comes to the "landscape" of string theory. Some theorists (for example, Andrei Linde and Leonard Susskind) argue that the landscape expresses the distribution of the inflationary multiverse. Others (Stephen Hawking and Thomas Hertog in particular) argue that it maps onto the Many-Worlds Interpretation of quantum mechanics. And others still (most notably, Laura Mersini-Houghton) argue that all three models can be reconciled with one another. See Linde, "Eternally Existing," 399; Susskind, *Cosmic Landscape*, 12; Stephen Hawking and Thomas Hertog, "Populating the Landscape: A Top-Down Approach," *Physical Review D* 73 (2006): 1–9; and Laura Mersini-Houghton, "Thoughts on Defining the Multiverse," April 27, 2008, available only through arXiv/0804.4280.

22. Greene calls the two configurations of this type the "Quilted" and "Inflationary" multiverses, respectively, in *Hidden Reality*, 10–71.

23. Examples of works on contemporary temporal or cyclic models include Lee Smolin, *The Life of the Cosmos* (New York: Oxford University Press, 1997); Baum and Frampton, "Turnaround in Cyclic Cosmology"; Steinhardt and Turok, *Endless Universe*; Martin Bojowald, *Once Before Time: A Whole Story of the Universe* (New York: Vintage, 2010); and Roger Penrose, *Cycles of Time: An Extraordinary New View of the Universe* (New York: Knopf, 2011).

24. David Deutsch, *The Fabric of Reality* (New York: Penguin, 1997); Colin Bruce, *Schrödinger's Rabbits: The Many-Worlds of Quantum* (Washington, D.C.: Joseph Henry, 2004).

25. R. A. Montgomery, *Space and Beyond*, Choose Your Own Adventure (Waitsfield, Vt.: Chooseco, 2006); Jeff Melman, dir., and Chris McKenna, writer, "Remedial Chaos Theory," *Community*, season 3, episode 4, NBC, October 13, 2011.

26. Tegmark, "Parallel Universes"; Max Tegmark, "The Multiverse Hierarchy," in *Universe or Multiverse?* ed. Carr, 99–125. Some direct and very recent philosophical precedents include Robert Nozick, *Philosophical Explanations* (Cambridge, Mass.: Harvard University Press, 1981); and David Lewis, *On the Plurality of Worlds* (Oxford: Oxford University Press, 1986).

27. Greene, *Hidden Reality*, 294; Paul Davies, *Cosmic Jackpot: Why Our Universe Is Just Right for Life* (New York: Houghton Mifflin, 2007), 210.

28. G. W. Leibniz, *Theodicy: Essays on the Goodness of God, the Freedom of Man, and the Origin of Evil*, trans. E. M. Huggard (Eugene, Ore.: Wipf and Stock, 2001), 377–88. Voltaire parodically represents this position through Dr. Pangloss in *Candide* (1759).

29. Tegmark, "Parallel Universes," 49.

30. Martin Rees, "Concluding Perspective," January 16, 2001, in *New Cosmological Data and the Values of the Fundamental Parameters: Proceedings of the 201st Symposium of the International Astronomical Union Held During the IAU General Assembly XXIV, the Victoria University of Manchester, United Kingdom, 7–11 August, 2000*, ed. Anthony Lasenby and Althea Wilkinson (San Francisco: Astronomical Society of the Pacific, 2005), 421, arXiv:astro-ph/0101268v1.

31. Hugh Everett, "'Relative State' Formulation of Quantum Mechanics," *Review of Modern Physics* 29 (1957): 454–62. Bryce DeWitt developed and popularized this thesis in "Quantum Mechanics and Reality," *Physics Today* 23, no. 9 (1970): 155–65, and Bryce DeWitt and Neill Graham, eds., *The Many-Worlds Interpretation of Quantum Mechanics* (Princeton, N.J.: Princeton University Press, 1973).

32. Vilenkin, "Birth of Inflationary Universes"; Linde, "Eternally Existing"; Andreas Albrecht and Paul J. Steinhardt, "Cosmology for Grand Unified Theories with Radiatively Induced Symmetry Breaking," *Physical Review Letters* 48 (1982): 1220–23; Alan H. Guth and Paul J. Steinhardt, "The Inflationary Universe," *Scientific American*, May 1984, 116–28.

33. Vilenkin, *Many Worlds in One*, 86.

34. Alan P. Lightman, "The Accidental Universe: Science's Crisis of Faith," *Harper's Magazine*, December 22, 2011.

35. Rees, "Cosmology and the Multiverse," 58.

36. Quoted in Tim Folger, "Science's Alternative to an Intelligent Creator: The Multiverse Theory" [interview with Andrei Linde], *Discover*, November 10, 2008.

37. The classic elaborations of these various fine-tunings can be found in Bernard Carr and Martin Rees, "The Antrophic Principle and the Structure of the Physical World," *Nature*, April 12, 1979, 605–12; Paul C. W. Davies, *The Accidental Universe* (Cambridge: Cambridge University Press, 1982); John D. Barrow and Frank Tipler, *The Anthropic Cosmological Principle* (Oxford: Clarendon Press, 1986); Martin Rees, *Just Six Numbers: The Deep Forces That Shape the Universe* (New York: Basic Books, 2000); John D. Barrow, *The Constants of Nature* (New York: Pantheon, 2002); and Bernard Carr, "The Anthropic Principle Revisited," in *Universe or Multiverse?* ed. Carr, 77–89.

38. Most experimental physicists, by contrast, are happy just to work with the constants they are given—as a common expression puts it, to "shut up and calculate." See David I. Kaiser, *How the Hippies Saved Physics: Science, Counterculture, and the Quantum Revival* (New York: Norton, 2012), 1–24.

39. "We see that things which lack intelligence, such as natural bodies, act for an end . . . so as to obtain the best result. Hence it is plain that not fortuitously, but designedly, do they achieve their end. Now whatever lacks intelligence cannot move towards an end, unless it be directed by some being endowed with knowledge and intelligence; as the arrow is shot to its mark by the archer. Therefore some intelligent being exists by whom all natural things are directed to their end; and this being we call God" (Thomas Aquinas, *Summa theologiae*, trans. Fathers of the English Dominican Province, 5 vols. [Allen,

Tex.: Christian Classics, 1981], 1.2.3). An earlier formulation of this argument can be found in Augustine of Hippo, *Concerning the City of God Against the Pagans*, trans. Henry Bettenson (New York: Penguin, 2003), 11.4.2.

40. David Hume, *Dialogues Concerning Natural Religion*, ed. Richard H. Popkin, 2nd ed. (Indianapolis: Hackett, 1998), 15 (subsequent references are cited in the text).

41. It is important to note that this God of natural religion is not the God of revealed religion, who is said to be "wholly other" from human beings and totally unknowable. Rather, the creator of natural religion is similar in kind to human beings but different in degree; just like human designers, the designer-God is intelligent, powerful, and benevolent, only *more* so. Hume half-heartedly represents the "revealed" viewpoint through the confused, fairly useless character of Demea, who claims that "intellect or understanding is not to be ascribed to the Deity, and that our most perfect worship of him consists, not in acts of veneration, reverence, gratitude, or love, but in a certain mysterious self-annihilation or total extinction of all our faculties." Demea's critique of natural religion is that "by representing the Deity as so intelligible and comprehensible, and so similar to a human mind, we are guilty of the grossest and most narrow partiality, and make ourselves the model of the whole universe" (ibid., 26–27).

42. This is David Hume's theory of causality as "constant conjunction," set forth in *An Enquiry Concerning Human Understanding*, ed. Eric Steinberg, 2nd ed. (Indianapolis: Hackett, 1993).

43. The resonance with the book of Job is striking: "Where were you when I laid the foundations of the earth?" God asks from out of the whirlwind. "Tell me, if you have understanding. Have the gates of death been revealed to you? Have you comprehended the expanse of the earth? Declare, if you know all this" (Job 38:4, 17–18, New Revised Standard Version [NRSV]).

44. For a careful roundup and scathing critique of such arguments, see Kenneth Surin, *Theology and the Problem of Evil* (Eugene, Ore.: Wipf and Stock, 1986).

45. William Paley, *Natural Theology; or, Evidences of the Existence and Attributes of the Deity. Collected from the Appearances of Nature* (London: Faulder, 1802), 1–2, 19.

46. Ibid., 68.

47. Ibid., 261, 64, 65, 67, 68, 352, 57, 579 (emphasis in original).

48. Ibid., 474.

49. Quoted in Lightman, "Accidental Universe," 2 (emphasis added).

50. Paul C. W. Davies, "Universes Galore: Where Will It All End?" in *Universe or Multiverse?* ed. Carr, 487.

51. Davies, *Cosmic Jackpot*, 15.

52. Brandon Carter, "Large Number Coincidences and the Anthropic Principle in Cosmology," in *Physical Cosmology and Philosophy*, ed. John Leslie and Paul Edwards (Cambridge: Cambridge University Press, 1990), 126.

53. On the distinction between strong and weak arguments, see ibid.; and Ernan Mc-Mullin, "Indifference Principle and Anthropic Principle in Cosmology," *Studies in History and Philosophy of Science* 24 (1993): 359–89. The philosopher Nick Bostrom counts thirty versions of the anthropic argument in all, in *Anthropic Bias: Observation Selection Effects in Science and Philosophy* (New York: Routledge, 2002). For a deeply critical assessment of the four major strands, see Martin Gardner, "WAP, SAP, PAP, & FAP," *New York Review of Books*, May 8, 1986, 22–25, in which Gardner concludes that WAP, SAP, PAP, and FAP amount to so much "CRAP."

54. William R. Stoeger, S.J., "Are Anthropic Arguments, Involving Multiverses and Beyond, Legitimate?" in *Universe or Multiverse?* ed. Carr, 446.

55. Susskind, *Cosmic Landscape*, 7. Susskind is just explaining the SAP in this passage; he is not affirming it.

56. Andrei Linde, "The Inflationary Multiverse," in *Universe or Multiverse?* ed. Carr, 129.

57. See, for example, Richard Swinburne, "Argument from the Fine-tuning of the Universe," in *Physical Cosmology and Philosophy*, ed. Leslie and Edwards, 154–73.

58. John Polkinghorne, *Beyond Science: The Wider Human Context* (Cambridge: Cambridge University Press, 1996), 92.

59. William Lane Craig, "Barrow and Tipler on the Anthropic Principle vs. Divine Design," *British Journal for the Philosophy of Science* 38 (1988): 393. See also William Lane Craig, "Design and the Anthropic Fine-tuning of the Universe," in *God and Design: The Teleological Argument and Modern Science*, ed. Neil A. Manson (New York: Routledge, 2003), 155; and Barrow and Tipler, *Anthropic Cosmological Principle*.

60. Brandon Carter, "Anthropic Principle in Cosmology," June 27, 2006, in *Current Issues in Cosmology*, ed. Jean-Claude Pecker and Jayant Narlikar (Cambridge: Cambridge University Press, 2011), 174, arXiv:gr-qc/0606117v1. For other theological appropriations of the anthropic principle, see Michael J. Denton, *Nature's Destiny* (New York: Free Press, 1998); Robin Collins, "God, Design, and Fine-tuning," in *God Matters: Readings in the Philosophy of Religion*, ed. Raymond Martin and Christopher Bernard (New York: Longman, Pearson, 2002), 119–35; Stephen Barr, *Modern Physics and Ancient Faith* (Notre Dame, Ind.: Notre Dame University Press, 2003); Guillermo Gonzalez and Jay W. Richards, *The Privileged Planet* (Washington, D.C.: Regnery, 2004); and Hugh Ross, *Why the Universe Is the Way It Is* (Grand Rapids, Mich.: Baker, 2008).

61. Swinburne, "Argument from the Fine-tuning of the Universe," 165.

62. It is, however, the preferred "answer" among the majority of working physicists, who tend to refuse the question in the first place. See Davies, *Cosmic Jackpot*, 261.

63. Saul Perlmutter, Brian Schmidt, and Adam Reiss were awarded the Nobel Prize in Physics for this discovery in 2011. See http://www.nobelprize.org/nobel_prizes/physics/laureates/2011/press.html/ (accessed May 29, 2013).

64. Carroll, *From Eternity to Here*, 59. Now that the Higgs boson (or a "Higgslike particle") has been found by the Large Hadron Collider in Geneva, "empty" space is also said to be filled with the Higgs field, which gives elementary particles their mass. See ATLAS Collaboration, "Observation of a New Particle in the Search for the Standard Model Higgs Boson with the ATLAS Detector at the LHC," *Physics Letters B* 716 (2012): 1–29, http://www.sciencedirect.com/science/article/pii/S037026931200857X (accessed September 11, 2013).

65. High Energy Physics Advisory Panel, "Quantum Universe: The Revolution in 21st Century Particle Physics," October 22, 2003, http://www.interactions.org/pdf/Quantum_Universe.pdf (accessed May 29, 2013). For two reader-friendly explanations of this calculation, see Barrow, *Book of Universes*, 291; and Carroll, *From Eternity to Here*, 20.

66. Quoted in Michael D. Lemonick, "The End," *Time*, June 25, 2001; Carroll, *From Eternity to Here*, 20; Lee Smolin, *The Trouble with Physics: The Rise of String Theory, the Fall of a Science, and What Comes Next* (New York: First Mariner, 2007), 153.

67. Einstein had introduced lambda in his gravitational equations in order to keep the cosmos static, but he revoked it when Edwin Hubble discovered the universe was expanding (see chap. 5, sec. "Let There Be Light").

68. Lightman, "Accidental Universe," 5.

69. This is an overstatement, but only a slight one: "[A]t present, the anthropic upper limit on the vacuum energy is larger than the present mass density, but not many orders of magnitude greater" (Steven Weinberg, "Living in the Multiverse," in *Universe or Multiverse?* ed. Carr, 32).

70. Susskind, *Cosmic Landscape*, 83.

71. Ibid., x, 22.

72. "An anthropic explanation of the value of ρv [the energy density of the cosmological constant] makes sense if and only if there is a very large number of big bangs, with different values for v" (Steven Weinberg, "The Cosmological Constant Problems" [lecture given at the Dark Matter 2000 Conference, Marina del Ray, Calif., February 22–24, 2000], arXiv:astro-ph/0005265v1). Weinberg's paper built on an argument he had set forth more than a decade earlier in "Anthropic Bound on the Cosmological Constant," *Physical Review Letters* 59 (1987): 2607–10. According to string theorist Brian Wecht, "[M]any people cite [this earlier] paper . . . as the first compelling evidence they'd seen about there being some merit to the anthropic principle" (personal communication with the author, August 21, 2012).

73. For the initial unpopularity of these accounts as well as their sudden fame after the discovery of dark energy, see Alex Vilenkin and Jaume Garriga, "Many Worlds in One," *Physical Review D* 64 (2001): 1–5. For other accounts of the connection between dark energy and the revival of eternal inflation, see Gribbin, *In Search of the Multiverse*, 135; Mersini-Houghton, "Thoughts on Defining the Multiverse"; and Greene, *Hidden Reality*, 7.

74. The foundational paper in this regard is Raphael Bousso and Joseph Polchinski, "Quantization of Four-Form Fluxes and Dynamical Neutralization of the Cosmological Constant," *Journal of High Energy Physics*, no. 6 (2000): 1–25. Developments can be found in Leonard Susskind, "The Anthropic Landscape of String Theory," in *Universe or Multiverse*, ed. Carr, 247–66, arXiv:hep-th/0302219v1; and Sujay K. Ashok and Michael R. Douglas, "Counting String Vacua," *Journal of High Energy Physics*, no. 1 (2004): 1–35. Some theorists have argued that the number of vacua is actually infinite. See Jessie Shelton, Washington Taylor, and Brian Wecht, "Generalized Flux Vacua," *Journal of High Energy Physics*, no. 2 (2007): 1–27, arXiv:hep-th/0607015v2.

75. Although most theorists agree that the sources of the modern turn to the multiverse are, on the one hand, developments within cosmology and particle physics and, on the other, philosophical expedience, they disagree as to which came first. Bernard Carr, for example, argues that "these multiverse proposals have not generally been motivated by an attempt to explain the anthropic fine-tunings; most of them have arisen independently out of developments in cosmology and particle physics" ("Introduction and Overview," in *Universe or Multiverse?* ed. Carr, 4; compare Sean M. Carroll, "Does This Ontological Commitment Make Me Look Fat?" *Discover*, June 4, 2012, http://blogs.discovermagazine .com/cosmicvariance/2012/06/04/does-this-ontological-commitment-make-me-look-fat / [accessed May 29, 2013]). Martin Rees, by contrast, has written that the multiverse hypothesis "was *originally* just a conjecture, motivated by a wish to explain the apparent fine-tuning in our universe—and incidentally a way to undercut the so-called theological design argument, which said that there was something special about these laws" ("In the Matrix," *Edge*, May 19, 2003 [emphasis added]; compare Martin Gardner, *Are Universes Thicker Than Blackberries? Discourses on Gödel, Magic Hexagrams, Little Red Riding Hood, and Other Mathematical and Pseudoscientific Topics* [New York: Norton, 2003], 6).

In the face of this conflict, it is probably safest to say with Davies that the philosophical commitments condition the models in the first place and vice versa: "[A]ll cosmological models are constructed by augmenting the results of observations by some sort of philosophical principle" ("Universes Galore," 487).

76. Quoted in George Brumfiel, "Outrageous Fortune," *Nature*, January 5, 2006, 10.

77. Hawking and Mlodinow, *Grand Design*, 165.

78. For some of the most dramatic expressions of what I would call "cosmological antitheism," see Susskind, *Cosmic Landscape*; Hawking and Mlodinow, *Grand Design*; and Lawrence Krauss, *A Universe from Nothing: Why There Is Something Rather Than Nothing* (New York: Free Press, 2012).

79. Quoted in Folger, "Science's Alternative."

80. See, for example, Don N. Page, "Does God So Love the Multiverse?" in *Science and Religion in Dialogue*, ed. Melville Y. Stewart (Malden, Mass.: Wiley-Blackwell, 2010), 1:380–95, arXiv/0801.0246; and Jeffrey A. Zweernick, *Who's Afraid of the Multiverse?* (Pasadena, Calif.: Reasons to Believe, 2008).

81. Homer, *The Iliad*, trans. Richmond Lattimore (Chicago: University of Chicago Press, 1961), 12.225, and *The Odyssey*, trans. Richmond Lattimore (New York: Harper, 2007), 13.77.

82. Edward Adams, "Graeco-Roman and Ancient Jewish Cosmology," in *Cosmology and New Testament Theology*, ed. Jonathan T. Pennington and Sean M. McDonald (New York: Clark, 2008), 6.

83. Ibid.

84. David J. Furley, *The Greek Cosmologists: The Formation of the Atomic Theory and Its Earliest Critics* (Cambridge: Cambridge University Press, 1987), 58; Diogenes Laertius, *Lives of Eminent Philosophers*, trans. R. D. Hicks, vol. 2, Loeb Classical Library 185 (Cambridge, Mass.: Harvard University Press, 1942), 8.1.48.

85. Adams, "Graeco-Roman and Ancient Jewish Cosmology," 6.

86. What I am calling "the multiple" is therefore akin to what James calls "the plural." It is even closer to what Jean-Luc Nancy calls "singular plurality"—an equiprimordiality of the many and the one by virtue of which existents are irreducibly "with" one another. "Multiplicity" signals, furthermore, that this "withness" not only constitutes the multiple but deconstitutes it as well; as Judith Butler has shown in her work on vulnerability, the very relations that compose a self or city or world also threaten to undo it. (It should be said that in her most recent monograph on Israel–Palestine, Butler calls this interdetermination "plurality," rather than multiplicity.) And, perhaps most noticeably, my use of this term conjures Gilles Deleuze's understanding of multiplicity not as unmediable difference, but as "the affirmation of unity." There is an order to multiplicity's disorder, an irreducible web of relations that distinguishes multiplicity from sheer plurality. See Jean-Luc Nancy, *Being Singular Plural*, trans. Robert D. Richardson and Anne E. O'Byrne (Stanford, Calif.: Stanford University Press, 2000); Mary-Jane Rubenstein, *Strange Wonder: The Closure of Metaphysics and the Opening of Awe* (New York: Columbia University Press, 2009), 99–131, and "Undone by Each Other: Interrupted Sovereignty in Augustine's *Confessions*," in *Polydoxy*, ed. Catherine Keller and Laurel Schneider (New York: Routledge, 2010), 105–25; Judith Butler, *Precarious Life: The Powers of Mourning and Violence* (New York: Verso, 2004), and *Parting Ways: Jewishness and the Critique of Zionism* (New York: Columbia University Press, 2012); and Gilles Deleuze, *Nietzsche and Philosophy*, trans. Hugh Tomlinson (New York: Columbia University Press, 1983), 23–24.

87. Quoted in both Pierre Duhem, *Medieval Cosmology: Theories of Infinity, Place, Time, Void, and the Plurality of Worlds*, trans. Roger Ariew (Chicago: University of Chicago Press, 1985), 450; and Steven J. Dick, *Plurality of Worlds: The Origins of the Extraterrestrial Life Debate from Democritus to Kant* (Cambridge: Cambridge University Press, 1982), 28.

1. A Single, Complete Whole

1. Hesiod, *Theogony*, in *Theogony, Works and Days, Theognis, Elegies* (New York: Penguin, 1973), 27.

2. The Ionians were "hylozoists," meaning that they believed "matter as such has the property of life and growth" (David J. Furley, *The Greek Cosmologists: The Formation of the Atomic Theory and Its Earliest Critics* [Cambridge: Cambridge University Press, 1987], 18).

3. "[Thales's] doctrine was that water (*hudor*) is the universal primary substance, and that the world is animate and full of divinities" (Diogenes Laertius, *Lives of Eminent Philosophers*, trans. R. D. Hicks, vol. 1, Loeb Classical Library 184 [Cambridge, Mass.: Harvard University Press, 1942], 1.1.27).

4. "[Anaximenes] took for his first principle air (*aera*) or that which is unlimited" (ibid., 2.2.3).

5. "All things are composed of fire, and into fire they are again resolved" (Diogenes Laertius, *Lives of Eminent Philosophers*, trans. R. D. Hicks, vol. 2, Loeb Classical Library 185 [Cambridge, Mass.: Harvard University Press, 1942], 9.1.7). Although fire is the first of the four elements for Heraclitus, it is not the underlying substance that his predecessors considered it. In fact, Heraclitus rejects the notion of "substance" beneath change *tout court*. See Edward Adams, "Graeco-Roman and Ancient Jewish Cosmology," in *Cosmology and New Testament Theology*, ed. Jonathan T. Pennington and Sean M. McDonald (New York: Clark, 2008), 10; and Furley, *Greek Cosmologists*, 34–36.

6. "[Aniximander] laid down as his principle and element that which is unlimited (*apeiron*) without defining it as air or water or anything else" (Diogenes Laertius, *Lives*, 1:2.1.1). Aristotle's rundown of the pre-Socratics is as follows:

Thales, the introducer of this sort of philosophy, said that [the principle of all things] was water . . . perhaps drawing this supposition from seeing that the nourishment of all creatures is moist. . . . Anaximenes, however, assumed that air was prior to water . . . and Diogenes [of Apollonia] thought the same, while Hippasus of Metapontum and Heraclitus of Ephesus thought it was fire. Empedocles thought that there were four elements, adding to those mentioned earth as a fourth. . . . Anaxagoras of Clazomenae was earlier than he in date but later in his works, and he said that the number of principles was infinite. (*The Metaphysics*, trans. Hugh Lawson-Tancred [New York: Penguin, 1998], 983b–84a)

7. Diogenes Laertius, *Lives*, 2:8.2.76. See also Friedrich Solmsen, "Love and Strife in Empedocles' Cosmology," *Phronesis* 10, no. 2 (1965): 109–48.

8. "[Democritus's] opinions are these. The first principles of the universe are atoms and empty space. . . . The worlds are unlimited; they come into being and perish. . . . Further, the atoms are unlimited in size and number, and they are borne along in the whole universe in a vortex, and thereby generate all composite things—fire, water, air,

earth; for even these are conglomerations of given atoms" (Diogenes Laertius, *Lives*, 2:9.7.44). See also Edward Adams, *The Stars Will Fall from Heaven: Cosmic Catastrophes in the New Testament and Its World* (New York: Clark, 2007), 106–7.

9. According to Aristocles, "The element of all things is fire. . . . At certain fated times the whole universe will be converted into fire; next, it is again made into an ordered universe. The primal fire (*proton pyr*) is so to speak a kind of seed, containing the *logoi* of all things that have become, do become, and will become" (quoted in J. Mansfield, "Providence and the Destruction of the Universe in Early Stoic Thought: With Some Remarks on the 'Mysteries of Philosophy,'" in *Studies in Hellenistic Religion*, ed. M. J. Vermaseren [Leiden: Brill, 1979], 145).

10. Diogenes Laertius is careful to point out that "Plato, who mentions almost all the early philosophers, never once alludes to Democritus, not even where it would be necessary to controvert him," and that Plato is said to have wanted to burn all of Democritus's books (*Lives*, 2:9.7.40). The analysis in this chapter makes clear just how threatening Atomist cosmology was to the singularity and imperishability of the Platonic universe.

11. Plato, *Timaeus*, in *Timaeus and Critias*, trans. Desmond Lee (New York: Penguin, 1977), 30a (subsequent references are cited in the text). Although I use the word *cosmos* (*kosmos*) more than *universe* (*to pan* [literally, "the all"]) in this reading of the *Timaeus*, Plato uses these terms interchangeably, along with *sky/heavens* (*ouranos*). See John Sallis, *Chorology: On Beginning in Plato's "Timaeus*," Studies in Continental Thought (Bloomington: Indiana University Press, 1999), 53.

12. This is not to say that creation stories did not exist elsewhere. It is common to creation myths of the Near East, India, and the Far East to see a (usually male) "god of order" overcome a (usually female) principle of chaos. That said, these gods tend to come *from* the chaos in the first place, whereas Timaeus's demiurge exists outside it. See Richard Clifford, *Creation Accounts in the Ancient Near East and in the Bible* (Washington, D.C.: Catholic Biblical Association of America, 1994); Mary K. Wakeman, *God's Battle with the Monster* (Leiden: Brill, 1973); and Paul Ricoeur, *The Symbolism of Evil* (New York: Harper & Row, 1967), 175–210.

13. Francis M. Cornford reminds us that this god is not the omnipotent creator of the Abrahamic traditions—at least in their orthodox configurations. This god does not create the material he uses but rather works within the constraints of necessity. As such, he is not omnipotent, nor does Timaeus suggest that he is an object of worship. Moreover, this demiurge has very little personality, suggesting to Cornford that he is not so much a god as a symbol of reason itself, "working for ends that are good. The whole purpose of the *Timaeus* is to teach men to regard the universe as revealing the operation of such a Reason, not the fortuitous outcome of blind and aimless bodily motions" (*Plato's Cosmology: The "Timaeus" of Plato*, trans. Francis MacDonald Cornford, Library of Liberal Arts [New York: Bobbs-Merrill, 1957], 38; compare 15, 48).

14. Plato, *Republic*, trans. G. M. A. Grube, ed. C. D. C. Reeve (Indianapolis: Hackett, 1992), 540a–b.

15. This assurance should be contrasted with a passage in Plato's *Statesman* in which the alternating cosmic periods of order and chaos are said to be beyond the control of the god. See Plato, *Statesman*, ed. Julia Annas and Robin Waterfield, trans. Robin Waterfield, Cambridge Texts in the History of Political Thought (Cambridge: Cambridge University Press, 1995), 269c.

16. Plato, *Plato's Cosmology*, 43.

17. Timaeus also refers to his story as an *eikos logos*, seemingly flouting the mythos–logos distinction on which Critias insists at the beginning of the dialogue. On this puzzling conflation, see T. K. Johansen, *Plato's Natural Philosophy: A Study of the "Timaeus–Critias"* (Cambridge: Cambridge University Press, 2004), 50–68.

18. In relation to the sudden appearance of a more primordial beginning (*khôra*) in the middle of the text, Sallis argues that the *Timaeus* is by necessity "a badly told story, one that violates the very injunction it issues about how to begin" (*Chorology*, 5).

19. Michel Serres, *Genesis*, trans. Genevieve James and James Nielson (Ann Arbor: University of Michigan Press, 1995), 112 (ellipses in original), 133.

20. Ibid., 111.

21. Adams, "Graeco-Roman and Ancient Jewish Cosmology," 13.

22. Gerhard May, *Creatio ex Nihilo: The Doctrine of "Creation out of Nothing" in Early Christian Thought*, trans. A. S. Worrall (Edinburgh: Clark, 1994), 4, 15–16; Plato, *Plato's Cosmology*, 37.

23. These groups are represented by Philo of Alexandria, Plutarch, and Atticus, respectively. See David Sedley, *Creationism and Its Critics in Antiquity* (Berkeley: University of California Press, 2007), 133.

24. Plato, *Plato's Cosmology*, 203.

25. Following James Joyce, Catherine Keller has called this particular interdetermination "chaosmos" (*Face of the Deep: A Theology of Becoming* [New York: Routledge, 2003], 12).

26. Plato, *Plato's Cosmology*, 37.

27. Although I do not engage their landmark studies directly here, my own reading of this part of the *Timaeus* has been influenced in countless subtle ways by the groundbreaking works of Jacques Derrida, "Khôra," in *On the Name*, ed. Thomas Dutoit (Stanford, Calif.: Stanford University Press, 1995), 89–130; and Sallis, *Chorology*.

28. For Luce Irigaray, excluding the women they rely on is what philosophy, politics, language, and culture do best: women provide the raw materials for the systems that exclude them and can at best be said to be "envelopes" or "receptacles" in which great men became great—their "point of departure" ("Sexual Difference," in *An Ethics of Sexual Difference*, trans. Carolyn Burke and Gillian C. Gill [Ithaca, N.Y.: Cornell University Press, 1993], 5–19, and "Plato's Hystera," in *Speculum of the Other Woman*, trans. Gillian C. Gill [Ithaca, N.Y.: Cornell University Press, 1985], 243–365).

29. In line with Sallis's critiques of the usual translations of χωρα, I have followed his lead in leaving the term in the (transliterated) Greek, *khôra*. The difficulty, as he explains it, is that

> if, following Cornford and A. E. Taylor, one proposed to translate χωρα as *space*, then one would have to set about immediately withdrawing from the word much that we cannot but hear in it. For clearly the χωρα is not the isotropic space of post-Cartesian physics. Nor is it even empty space, the void, as discussed in Greek atomism, for this is called το κενου and is in fact discussed as such later in the *Timaeus* (58b). It would hardly be otherwise if one were to translate χωρα as *place* . . . for one would then have conflated the difference between χωρα and τοπος and would risk assimilating Plato's chorology to the topology of Aristotle's *Physics*. (*Chorology*, 115)

It remains to be seen whether *khôra* might be construed more productively in relation to Leibniz's or Einstein's conceptions of space(-time), according to which bodies shape the space that sets them in motion.

30. Here I retain Timaeus's use of the past tense "for convenience," bearing in mind the strong possibility that the tale is a linear mapping of an eternal process.

31. This view of precosmic forces constitutes a remarkable difference from Aristotle, who assigns different regions of the cosmos to the five elements (fire, water, earth, air, and aether). See Aristotle, *On the Heavens* (*De caelo*), trans. J. L. Stocks, in *The Complete Works of Aristotle: The Revised Oxford Translation*, ed. Jonathan Barnes, vol. 1 (Princeton, N.J.: Princeton University Press, 1971); and Friedrich Solmsen, *Aristotle's System of the Physical World: A Comparison with His Predecessors* (Ithaca, N.Y.: Cornell University Press, 1960).

32. Timaeus illustrates this interdependence by proposing we begin with water, "solidifying into stones and earth, and again dissolving and evaporating into wind and air; air by combustion becomes fire, and fire in turn when extinguished and condensed takes the form of air again" (Plato, *Timaeus*, 49c).

33. This cosmic mixing stands in stark tension with Socrates's, Critias's, and presumably Timaeus's insistence that the perfect city is one in which the various classes are kept totally separate (ibid., 24a).

34. In his primordial scene of unrelated differences, Timaeus describes the realm of the Forms as "admitting no modification and entering no combination" (ibid., 52a). In mixing the world, the demiurge therefore brings into combination that which enters no combination.

35. Serres, *Genesis*, 112, 33.

36. Plato does not mention the Atomists by name here—or, indeed, in any of his works—but Diogenes Laertius tells us that "Aristoxenus in his *Historical Notes* affirms that Plato wished to burn all the writings of Democritus that he could collect" (*Lives*, 2:9.7.40). Whether this story is true or not, it gives voice to the profound antinomy between Platonists and Atomists.

37. Aristotle, *On the Heavens*, 277b28 (subsequent references are cited in the text).

38. In fact, Aristotle begins to speak of these qualities as if one necessarily entailed the other, resolving to demonstrate "that the heaven as a whole neither came into being nor admits of destruction . . . but is *one* and *eternal*" (ibid., 283b27–28).

39. Aristotle offers two proofs in the *De caelo*, one based on natural motion and one based on "natural place" (ibid., 278a26–79a6). I treat only the former here, partly because the two proofs are so similar, but mainly because the proof from natural motion is the one to which subsequent thinkers primarily respond. For a very capable account of the argument from natural place, see Dana Miller, "Plutarch's Argument for a Plurality of Worlds in *De defectu oraculorum* 424c10–425e7," *Ancient Philosophy* 17, no. 2 (1997): 377n.7.

40. Aristotle, *Physics*, trans. R. P. Hardie and R. K. Gaye, in *The Complete Works of Aristotle*, ed. Jonathan Barnes (Princeton, N.J.: Princeton University Press, 1984), 1:255b14–17; Aristotle, *On the Heavens*, 269a31–35, 308b12–15.

41. "The shape of the heaven is of necessity spherical; for that is the shape most appropriate to its substance and also by nature primary" (Aristotle, *On the Heavens*, 286b10). By "primary," Aristotle means that, unlike all other shapes, a sphere is indivisible. He goes on to add the evidence that the heavens must move in the "swiftest" and "shortest"

path possible, which is to say in circular paths: "Therefore, if the heaven moves in a circle and moves more swiftly than anything else, it must necessarily be spherical" (ibid., 287a28–29).

42. As Miller points out, "This argument requires that one accept the claim—which Aristotle omits . . . that given a plurality of center places, bodies within a world will move relative to *any* center place." Against this claim, Plutarch argued that the notion of "center" is relative to a particular world, so elements from one world would move only relative to its own center, and there is no obstacle to a plurality of worlds ("Plutarch's Argument for a Plurality of Worlds," 380).

43. Aristotle, *Metaphysics*, trans. Lawson-Tancred, 1072a (subsequent references to this translation are cited in the text).

44. Aristotle uses the terms *planeton* and *astron* more or less interchangeably.

45. Hugh Lawson-Tancred, introduction to the text of Book Lambda, in Aristotle, *Metaphysics*, trans. Lawson-Tancred, 376.

46. Aristotle, *Metaphysics*, trans. Hugh Tredennick, Loeb Classical Library 271 (Cambridge, Mass.: Harvard University Press, 1961), 1074a–b.

47. Western philosophy would have to wait to hear such an argument until the early fifteenth century, when Nicholas of Cusa held the singularity of God together with a plurality of worlds by means of the Trinity, a singular-plurality. Cusa posited the earth as just another star and stipulated that each star was in motion (see chap. 3, sec. "End Without End"): "The center of the world, therefore, coincides with the circumference. And therefore, the world has no circumference" (*On Learned Ignorance*, in *Selected Spiritual Writings*, trans. H. Lawrence Bond, Classics of Western Spirituality [New York: Paulist Press, 1997], 158; see also Steven J. Dick, *Plurality of Worlds: The Origins of the Extraterrestrial Life Debate from Democritus to Kant* [Cambridge: Cambridge University Press, 1982], 40–43).

48. Lawson-Tancred, introduction to the text of Book Lambda in Aristotle, *Metaphysics*, trans. Lawson-Tancred, 376. For one such "compatibilist" attempt, see G. E. R. Lloyd, "Metaphysics Lambda 8," in *Aristotle's "Metaphysics" Lambda: Symposium Aristotelicum*, ed. Michael Frede and David Charles (Oxford: Clarendon Press, 2000), 245–73.

49. Serres, *Genesis*, 133.

50. Aristarchus of Samos had posited a heliocentric model around 320 B.C.E., but his theory, in Adams's words, "was almost universally rejected" ("Graeco-Roman and Ancient Jewish Cosmology," 7).

51. Nicolaus Copernicus, *On the Revolutions of the Heavenly Spheres*, trans. A. M. Duncan (New York: Barnes and Noble Books, 1976), 46–51. On the extent to which Copernicus stops short of the "infinite universe," see Alexander Koyre, *From the Closed World to the Infinite Universe* (Baltimore: Johns Hopkins University Press, 1957), 25.

52. Johannes Kepler, *Kepler's Conversation with Galileo's Sidereal Messenger*, ed. Edward Rosen (New York: Johnson Reprint, 1965), 35–36. See also chap. 4, sec. "From Infinity to Pluralism."

53. On this development, see Edward R. Harrison, "Newton and the Infinite Universe," *Physics Today* 39, no. 2 (1986): 24–32. See also chap. 4, sec. "From Infinity to Pluralism."

54. Plato, *Timaeus*, 31a.

2. Ancient Openings of Multiplicity

1. Although Democritus seems to have been "one of the most prolific of all ancient authors," none of his works survives intact. G. S. Kirk and J. E. Raven tell us that of the fragments that have been preserved, "nearly all [are] taken from the ethical works" rather than from his cosmological treatises (*The Presocratic Philosophers: A Critical History with a Selection of Texts* [Cambridge: Cambridge University Press, 1957], 404). The exceptions are three letters preserved in book 10 of Diogenes Laertius's *Lives of Eminent Philosophers*. For the most part, however, the teachings of the Atomists must be reconstructed by means of treatises written *against* them by Aristotle and his followers. As David Furley explains it, the catastrophic loss of every Atomist text was not exactly accidental: "In the era before printing, copies were made very selectively, and the learned world vastly preferred the 'Closed World' cosmology of Plato and Aristotle. This was especially so in the centuries when learning was concentrated among Christians, but in Simplicius' neglect or ignorance of Democritus' writings we can see clear evidence that the judgment against Atomism was made independently of Christianity" (*The Greek Cosmologists: The Formation of the Atomic Theory and Its Earliest Critics* [Cambridge: Cambridge University Press, 1987], 116–17; compare 15, 3).

2. Furley tells us that the only pre-Democritean use of the term *atomos* "is in Sophocles' *Trachiniae* 200, where it is used of grass and means 'unmown'" (*Greek Cosmologists*, 123). Whereas Leucippus seems to have thought atoms were indivisible because they were too small to be reduced in size, Democritus probably held that an atom's indivisibility came from its having "no void and no interstices" (Kirk and Raven, *Presocratic Philosophers*, 408).

3. Diogenes Laertius, *Lives of Eminent Philosophers*, trans. R. D. Hicks, vol. 2, Loeb Classical Library 185 (Cambridge, Mass.: Harvard University Press, 1942), 9.6.31 (translation altered slightly). Aristotle describes this formation of the cosmic vortex as a "spontaneous" occurrence, happening by "chance" (*Physics*, trans. R. P. Hardie and R. K. Gaye, in *The Complete Works of Aristotle*, ed. Jonathan Barnes [Princeton, N.J.: Princeton University Press, 1984], 1:196a26).

4. Diogenes Laertius, *Lives*, 9.6.31.

5. Ibid., 9.6.30. David Furley argues that this *hymen* (in other contexts the word for "the caul that surrounds the embryo") is an indication that the Atomist cosmos is not mechanistically but, rather, organically construed: "Cosmogony is the story of a birth, even for an Atomist" (*Cosmic Problems: Essays on Greek and Roman Philosophy of Nature* [Cambridge: Cambridge University Press, 1989], 230).

6. Diogenes Laertius, *Lives*, 9.6.31–33. It is important to emphasize that it is "the all" that is unlimited, not the cosmos. Furley explains that "no one in classical antiquity believed that the *world* is infinite. The controversy was not about the *existence* of a closed world, but about its status: [I]s that all there is, or is there something else too?" (*Cosmic Problems*, 2, emphasis in original).

7. "Apeirous te einai kosmous" (Diogenes Laertius, *Lives*, 9.7.44).

8. Kirk and Raven, *Presocratic Philosophers*, 412. By "the first," Kirk and Raven mean the first among the Greek philosophers; although the theory of multiple worlds has occasionally been attributed to Anadimander, Anaxagoras, or Anaximenes, the evidence is inconclusive. See James Warren, "Ancient Atomists and the Plurality of Worlds," *Classical Quarterly* 54, no. 2 (2004): 354n.2; and Furley, *Cosmic Problems*, 4, and *Greek Cosmolo-*

gists, 29–30, 70–71. Kirk and Raven do not take into account Atomism's similarity to pre-Mahayana Buddhist cosmologies, many of which set forth atomic theories of the universe and posit *vast quantities* of worlds. That having been said, the Buddhist philosopher Akira Sadakata has suggested that there is a "strong possibility that Greek philosophy influenced Indian atomic theories" and that although the Buddhists were most likely "ahead" of the Greeks in terms of elemental theories, the Greeks seem to have been the first to set forth a theory of Atomism (*Buddhist Cosmology: Philosophy and Origins* [Tokyo: Kosei, 2004], 20–23).

9. "Plato, who mentions almost all the early philosophers, never once alludes to Democritus, not even where it would be necessary to controvert him, obviously because he knew that he would have to match himself against the prince of philosophers, for whom, to be sure, Timon has this meed of praise: 'Such is the wise Democritus, the guardian of discourse, keen-witted disputant, among the best I ever read'" (Diogenes Laertius, *Lives*, 9.7.40).

10. For one such reference to Lucretius and Democritus, see Aristotle, *On the Heavens* (*De caelo*), trans. J. L. Stocks, in *The Complete Works of Aristotle: The Revised Oxford Translation*, ed. Jonathan Barnes (Princeton, N.J.: Princeton University Press, 1971), 1:275b30. On Aristotle's refutation of the existence of a void, see Furley, *Cosmic Problems*, 77–90, and *Greek Cosmologists*, 189–93. On his refutation of the existence of an actual infinity and on that argument's reception, see Furley, *Cosmic Problems*, 103–14; and Pierre Duhem, *Medieval Cosmology: Theories of Infinity, Place, Time, Void, and the Plurality of Worlds*, trans. Roger Ariew (Chicago: University of Chicago Press, 1985), 3–138.

11. Plato, *Timaeus*, in *Timaeus and Critias*, trans. Desmond Lee (New York: Penguin, 1977), 31a; Aristotle, *On the Heavens*, 280a24–28.

12. Quoted in Furley, *Greek Cosmologists*, 140.

13. In his highly unsympathetic portrayal of Epicurean philosophy, Cicero's Cotta suggests that Epicurus's claim to be self-taught is no surprise: "It is like the owner of a jerry-built house boasting that he has not employed an architect" (Cicero, *On the Nature of the Gods*, trans. Horace C. P. McGregor [New York: Penguin, 1972], 1.25.72). On the likely teachers of Epicurus, see Furley, *Cosmic Problems*, 77; and Howard Jones, *The Epicurean Tradition* (New York: Routledge, 1989), 15–18.

14. David Sedley, *Creationism and Its Critics in Antiquity* (Berkeley: University of California Press, 2007), 133, 40.

15. Lucretius, *De rerum natura*, trans. W. H. D. Rouse, rev. Martin F. Smith, Loeb Classical Library 181 (Cambridge, Mass.: Harvard University Press, 1975), 5.186–94.

16. In *Letter to Herodotus*, Epicurus writes, "There is an infinite number of worlds, some like this world, others unlike it" (quoted in Diogenes Laertius, *Lives*, 10.45).

17. In the *Academica II*, Cicero writes that Democritus "claims that there are infinitely many worlds, some of them not only similar to each other but in every respect so utterly alike that there is no difference whatsoever between them, and indeed that there are infinitely many of *those*, and likewise the people in them" (quoted in Sedley, *Creationism and Its Critics*, 137, emphasis added in Sedley).

18. Aristotle, *On the Heavens*, 303a3–b4.

19. According to Leucippus and Democritus, Aristotle tells us, "[T]he primary masses . . . are infinite in number . . . and there is an infinity of shapes" (ibid., 303a3–11). Yet this does not seem to be what Democritus actually taught. In *Letter to Herodotus*, Epicurus wrote, "[T]he atoms . . . vary indefinitely in their shapes . . . but the variety of

shapes, though indefinitely large, *is not absolutely infinite*" (quoted in Diogenes Laertius, *Lives*, 10.42, emphasis added; see also 10.55).

20. Diogenes Laertius, *Lives*, 10.42–43. For an account of the difference between Democritus and Epicurus on this matter, see Sedley, *Creationism and Its Critics*, 160–61.

21. Lucretius, *De rerum natura*, 2.1130. See also Friedrich Solmsen, "Epicurus on the Growth and Decline of the Cosmos," *American Journal of Philology* 74, no. 1 (1953): 42–44.

22. Solmsen, "Epicurus on the Growth and Decline of the Cosmos," 34.

23. On Lucretius's stance against the Platonists, see Sedley, *Creationism and Its Critics*, 139–55. On his positions against Aristotle, see Furley, *Cosmic Problems*, 223–35.

24. Lucretius, *De rerum natura*, 1.63–79 (translation altered slightly; subsequent references are cited in the text).

25. "Extra processit longe flammantia moenia mundi atque omne immensum peragravit." Here I am using Furley's translation of "longe flammantia moenia mundi" (*Cosmic Problems*, 2).

26. Throughout the text, Lucretius insists that that the "first beginnings" had nothing to do with "design," show no sign of "intelligence," and took place without "the help of the gods" (*De rerum natura*, 1.1021–23, 2.1093, 5.420–22).

27. Lucretius, *The Nature of Things*, trans. A. E. Stallings (New York: Penguin, 2007), 2.181.

28. Friedrich Solmsen, "Epicurus and Cosmological Heresies," *American Journal of Philology* 72, no. 1 (1951): 14.

29. David Hume, *Dialogues Concerning Natural Religion*, ed. Richard H. Popkin, 2nd ed. (Indianapolis: Hackett, 1998), 59.

30. Lucretius, *Nature of Things*, 1.968–69. This illustration can be traced back to a fourth-century Pythagorean named Archytas, whom Simplicius records as having asked, "If I were at the edge of the world, as it might be in the region of the fixed stars, could I stretch out my hand or a stick into the outer region or not?" (quoted in Furley, *Cosmic Problems*, 7). It also relies on Epicurus's argument in *Letter to Herodotus*: "Again, the sum of things is infinite. For what is finite has an extremity, and the extremity of anything is discerned only by comparison with something else. Now the sum of things is not discerned by comparison with anything else: hence, since it has no extremity, it has no limit; and, since it has no limit, it must be unlimited or infinite" (quoted in Diogenes Laertius, *Lives*, 10.41–42).

31. On the distinction between "world" and "universe," see Furley, *Cosmic Problems*, 2.

32. For Arthur O. Lovejoy, the "principle of plenitude" extends to "deductions from the assumption that no genuine potentiality of being can remain unfulfilled, that the extent and abundance of the creation must be as great as the possibility of existence and commensurate with the productive capacity of a 'perfect' and inexhaustible Source, and that the world is the better, the more things it contains" (*The Great Chain of Being: A Study of the History of an Idea* [Cambridge, Mass.: Harvard University Press, 1976], 52).

33. "Probably a crab would be filled with a sense of personal outrage if it could hear us class it without ado or apology as a crustacean, and thus dispose of it. 'I am no such thing,' it would say, 'I am MYSELF, MYSELF alone'" (William James, *The Varieties of Religious Experience* [New York: Penguin, 1982], 9).

34. Plato, *Timaeus*, 52d.

35. Genesis 1:2 reads, "The earth was a formless void and darkness covered the face of the deep, while a wind from God swept over the face of the waters." Catherine Keller

calls on this verse to reconfigure our understanding of the Trinity, in *Face of the Deep: A Theology of Becoming* (New York: Routledge, 2003), 229–38. Genesis 2:5–7a reads, "[N]o plant of the field was yet in the earth and no herb of the field had yet spring up—for the Lord God had not caused it to rain upon the earth, and there was no one to till the ground; but a stream would rise from the earth, and water the whole face of the ground—then the Lord God formed man from the dust of the ground" (NRSV).

36. Gerhard May, *Creatio ex Nihilo: The Doctrine of "Creation out of Nothing" in Early Christian Thought*, trans. A. S. Worrall (Edinburgh: Clark, 1994).

37. For a critical analysis of this logic, see Keller, *Face of the Deep*, 43–64; Whitney Bauman, "*Creatio ex Nihilo, Terra Nullius*, and the Erasure of Presence," in *Ecospirit: Religions and Philosophies for the Earth*, ed. Laurel Kearns and Catherine Keller (New York: Fordham University Press, 2007), 353–72; and Mary-Jane Rubenstein, "Cosmic Singularities: On the Nothing and the Sovereign," *Journal of the American Association of Religion* 80, no. 2 (2012): 485–517, and "Myth and Modern Physics: On the Power of Nothing," in *Creation Options: Rethinking Initial Creation*, ed. Thomas Oord and Richard Livingston (New York: Routledge, forthcoming).

38. Quoted in Diogenes Laertius, *Lives*, 10.39.

39. Edward Adams, "Graeco-Roman and Ancient Jewish Cosmology," in *Cosmology and New Testament Theology*, ed. Jonathan T. Pennington and Sean M. McDonald (New York: Clark, 2008), 11.

40. Diogenes Laertius, *Lives*, 9.31, 9.44, 10.90.

41. Furley, *Greek Cosmologists*, 141–42; Michel Serres, *The Birth of Physics*, trans. Jack Hawkes (Manchester, Eng.: Clinamen Press, 2000), 6.

42. Serres, *Birth of Physics*, 30.

43. Democritus, "Fragment 164," quoted in Furley, *Greek Cosmologists*, 142.

44. Serres, *Birth of Physics*, 31.

45. Ibid., 5–6.

46. Rod Nave, "Laminar Flow," HyperPhysics, Georgia State University, http://hyperphysics.phy-astr.gsu.edu/hbase/pfric.html (accessed April 13, 2013).

47. Serres, *Birth of Physics*, 6.

48. Ibid., 11.

49. Ibid., 136.

50. Ibid., 144.

51. Ibid., 145.

52. Plato, *Timaeus*, 35a.

53. For differing interpretations on the role of divinity in Lucretius, see William S. Anderson, "Discontinuity in Lucretian Symbolism," *Transactions and Proceedings of the American Philological Association* 91 (1960): 1–29; Elizabeth Asmis, "Lucretius' Venus and Stoic Zeus," *Hermes* 110 (1982): 458–70; and Bonnie A. Catto, "Venus and Natura in Lucretius' *De rerum natura* 1.1–23 and 2.167–74," *Classical Journal* 84 (1989): 97–104. My thanks to Tushar Irani and Alex Ray for having suggested these sources.

54. This distinction was ridiculed by Epicurus's numerous rivals, including Cicero's Balbus (the Stoic), who suggests that gods who do not govern the cosmos might as well not exist. See Cicero, *On the Nature of the Gods*, 2.17.44.

55. Quoted in Warren, "Ancient Atomists and the Plurality of Worlds," 357. See also Walter Charleton, *Physiologia Epicuro-Gassendo-Charltoniana: Or a Fabrick of Science Natural Upon the Hypothesis of Atoms* (London: Tho. Newcomb for Thomas Heath, 1654), 9.

56. Judith Butler, *Precarious Life: The Powers of Mourning and Violence* (New York: Verso, 2004), 19–49, and *Undoing Gender* (New York: Routledge, 2004), 17–39.

57. J. M. Ross, "Introduction," in *Cicero: The Nature of the Gods* (New York: Penguin, 1972), 41.

58. "Outside of the world is diffused the infinite void, which is incorporeal" (Diogenes Laertius, *Lives*, 7.140).

59. Michael Lapidge, "Stoic Cosmology," in *The Stoics*, ed. J. M. Rist (Berkeley: University of California Press, 1978), 177.

60. Diogenes Laertius, *Lives*, 7:140 (emphasis added).

61. Ibid., 7.141.

62. A fragment from Zeno reads: "The universe (*to pan*) will be totally destroyed. Everything which burns [something], having [what] it burns, shall burn up the whole of it. The sun is *fire*—shall it not, then, burn up what it has?" (quoted in J. Mansfield, "Providence and the Destruction of the Universe in Early Stoic Thought: With Some Remarks on the 'Mysteries of Philosophy,'" in *Studies in Hellenistic Religion*, ed. M. J. Vermaseren [Leiden: Brill, 1979], 148).

63. For an account of the history of interpretation of these cycles, see Richard Parry, "Empedocles," in *Stanford Encyclopedia of Philosophy* (2005), http://plato.stanford.edu/entries/empedocles/ (accessed May 31, 2013). See also Friedrich Solmsen, "Love and Strife in Empedocles' Cosmology," *Phronesis* 10, no. 2 (1965): 109–48.

64. Helge Kragh, "Ancient Greek–Roman Cosmology: Infinite, Eternal, Finite, Cyclic, and Multiple Universes," *Journal of Cosmology* 9 (2010): 2172–78.

65. Chief among these treatises is Plutarch's *De Stoicorum repugnantiis*. See Ricardo Salles, "Chrysippus on Conflagration and the Indestructibility of the Cosmos," in *God and Cosmos in Stoicism*, ed. Ricardo Salles (Oxford: Oxford University Press, 2009), 118–34.

66. Cicero, *On the Nature of the Gods*, 2.46.118.

67. One passing but tantalizing reference can be found in Seneca's *Ad Marciam: De consolatione*. See Seneca, *To Marcia on Consolation*, in *Seneca: Moral Essays*, trans. John W. Basore, Loeb Classical Library 254 (New York: Putnam, 1932), 2–97. On other cosmic catastrophes in Seneca's work, see Edward Adams, *The Stars Will Fall from Heaven: Cosmic Catastrophes in the New Testament and Its World* (New York: Clark, 2007), 122.

68. Mircea Eliade, *The Myth of the Eternal Return: Cosmos and History*, trans. Willard R. Trask (Princeton, N.J.: Princeton University Press, 2005), 135, 57, 136. As Eliade points out, the phrase "empire without end" comes from Jupiter's promise in Virgil's *Aeneid*.

69. Ibid., 135.

70. Diogenes Laertius, *Lives*, 7.139.

71. Ibid., 7:147.

72. Ross, "Introduction," 42.

73. Lapidge, "Stoic Cosmology," 163.

74. Diogenes Laertius, *Lives*, 7.134.

75. Lapidge, "Stoic Cosmology," 164.

76. Hans Frederich August von Anim, ed., *Zeno et Zenonis discipuli: Exemplar anastatice iteratum*, vol. 1 of *Stoicorum veterum fragmenta* (Leipzig: Teubneri, 1921), 1.125–29 (hereafter *SVF* I). See also Lapidge, "Stoic Cosmology," 166.

77. Hans Frederich August von Anim, ed., *Chryssipi fragmenta logica et physica*, vol. 2 of *Stoicorum veterum fragmenta* (Leipzig: Teubneri, 1903), 2.622 (hereafter *SVF* II). See

also Pheme Perkins, "On the Origin of the World: A Gnostic Physics," *Vigiliae Christianae* 34, no. 1 (1980): 38.

78. Anim, *SVF* II, 2.1071. See also Mansfield, "Providence," 181.

79. Lapidge, "Stoic Cosmology," 164.

80. David E. Hahm, *The Origins of Stoic Cosmology* (Columbus: Ohio State University Press, 1977), 31–33. An excerpt from Chrysippus's *Physics* in Plutarch reads: "The change of fire is as follows: it is changed through air into water. And from this, when earth has settled down, air is evaporated. Then, when the air has been thinned, the *aether* is poured around in a circle" (quoted in ibid., 31).

81. Quoted in Adams, "Graeco-Roman and Ancient Jewish Cosmology," 10. On the scope of the relationship between Heraclitus and the Stoics, see Adams, *Stars Will Fall*, 105–6; and Lapidge, "Stoic Cosmology," 162–63, 80–81.

82. Hahm, *Origins of Stoic Cosmology*, 31.

83. Alexander of Aphrodisias accounts for this indeterminacy by saying that fire itself contains both *archai*, which he calls "god" and "matter," respectively (quoted in ibid., 33).

84. Jean-Baptiste Gourinat, "The Stoics on Matter and Prime Matter: 'Coporealism' and the Imprint of Plato's *Timaeus*," in *God and Cosmos in Stoicism*, ed. Ricardo Salles (Oxford: Oxford University Press, 2009), 68 (emphasis added).

85. Michel Serres, *Genesis*, trans. Genevieve James and James Nielson (Ann Arbor: University of Michigan Press, 1995), 112.

86. Gourinat, "Stoics on Matter and Prime Matter," 68.

87. Diogenes Laertius, *Lives*, 7.139.

88. Ibid., 7.137.

89. Adams, "Graeco-Roman and Ancient Jewish Cosmology," 16.

90. Diogenes Laertius, *Lives*, 7.147.

91. This quotation is from a fragment not included in the *SVF*. See Mansfield, "Providence," 148; compare Anim, *SVF* II, 2.593; and Lapidge, "Stoic Cosmology," 181.

92. This moment of destruction and renewal is traditionally called the "Great Year." See Anim, *SVF* II, 2.625; Lapidge, "Stoic Cosmology," 181; Mansfield, "Providence," 145; Henri-Charles Puech, "Gnosis and Time," in *Man and Time: Papers from the Eranos Yearbooks*, ed. Joseph Campbell and R. F. Hull (New York: Pantheon, 1958), 41; and Eliade, *Myth of the Eternal Return*, 87. Plato discusses the Great Year briefly in the *Timaeus* (39d) but imagines that the destruction and renewal will take place within parts of the cosmos (cities, for example) rather than happening to the cosmos itself.

93. Anim, *SVF* II, 2.604–5; Lapidge, "Stoic Cosmology," 181.

94. Of this process, Diogenes Laertius simply says, "[The cosmos] is first dried up completely and then made watery." Plutarch explains in a bit more depth that "when *ekpyrosis* takes place, [Chrysippus] says that the universe is totally alive and is a living being, but thereafter, as it is quenched and becomes concentrated, it turns into water and earth and all things substantial" (Anim, *SVF* II, 2.589, 2.605; see also Lapidge, "Stoic Cosmology," 183).

95. "No resolution of this patent contradiction seems ever to have been proposed," says Lapidge, "nor can I propose one" ("Stoic Cosmology," 181).

96. Mansfield, "Providence," 160.

97. Adams, "Graeco-Roman and Ancient Jewish Cosmology," 17.

98. Sedley, *Creationism and Its Critics*, 208; Adams, *Stars Will Fall*, 119.

99. Anim, *SVF* I, 1.109, and *SVF* II, 2.625–26; Hahm, *Origins of Stoic Cosmology*, 185.

100. Anim, *SVF* II, 2.624.

101. Friedrich Nietzsche, *Ecce Homo*, trans. Walter Kaufman (New York: Vintage, 1967), "Birth of Tragedy," para. 3.

102. Alexander Nehemas, "The Eternal Recurrence," in *Nietzsche*, ed. John Richardson and Brian Leiter (Oxford: Oxford University Press, 2001), 119. Although Nietzsche never gives a satisfying physical proof of the eternal return, Sean Carroll argues that Henri Poincaré does. In 1890, Poincaré posited his "recurrence theorem," according to which any system will eventually return to its original configuration if one waits long enough. As Carroll explains, the recurrence time for a "typical macroscopic-sized object" is $10^{1,000,000,000,000,000,000,000,000}$ seconds (by comparison, he reminds us, the universe is only 10^{18} seconds old). So, Carroll concludes, "Nietzsche's Demon isn't wrong; it's just thinking long-term" (*From Eternity to Here: The Quest for the Ultimate Theory of Time* [New York: Dutton, 2010], 206).

103. Friedrich Nietzsche, *The Gay Science*, trans. Walter Kaufman (New York: Vintage, 1974), para. 341.

104. "A new will I teach men: to will this way which man has walked blindly, and to affirm it!" (Friedrich Nietzsche, *Thus Spoke Zarathustra*, trans. Walter Kaufman [New York: Penguin, 1978], 32).

105. Mansfield, "Providence," 132.

106. Anim, *SVF* II, 2.975.

107. Friedrich Nietzsche, *On the Genealogy of Morals*, trans. Walter Kaufmann (New York: Vintage, 1989), 1.13; Nehemas, "Eternal Recurrence," 128.

108. Nehemas, "Eternal Recurrence," 123.

109. Nietzsche, *Thus Spoke Zarathustra*, 323.

110. Gilles Deleuze, *Nietzsche and Philosophy*, trans. Hugh Tomlinson (New York: Columbia University Press, 1983), 46.

111. Marcus Aurelius Antoninus, *Meditations of Marcus Aurelius Antoninus* (2006), 1.17, http://www.felix.org/node/34464 (accessed June 3, 2013).

112. Marcus Aurelius Antoninus, *Meditations*, trans. Maxwell Staniforth (New York: Penguin, 1964), 7.9.

113. Origen, *De principiis*, in *The Ante-Nicene Fathers: Translations of the Writings of the Fathers Down to A.D. 325*, ed. Alexander Roberts and James Donaldson (Buffalo, N.Y.: Christian Literature, 1885), 3.5.3.

114. Ibid., 3.5.1, 3.5.3. For similar passages, see Isaiah 66:22 and 51:6, not to mention Revelation 21, which is said to fulfill the Isaianic promise. But these later chapters, which compose "second Isaiah," are a fairly late set of writings with respect to the rest of the Hebrew Bible; they were written either during or after the Babylonian Exile. And so the idea of a new heaven and a new earth is, as it were, *new*. In fact, the oldest biblical sources seem to teach that creation will endure forever, even "forever and ever" (Psalm 148:6), promising that because "[God] has established the world, it shall never be moved" (Psalm 93:1). According to Adams, "[T]he conviction that creation will perish and wear out (esp. Ps 102:25–27 and Isa. 51:6) [therefore] represents a development of and departure from this cosmological tradition." This departure seems to stop short of a full-fledged proto-Stoicism insofar as it is unclear "whether Isaiah 65:17–25 envisions the destruction and recreation of the world or a non-destructive renewal of the cosmos" (*Stars Will Fall*, 50).

115. Origen, *De principiis*, 3.5.3.

116. Ibid., 2.3.4.

117. Ibid., preface.4, 2.2.5.

118. Ibid., 1.1.5.

119. Ibid., 1.1.1–4.

120. Ibid., 2.3.4.

121. Ibid., 3.5.3. As Mark Worthing has observed, "[W]hat is not entirely clear is whether Origen intends the *creatio ex nihilo* to apply to the creation of this present cycle or to the very first of the sequence" ("Christian Theism and the Idea of an Oscillating Universe," in *God, Life, and the Cosmos: Christian and Islamic Perspectives* [Burlington, Vt.: Ashgate, 2002], 293).

122. Origen, *De principiis*, 2.1.3.

123. In fact, the theory was condemned in 553 at the second Council of Constantinople. On the sources and misunderstandings of the "Origenist controversy," see Augustine of Hippo, *Arianism and Other Heresies*, ed. Roland J. Teske, vol. 1 of *The Works of St. Augustine: A Translation for the 21st Century*, ed. John E. Rotelle (Hyde Park, N.Y.: New City Press, 1995), 86–93.

124. Augustine of Hippo, *Concerning the City of God Against the Pagans*, trans. Henry Bettenson (New York: Penguin, 2003), 12.10–12.

125. Ibid., 11.4.

126. Ibid., 12.12.

127. Plato, *Timaeus*, 39d.

128. Augustine of Hippo, *Concerning the City of God*, 12.14.

129. Ibid.

130. Seneca, *To Marcia on Consolation*, 26.6.

131. "A quo ludibrio prorsus inmortalem animam, etiam cum sapientiam pereperit, liberare non possunt" (Augustine of Hippo, *Concerning the City of God*, 12.14).

132. Ibid., 12.21. I refer the reader to the full passage in Latin; the rhythm and force of it is lost in translation. See Augustine of Hippo, *The City of God*, vol. 4, trans. Philip Levine, Loeb Classical Library 414 (Cambridge, Mass.: Harvard University Press, 1966).

133. "Quis haec audiat? Quis credit? Quis ferat?" (Augustine of Hippo, *City of God*, 12.21, my translation).

134. "Ut quo modo valeo dicam quod volo" (ibid., 12.21), translation from Augustine of Hippo, *Concerning the City of God*, 12.21 (emphasis added).

135. Augustine of Hippo, *Concerning the City of God*, 11.22.

136. Ibid., 12.21.

137. Augustine of Hippo, *Confessions*, trans. Owen Chadwick (Oxford: Oxford University Press, 1991), 2.4.9, 1.20.31.

138. Augustine of Hippo, *Concerning the City of God*, 12.21.

139. Ibid., 12.14. The original reads: "[S]ine cessation ad falsam beatitudinem et ad veram miseriam sine cessation redeuntem" (Augustine of Hippo, *City of God*, 12.14).

140. Augustine of Hippo, *Concerning the City of God*, 1.23.

141. Ibid., 12.14.

142. Origen foresaw and avoided this pitfall (*De principiis*, 2.3.4–5), but Augustine seems not to have taken notice. The question of Christ's singularity in the face of the possibility of many worlds will be debated vigorously in the seventeenth century (see chap. 4, sec. "From Infinity to Pluralism").

143. Augustine of Hippo, *Concerning the City of God*, 12.4.

144. Ibid., 12.21.

145. Ibid.

146. Ibid., 12.15.

147. This notion of an "oscillating universe" gained some popularity in the mid-1930s, resurged in the mid-1960s, and has found a new set of twenty-first-century advocates—all among theorists who have sought to escape the notion of an absolute cosmic beginning (see chap. 5, secs. "Let There Be Light" and "Other Answers").

148. The observational data confirming the overwhelming likelihood that we inhabit a "flat" universe were released in 2003, following the 2001 mission of the Wilkinson Microwave Anisotropy Probe (WMAP) satellite.

149. Arthur Eddington, "The Arrow of Time, Entropy, and the Expansion of the Universe," in *The Concepts of Space and Time*, ed. Milic Capec (Dordrecht: Reidel, 1976), 466 (emphasis added).

150. Ibid.

151. Arthur Eddington, *The Nature of the Physical World* (Cambridge: Cambridge University Press, 1935), 86. Perhaps unsurprisingly, Eddington's preferred solution to Einstein's equations produced an early formulation of the big bang hypothesis known as the "Eddington–Lemaitre model."

152. Norman Martin, "The Song That Doesn't End" (1988); lyrics changed slightly by Shari Lewis in the television series for children *Lamb Chop's Play-Along* (PBS).

153. Poggio Bracciolini rediscovered Lucretius's *De rerum natura* in 1417. Concerning both this discovery and the readiness of early-fifteenth-century Europe to re-receive it, see Jones, *Epicurean Tradition*, 142–50. For a not quite accurate but broadly compelling treatment of Bracciolini's discovery and its influence on the modern world more broadly, see Stephen Greenblatt, *The Swerve: How the World Became Modern* (New York: Norton, 2011). For a helpful, if overstated, correction, see Jim Hinch, "Why Stephen Greenblatt Is Wrong—and Why It Matters," *Los Angeles Review of Books*, December 1, 2012.

154. Velleius begins his short and poorly argued address by lambasting Plato's "artisan deity" as well as "that old hag of a fortune-teller, the *Pronoia* of the Stoics," and this point remains the only one he makes throughout his speech: that the world is not the contrivance of a deity. He offers no details of how the world *did* come into being (atoms, void, *clinamen*, vortex, etc.); all he says is that the gods had nothing to do with it. See Cicero, *De natura deorum. Academia*, trans. H. Rackham, Loeb Classical Library 268 (Cambridge, Mass.: Harvard University Press, 2005), 1.8.18.

155. Jones, *Epicurean Tradition*, 102.

156. Augustine, *Epistle* 118, quoted in ibid., 94.

3. Navigating the Infinite

1. "Videtur quod non sit unus mundus tantum, sed plures" (Thomas Aquinas, *Summa theologiae*, trans. Fathers of the English Dominican Province, 5 vols. [Allen, Tex.: Christian Classics, 1981], 1.47.3 (subsequent references are cited in the text).

2. Charles Homer Haskings, *The Renaissance of the Twelfth Century* (Cambridge, Mass.: Harvard University Press, 1971); Fernand Van Steenberghen, *Aristotle in the West: The Origins of Latin Aristotelianism* (Louvain, Belgium: Nauwelaerts, 1970).

3. Aristotle, *On the Heavens* (*De caelo*), trans. J. L. Stocks, in *The Complete Works of Aristotle: The Revised Oxford Translation*, ed. Jonathan Barnes (Princeton, N.J.: Princeton University Press, 1971), 1:276b12–17. See also chap. 1, sec. "Reflecting Singularity."

4. Aristotle, *On the Heavens*, 276b21.

5. Ibid., 280a24–28.

6. Steven J. Dick, *Plurality of Worlds: The Origins of the Extraterrestrial Life Debate from Democritus to Kant* (Cambridge: Cambridge University Press, 1982), 26.

7. Ibid., 23.

8. The first neo-Aristotelian to engage this question seems to have been the astronomer-philosopher Michael Scot (1175–1232), who sought to reaffirm Aristotle's stance against the plurality of worlds. Michael considers the position of "those who maintain that God, who is omnipotent, could and can still create, in addition to this world, another world, or several other worlds, or even an infinity of worlds," and he even concedes that God *could*, in fact, create other worlds. But he insists that God *does* not do so because "nature cannot withstand it" (*Super auctorem spherae*, 2.146, quoted in Pierre Duhem, *Medieval Cosmology: Theories of Infinity, Place, Time, Void, and the Plurality of Worlds*, trans. Roger Ariew [Chicago: University of Chicago Press, 1985], 443). To support this position, Michael appeals not to the theory of natural motion, but to Aristotle's insistence that there is no such thing as a void (*Physics*, trans. R. P. Hardie and R. K. Gaye, in *The Complete Works of Aristotle*, ed. Jonathan Barnes [Princeton, N.J.: Princeton University Press, 1984], 1:213a12–17b28). After all, if there is no extracosmic space, then there is no *place* for any other worlds, let alone any stuff with which to make more of them. Following Michael's lead, William of Auvergne (1180/90–1249) will conclude that although God cannot create more than one universe, "this impossibility is not a defect in God, nor a defect issuing from God, rather it is a defect on the part of the world, which cannot exist in multiples" (*De universo*, 16.100b, quoted in Duhem, *Medieval Cosmology*, 444). In this manner, these authors likened the creator-God to Plato's demiurge, who must work within the constraints of his materials. God's power is limitless *in itself* but limited with respect to creation. Therefore, even an omnipotent God can create only one world. This solution to the threat of multiple worlds will clearly not work for Thomas Aquinas, whose God can create the stuff of other worlds ex nihilo and who must therefore find a different argument against the plurality of worlds.

9. Compare Aristotle, *On the Heavens*, 278a26.

10. "Quaecumque autem a Deo sunt, ordinem habent ad invicem et ad ipsum Deum" (Thomas Aquinas, *Summa theologiae*, 1.47.3),

11. Thomas refers to the specific order of creation in Article 2 of the same question. "In natural things," he argues, "species seem to be arranged in degrees; as the mixed things are more perfect than the elements, and plants than minerals, and animals than plants, and men than other animals; and in each of these one species is more perfect than the other" (*Summa theologiae*, 1.47.2). For accounts of the historical evolution of this cosmic hierarchy, see Ernst Cassirer, *The Individual and the Cosmos in Renaissance Philosophy*, trans. Mario Domandi (New York: Harper Torchbooks, 1964), 9; and Arthur O. Lovejoy, *The Great Chain of Being: A Study of the History of an Idea* (Cambridge, Mass.: Harvard University Press, 1976).

12. Of course, Augustine refuted the Stoics' temporal infinity, whereas Thomas is concerned with the Atomists' more spatial infinity. See chap. 2, notes 123–46.

13. Aristotle, *Metaphysics*, in *The Complete Works of Aristotle: The Revised Oxford Translation*, ed. Barnes, 2:1074a. See also chap. 1, sec. "Reflecting Singularity."

14. "The first effect of unity is equality; and then comes multiplicity; and therefore from the Father, to Whom, according to Augustine, is appropriated unity, the Son

proceeds . . . and then from Him the creature proceeds" (Thomas Aquinas, *Summa theologiae*, 1.47.2).

15. This overvaluation of unity over difference in Aquinas has been the source of feminist and Eastern Orthodox critiques alike, which seek to retrieve a more radically relational God beneath the Thomistic insistence on oneness. See Catherine La Cugna, *God for Us: The Trinity and Christian Life* (San Francisco: HarperSanFrancisco, 1991); Laurel Schneider, *Beyond Monotheism: A Theology of Multiplicity* (New York: Routledge, 2008); and John Zizioulas, *Being as Communion* (Crestwood, N.Y.: SVS Press, 1985).

16. Thomas Aquinas, *Exposition of Aristotle's Treatise "On the Heavens" (Unpublished)*, trans. R. F. Larcher and Pierre H. Conway (Columbus, Ohio: College of St. Mary of the Springs, 1963), chap. 195.

17. Duhem, *Medieval Cosmology*, 450; Dick, *Plurality of Worlds*, 28.

18. Dana Miller, "Plutarch's Argument for a Plurality of Worlds in *De defectu oraculorum* 424c10–425e7," *Ancient Philosophy* 17, no. 2 (1997): 375–76.

19. Duhem, *Medieval Cosmology*, 4.

20. The first Scholastic to adopt this strategy was Giles of Rome (1243–1316), who like Thomas Aquinas adhered to the Aristotelian teaching that "this world contains the entirety of its matter" (Aristotle, *On the Heavens*, 278a26). Because there is no matter from which another world might be formed, Giles insisted that the creation of another world would be impossible "by natural means," while granting that such a thing could be possible "by divine power" (*Quodlibet domini*, quoted in Duhem, *Medieval Cosmology*, 454). John Buridan (1300–1358) deployed a similar tactic in relation to Aristotle's infamous proof from natural motion, arguing that although earth indeed moves "naturally" down and fire moves "naturally" up, God could theoretically compose a different world out of different substances, with different sorts of motion. Or, Buridan imagined, God could "sidestep" the order of nature by orienting the elements of another world to its center alone rather than to ours (*Quaestiones super libros caelo et mundo magistri*, quoted in Duhem, *Medieval Cosmology*, 467). God might even go so far as to destroy a world if he wished it—even though celestial bodies are by nature incorruptible. In any of these cases, however, divine intervention would *interrupt* what Buridan held to be the order of the universe; God would be replacing natural causes with "voluntary and free" ones (quoted in Dick, *Plurality of Worlds*, 29). This voluntarist line arguably finds its culmination in Albert of Saxony (ca. 1316–1390), who reaffirmed each of Aristotle's arguments concerning the oneness of the cosmos, only to conclude with a divine exception. One can almost see him glancing over his shoulder as he wrote that, "following Aristotle's doctrine, we conclude that the existence of several . . . worlds is impossible naturally," only to add, "it is no less true that God could create many worlds, since He is omnipotent" (*In libros de caelo et mundo*, quoted in Duhem, *Medieval Cosmology*, 470).

21. This argument was set forth by Godfrey of Fontaines (1250–1306?) and William of Ware (d. 1306?). See Duhem, *Medieval Cosmology*, 458; and Dick, *Plurality of Worlds*, 28.

22. Anticipating one element of Newton's law of gravitation (that is, its reliance on distance), Godfrey of Fontaines wrote that "the earth of each world tends exclusively toward the center of the world it belongs to; it has no inclination driving it toward the center of another world" (quoted in Duhem, *Medieval Cosmology*, 460). In his commentary on Aristotle's *Physics*, Simplicius (490–560) had made a similar claim, arguing that

the movement of heavy bodies depends on their distance from one another. Averroës (1126–1198) refuted this position in his own commentary, arguing that the motion of bodies is absolute rather than relative. According to Duhem, Averroës was the "more faithful interpreter" of Aristotle (*Medieval Cosmology*, 442); indeed, in the hands of authors such as Godfrey, Simplicius's interpretation on this matter ended up unsettling part of the foundation of Aristotelian cosmology.

23. Quoted in Duhem, *Medieval Cosmology*, 452.

24. Quoted in Dick, *Plurality of Worlds*, 37. Oresme employed a similar strategy when he set forth arguments for the movement of the earth rather than the heavens, only to conclude with a characteristic "nevertheless" and that the earth must be at rest. See Norris S. Hetherington, "Introduction: A New Physics and a New Cosmology" [introduction to part 4], in *Cosmology: Historical, Literary, Philosophical, Religious, and Scientific Perspectives*, ed. Norris S. Hetherington (New York: Garland, 1993), 231–33. For Oresme's concentric hypothesis, see Edward Grant, *A History of Natural Philosophy: From the Ancient World to the Nineteenth Century* (Cambridge: Cambridge University Press, 2007), 228.

25. Dick, *Plurality of Worlds*, 37.

26. Cassirer, *Individual and the Cosmos*, 25.

27. On this connection, see Regine Kather, "'The Earth Is a Noble Star': Arguments for the Relativity of Motion in the Cosmology of Nicholaus Cusanus and Their Transformation in Einstein's Theory of Relativity," in *Cusanus: The Legacy of Learned Ignorance*, ed. Peter J. Casarella (Washington, D.C.: Catholic University of America Press, 2006), 226–50.

28. Nicholas of Cusa, *On Learned Ignorance*, in *Nicholas of Cusa: Selected Spiritual Writings*, trans. H. Lawrence Bond (New York: Paulist Press, 1997), 2.12.163–64 (subsequent references are cited in the text).

29. Alexander Koyré, *From the Closed World to the Infinite Universe* (Baltimore: Johns Hopkins University Press, 1957), 7–22; Cassirer, *Individual and the Cosmos*, 7–45. For more qualified evaluations of this claim, see Duhem, *Medieval Cosmology*, 505–9; Hans Blumenberg, *The Legitimacy of the Modern Age*, trans. Robert M. Wallace (Cambridge, Mass.: MIT Press, 1983), 503–10; and Elizabeth Brient, "Transitions to a Modern Cosmology: Meister Eckhart and Nicholas of Cusa on the Intensive Infinite," *Journal of the History of Philosophy* 37, no. 4 (1999): 575–600. For an argument that claims Cusa remained more or less entirely Aristotelian, see Rhys W. Roark, "Nicholas Cusanus, Linear Perspective, and the Finite Cosmos," *Viator* 41, no. 1 (2010): 315–66.

30. Albert Einstein, "On the Electrodynamics of Moving Bodies" (1905), Fourmilab Switzerland, http://www.fourmilab.ch/etexts/einstein/specrel/www/ (accessed June 3, 2013). On the similarities and differences between Cusa's and Einstein's relativities, see Kather, "'Earth Is a Noble Star.'" For a very clear introduction to the principles of special relativity, see Brian Greene, *The Elegant Universe: Superstrings, Hidden Dimensions, and the Quest for the Ultimate Theory* (New York: Norton, 1999), 23–52.

31. Koyré, *From the Closed World to the Infinite Universe*, 7.

32. Lucretius, *The Nature of Things*, trans. A. E. Stallings (New York: Penguin, 2007), 1.980. See also Mircea Eliade, *The Myth of the Eternal Return: Cosmos and History*, trans. Willard R. Trask (Princeton, N.J.: Princeton University Press, 2005), 135.

33. On the mechanics and significance of this rediscovery, see Stephen Greenblatt, *The Swerve: How the World Became Modern* (New York: Norton, 2011).

34. Dick, *Plurality of Worlds*, 42; the same point is made in Blumenberg, *Legitimacy of the Modern Age*, 511.

35. Thomas Aquinas, *Disputationes*, quoted in Tyrone Lai, "Nicholas of Cusa and the Finite Universe," *Journal of the History of Philosophy* 11, no. 2 (1973): 163.

36. "The universe cannot be negatively infinite, although it is boundless and thus privatively infinite, and in this respect neither finite nor infinite" (Nicholas of Cusa, *On Learned Ignorance*, 2.1.97).

37. Brient, "Transitions to a Modern Cosmology," 592.

38. "Deus est sphaera infinita, cuius centrum est ubique, circumferentia nusquam." This sentence can be traced back to the pseudo-Hermetic *Book of 24 Philosophers*. For an exhaustive history, see Karsten Harries, "The Infinite Sphere: Comments on the History of a Metaphor," *Journal of the History of Philosophy* 13, no. 1 (1975): 5–15.

39. Brient, "Transitions to a Modern Cosmology," 579.

40. See Tzinacán's ecstatic vision from prison in Jorge Luis Borges's "The Writing of the God":

> And at that . . . there occurred union with the deity, union with the universe (I do not know whether there is a difference between those two words). . . . I saw a wheel of enormous height, which was not before my eyes, or behind them, or to the sides, but everywhere at once. This Wheel was made of water, but also fire, and although I could see its boundaries, it was infinite. It was made of all things that shall be, that are, and that have been, all intertwined, and I was one of the strands within that all-encompassing fabric, and Pedro de Alvarado, who had tortured me, was another. In it were the causes and the effects, and the mere sight of that Wheel enabled me to understand all things, without end. (*Collected Fictions*, trans. Andrew Hurley [New York: Penguin, 1998], 253)

I am grateful to Marcelo Gleiser for calling my attention to this passage.

41. Brient, "Transitions to a Modern Cosmology," 592.

42. Cassirer, *Individual and the Cosmos*, 28.

43. For Catherine Keller, Cusa's tireless avoidance of pantheism is another means by which he continuously "complicates any notion of divine unity—whether of the classical theism of the divine One over and above the manifold world, or of an identification of that Unity with the Universe" ("The Cloud of the Impossible" [manuscript], chap. 3).

44. Nicholas of Cusa, *On the Vision of God*, in *Nicholas of Cusa*, trans. Bond, 13.53.

45. In this vein, Cusa writes that "the heavenly bodies are all particular, worldly parts of a single universe [partes particulares mundiales unius universi]" (quoted in Blumenberg, *Legitimacy of the Modern Age*, 516).

46. William James, *A Pluralistic Universe* (Lincoln: University of Nebraska Press, 1996), 325. See also introduction, sec. "The One and the Many."

47. Cassirer, *Individual and the Cosmos*, 37.

48. Jasper Hopkins, *Nicholas of Cusa's Debate with John Wenck: A Translation and an Appraisal of "De ignota litteratura" and "Apologia doctae ignorantiae"* (Minneapolis: Banning, 1981).

49. Quoted in Blumenberg, *Legitimacy of the Modern Age*, 494–509.

50. Quoted in ibid., 514.

51. On the relationship between these distinct parts of the Cusan corpus, see Mary-Jane Rubenstein, "End Without End: Cosmology and Infinity in Nicholas of Cusa," in *The Trials of Desire: A Festschrift for Denys A. Turner*, ed. Eric Bugyis and David Newheiser (Notre Dame, Ind.: University of Notre Dame Press, forthcoming).

52. Nicholas of Cusa, *De venatione sapientiae*, vol. 12 of *Nicolai de Cusa: Opera omnia*, ed. Ernest Hoffman and Raymond Klibansky (Leipzig: Felix Meiner, 1932). For a careful commentary on this work, see Clyde Lee Miller, *Reading Cusanus: Metaphor and Dialectic in a Conjectural Universe* (Washington, D.C.: Catholic University of America Press, 2003), chap. 6.

53. Nicholas of Cusa, *Directio speculantis seu de li non aliud*, vol. 13 of *Nicolai de Cusa: Opera omnia*, and *De possest*, vol. 11 of *Nicolai de Cusa: Opera omnia*.

54. As explained and quoted in Matthieu Herman van der Meer, "World Without End: Nicholas of Cusa's View of Time and Eternity," in *Christian Humanism*, ed. Alasdair A. Macdonald, R. W. M. von Martels, and Jan R. Veenstra Zweder (Leiden: Brill, 2009), 328.

55. Nicholas of Cusa, *De venatione sapientiae*, 38.109.17–23, quoted in ibid., 330.

56. Nicholas of Cusa, *De venatione sapientiae* 28.83.6–13, cited in ibid., 329.

57. Nicholas of Cusa, *De ludo globi/The Game of Spheres*, ed. Pauline Moffitt Watts (New York: Abaris Books, 1986), 63.

58. Arthur O. Lovejoy, quoted in Koyré, *From the Closed World to the Infinite Universe*, 31.

59. Ingrid D. Rowland, *Giordano Bruno: Philosopher/Heretic* (New York: Farrar, Straus and Giroux, 2008), 240.

60. Antonio Calcagno, *Giordano Bruno and the Logic of Coincidence: Unity and Multiplicity in the Philosophical Thought of Giordano Bruno*, Renaissance and Baroque Studies and Texts 23 (New York: Peter Lang, 1998), 29.

61. Giordano Bruno, *On the Infinite Universe and Worlds*, in *Giordano Bruno: His Life and Thought with Annotated Translation of His Work "On the Infinite Universe and Worlds,"* ed. Dorothea Singer (New York: Schuman, 1950), 280.

62. Giordano Bruno, *The Ash Wednesday Supper*, trans. Edward A. Gosselin and Lawrence S. Lerner, Renaissance Society of America Reprint Texts 4 (Toronto: University of Toronto Press, 1995), 86, 95.

63. Ibid., 87.

64. Aristotle, *On the Heavens*, 271b26. See also Miguel A. Granada, "Aristotle, Copernicus, Bruno: Centrality, the Principle of Movement, and the Extension of the Universe," *Studies in History and Philosophy of Science* 35 (2004): 92; and chap. 1, sec. "Reflecting Singularity."

65. The encyclopedia, *A Prognostication Everlasting*, had long been edited and published by Thomas's father, Leonard Digges. In 1576, Thomas issued a new version of the encyclopedia, adding a number of appendixes. They included the translation of Copernicus in the section "Perfit Description of the Caelestiall Orbes According to the Most Aunciente Doctrine of the Pythagoreans, Latelye Revived by Copernicus and by Geometricall Demonstrations Approved." See Hilary Gatti, "Giordano Bruno's Copernican Diagrams," *Filozofski Vestnik* 25, no. 2 (2004): 25–50.

66. Nicolaus Copernicus, *On the Revolutions of the Heavenly Spheres*, trans. A. M. Duncan (New York: Barnes and Noble Books, 1976), 1.8.

67. Bruno, *Ash Wednesday Supper*, 86.

68. Ibid., 89, 90.

69. Lucretius, *De rerum natura*, trans. W. H. D. Rouse, rev. Martin Ferguson Smith, Loeb Classical Library 181 (Cambridge, Mass.: Harvard University Press, 1975), 1.73.4. See also chap. 2, sec. "Accident and Infinity."

70. Bruno, *On the Infinite Universe*, 256.

71. "Nor is it necessary that these hypotheses should be true, nor indeed even probable, but it is sufficient if they merely produce calculations which agree with the observations" (Andreas Osiander, preface to Copernicus, *On the Revolutions*, 22).

72. Bruno, *Ash Wednesday Supper*, 137.

73. Gatti, "Giordano Bruno's Copernican Diagrams," 25.

74. According to Bruno, the void is not a vacuum, but the "spirit," "air," or "ether" between and within things. Bruno insists that this is the way Lucretius and Epicurus understood the void, too:

> [T]he ancients like ourselves regarded the Void as that in which a body may have its being, that which has containing power and contains atoms and bodies. Aristotle is alone in defining the void as that which is nothing, within which is nothing and which can be nothing but nothing. Giving to the Void a name and meaning accepted by none else, he raises castles in the air, and destroys his own Void, but not the Void discussed by all others who have used the term. (*On the Infinite Universe*, 274, translation altered slightly)

75. Ibid., 273 (translation altered slightly), 240.

76. Giordano Bruno, *Cause, Principle, and Unity*, trans. Richard J. Blackwell, in *Cause, Principle, and Unity and Essays on Magic*, ed. Richard J. Blackwell and Robert de Lucca, Cambridge Texts in the History of Philosophy (Cambridge: Cambridge University Press, 1998), 89 (spelling Americanized).

77. Bruno, *On the Infinite Universe*, 239.

78. Among these explanations, see, in particular, Miguel A. Granada, "L'infinité de l'univers et la conception du Système Solaire chez Giordano Bruno," *Revue des sciences philosophiques et théologiques* 82 (1998): 243–75.

79. Bruno, *Cause, Principle, and Unity*, 91, 44, 10.

80. Jorge Luis Borges, "The Aleph," in *Collected Fictions*, 281.

81. Bruno, *On the Infinite Universe*, 307.

82. Ibid., 304.

83. Miguel A. Granada, "Kepler and Bruno on the Infinity of the Universe and of Solar Systems," *Journal for the History of Astronomy* 39 (2008): 470.

84. Bruno, *On the Infinite Universe*, 328.

85. Bruno, *Ash Wednesday Supper*, 155–56.

86. Ibid., 267 (translation altered slightly).

87. Miguel Granada, personal correspondence to the author, July 26, 2011.

88. Bruno, *On the Infinite Universe*, 267 (translation altered slightly), 239.

89. Ibid., 256, 261.

90. Bruno, *Cause, Principle, and Unity*, 91. In the later *Articuli adversos mathematicos*, Bruno calls the universe an infinite sphere, attributing to it the infinite power and mo-

tion that Cusa reserved for God-in-the-universe. See Granada, "Aristotle, Copernicus, Bruno," 110.

91. We might note that in *De docta ignorantia*, Christ performs the function that the *posse fieri* will perform in Cusa's last works. For an extended treatment of Cusan Christology, see Bernard McGinn, "Maximum Contractum et Absolutum: The Motive for the Incarnation in Nicholas of Cusanus and His Predecessors," in *Nicholas of Cusa and His Age: Intellect and Spirituality: Essays Dedicated to the Memory of F. Edward Cranz, Thomas P. McTighe, and Charles Trinkaus*, ed. Thomas Izbicki and Christopher M. Bellitto, Studies in the History of Christian Traditions 105 (Boston: Brill, 2002), 149–74.

92. Blumenberg, *Legitimacy of the Modern Age*, 569.

93. Ibid., 550.

94. Ibid., 564–65; Miguel A. Granada, "Mersenne's Critique of Giordano Bruno's Conception of the Relation Between God and the Universe: A Reappraisal," *Perspectives on Science* 18, no. 1 (2010): 32, 40, and "Kepler and Bruno," 471.

95. Bruno, *On the Infinite Universe*, 265.

96. Granada, "Mersenne's Critique," 37.

97. Ibid., 39. As Granada points out, this is the orthodox distinction between *potentia absoluta* and *potentia ordinata*. Of course, Mersenne's God *does* act "necessarily and infinitely, but he does it on the *interior* level of the derivation of the Son" (ibid., 40, emphasis in original). For Bruno, by contrast, there *is* no "interior level" of divinity, no distinction between the "immanent" and "economic" Trinity.

98. Bruno, *On the Infinite Universe*, 260 (translation of first quote altered slightly).

99. Ibid., 246 (emphasis added).

100. Bruno, *Cause, Principle, and Unity*, 72 (subsequent references are cited in the text).

101. Aristotle, *The Metaphysics*, trans. Hugh Lawson-Tancred (New York: Penguin, 1998), 986a (emphasis added).

102. Grace Jantzen, *Becoming Divine: Towards a Feminist Philosophy of Religion* (Bloomington: Indiana University Press, 1999), 267.

103. Bruno, *On the Infinite Universe*, 229 (translation altered slightly).

104. As Calcagno points out, this passage must be read in light of the words that precede it: "I hate the mob, I loathe the vulgar herd." Especially if we read it alongside a similar statement in Bruno's *Ash Wednesday Supper* ("The mob . . . never contributes anything valuable and worthy. Things of perfection and worth are always to be found among the few" [100]), we can see Bruno's denigration of the "multitude" not to be an argument "against multiplicity itself, but against the uncritical mentality of the masses or, to use Nietzsche's language, 'the herd'" (Calcagno, *Giordano Bruno*, 162).

105. For alternative interpretations, see Calcagno, *Giordano Bruno*, 95–109; and Kevin Decker, "The Open System and Its Enemies: Bruno, the Idea of Infinity, and Speculation in Early Modern Philosophy of Science," *Catholic Philosophical Quarterly* 74, no. 4 (2000): 610.

106. Rowland, *Giordano Bruno*, 10, 274.

107. Maurice A. Finocchiaro, "Philosophy Versus Religion and Science Versus Religion: The Trials of Bruno and Galileo," in *Giordano Bruno: Philosopher of the Renaissance*, ed. Hilary Gatti (Burlington, Vt.: Ashgate, 2002), 56.

108. Bruno, *Cause, Principle, and Unity*, 90.

4. Measuring the Immeasurable

1. Alexander Koyré, *From the Closed World to the Infinite Universe* (Baltimore: Johns Hopkins University Press, 1957), 66–139; Steven J. Dick, *Plurality of Worlds: The Origins of the Extraterrestrial Life Debate from Democritus to Kant* (Cambridge: Cambridge University Press, 1982); Michael J. Crowe, *The Extraterrestrial Life Debate, 1750–1900: The Idea of a Plurality of Worlds from Kant to Lowell* (Cambridge: Cambridge University Press, 1986), 3–37; Karl S. Guthrie, *The Last Frontier: Imagining Other Worlds, from the Copernican Revolution to Modern Science Fiction*, trans. Helen Atkins (Ithaca, N.Y.: Cornell University Press, 1990), 43–198.

2. As early as 1601, Nicholas Hill of the Northumberland Circle (a group that the duke convened that included John Donne, Christopher Marlowe, and Walter Raleigh, among others) published *Philosophia epicurea*, which offers a rich description of the inhabitants of other planetary bodies. See Hugh Trevor-Roper, "Nicholas Hill, the English Atomist," in *Catholics, Anglicans, and Puritans: Seventeenth Century Essays* (Chicago: University of Chicago Press, 1987), 1–39.

3. Galileo Galilei, *Sidereus Nuncius, or, A Sidereal Message*, trans. William R. Shea (Sagamore Beach, Mass.: Science History, 2009).

4. As Crowe defines "pluralism" in this context, it is "the doctrine that the earth is but one of the inhabited planets of our solar system and the stars are suns surrounded by planets." Along this definition, he explains, "Copernicus, Kepler, [Tycho] Brahe, and Galileo cannot be called pluralists" (*Extraterrestrial Life Debate*, 10), even though their work reignited the pluralist project in the seventeenth century.

5. Guthrie, *Last Frontier*, 60.

6. Galileo Galilei, *Dialogue Concerning the Two Chief World Systems: Ptolemaic and Copernican*, trans. Stillman Drake, foreword by Albert Einstein (Berkeley: University of California Press, 1967). In addition to asserting the mobility of the earth, Galileo was charged with believing that there was water on the moon and that there were other worlds with human inhabitants, neither of which was a position he held. Nevertheless, Tomasso Campanella would defend these "Galilean" principles on scriptural and scientific grounds alike in *Defense of Galileo* (1622). See Guthrie, *Last Frontier*, 140–41.

7. Quoted in Maurice A. Finocchiaro, "Philosophy Versus Religion and Science Versus Religion: The Trials of Bruno and Galileo," in *Giordano Bruno: Philosopher of the Renaissance*, ed. Hilary Gatti (Burlington, Vt.: Ashgate, 2002), 74.

8. It is a good bet that Galileo did not say this insofar as the first source to suggest he did was published in 1757. See Guiseppe Marco Antonio Baretti, *The Italian Library*, electronic reproduction (Farmington Hills, Mich.: Thomson Gale, 2003), 52.

9. Robert Kargon, "Thomas Hariot, the Northumberland Circle, and Early Atomism in England," *Journal of the History of Ideas* 27, no. 1 (1966): 128–36.

10. An English translation of the *Syntagma* can be found in *The Selected Works of Pierre Gassendi*, trans. Craig B. Brush (New York: Johnson Reprint, 1972), 279–434.

11. Ibid., 399.

12. Dick, *Plurality of Worlds*, 55; Guthrie, *Last Frontier*, 172.

13. Dick, *Plurality of Worlds*, 59, 150.

14. Guthrie, *Last Frontier*, 172–74.

15. The natural philosopher and Anglican bishop John Wilkins made this distinction in Proposition 2 of *Discovery of a World in the Moone* (1638) in order to clarify what it was he was affirming when he affirmed a plurality of worlds:

> The term World may be taken in a double sense, more generally for the whole Universe, as it implies in it the elementary and aethereall bodies, the starres and the earth. Secondly, more particularly for an inferior World consisting of elements [like the Moon] . . . so that in the first sense I yeeld, that there is but one world, which is all that the arguments do prove; but understand it in the second sense, and so I affirm there may be more. (quoted in John D. Barrow, *The Infinite Book: A Short Guide to the Boundless, Timeless, and Endless* [New York: Vintage, 2005], 301n.9)

16. Nicholas of Cusa, *On Learned Ignorance*, in *Nicholas of Cusa: Selected Spiritual Writings*, trans. H. Lawrence Bond (New York: Paulist Press, 1997), 2.12.172. See also chap. 3, sec. "End Without End."

17. Finocchiaro, "Philosophy Versus Religion," 66.

18. Augustine of Hippo, *Concerning the City of God Against the Pagans*, trans. Henry Bettenson (New York: Penguin, 2003), 16.9. A century earlier Lactantius (ca. 240–320) had also ruled out the possibility of "antipodeans," largely by ridiculing the notion that there might be "men whose footsteps are higher than their heads," vegetables that "grow downwards," and a heaven that lies below the earth ("The Divine Institutes," in *The Ante-Nicene Fathers: Translations of the Writings of the Fathers Down to A.D. 325*, ed. Alexander Roberts and James Donaldson [Edinburgh: Clark, 1911], 7:bk. 3, chap. 24, Internet Sacred Text Archive, http://sacred-texts.com/chr/ecf/007/0070075.htm [accessed June 4, 2013]).

19. On the scope of this debate, see Rudolph Simek, *Heaven and Earth in the Middle Ages: The Physical World Before Columbus* (Woodbridge, Eng.: Boydell & Brewer, 1996), 48–55.

20. Pope Paul III, *Sublimus Dei* (1573), Papal Encyclicals Online, http://www.papalencyclicals.net/Paul03/p3subli.htm (accessed June 4, 2013).

21. Campanella's argument is loosely paraphrased from Guthrie, *Last Frontier*, 139. Similar comparisons between "antipodeans" and extraterrestrials were made by Edmund Spencer, Johannes Kepler, Henry More, Pierre Borel, and Bernard le Bovier de Fontenelle. The last compares his journey through different worlds with a journey across the globe, encountering increasingly strange, barbaric, and exotic races of "men" along the way—from "the English" to "the Iriquois" to "the women of Jesso" to "the Tartars" and "the beautiful Circassians." Later, he puts earthlings in the position not of Europeans, but of Native Americans, who "were so ignorant that they hadn't the slightest suspicion anyone could make roads across such vast seas." Someday, Fontenelle imagines, there will likely be "communication between the earth and the moon," but most earthlings are too unimaginative to foresee it (*Conversations on the Plurality of Worlds*, trans. H. A. Hargreaves [Berkeley: University of California Press, 1990], 21, 34).

22. John Locke suggested in 1689 that humans are in all likelihood "one of the lowest of all intellectual beings" (*An Essay Concerning Human Understanding: Complete and Unabridged* [Milwaukee: WLC Books, 2009], 4.3.23). By contrast, Christiaan Huygens

imagined in 1698 that all rational beings throughout the universe are likely to display the same "mixture of good with bad, of wise with fools, of war with peace" as earthlings do (*Cosmotheoros: Or, Conjectures Concerning the Planetary Worlds, and Their Inhabitants*, electronic reproduction [Glasgow: Robert Urie, 1762], 33).

23. Bernard le Bovier de Fontenelle imagined that the closer a planet was to the earth, the more humanlike its inhabitants would be, in "But One Little Family of the Universe," trans. Aphra Behn, in *The Book of the Cosmos: Imagining the Universe from Heraclitus to Hawking*, ed. Dennis Richard Danielson (Cambridge, Mass.: Perseus, 2000), 216.

24. Nicholas Hill thought that there were pygmies on the moon and giants on the sun; Christian Wolff calculated that the average height of Jupiterians was $13^{819}/_{1440}$ feet. See Guthrie, *Last Frontier*, 77; and Crowe, *Extraterrestrial Life Debate*, 30. Assuming that there would be astronomers on other planets, Huygens reasoned that in order to make use of their "tubes and engines," the extraterrestrials would have to be "larger than, or at least equal to, ourselves, especially in Jupiter and Saturn, which are so vastly bigger than the planet which we inhabit" (*Cosmotheoros*, 61).

25. A stark difference in opinion on this question can be seen between Fontenelle, who wrote that "men" on other planets were not men at all (*Conversations on the Plurality of Worlds*, 6), and Huygens, who believed that they were just like earthlings (*Cosmotheoros*, 60–65).

26. This set of questions was opened only seven years after the publication of Nicolaus Copernicus's *De revolutionibus*, when the Lutheran theologian Philip Melanchthon (1497–1560) used them to demonstrate the heretical nature of heliocentrism. In *Initia doctrina physicae* (1550), Melanchthon argued that if there were more than one world, Christ would have to die more than once, contrary to St. Paul's assurances. In response to this argument, Pierre Borel (1620–1671) countered in a 1657 treatise that "though we should certainly know, that those men in the stares have need of salvation, God hath so many means and wayes, to us unknown, for to save them . . . that we need not inform our selves about these things, but believe them in faith" (*A New Treatise Proving a Multiplicity of Worlds: That the Planets Are Regions Inhabited, and the Earth a Star, and That It Is out of the Center of the World in the Third Heaven, and Turns Round Before the Sun Which Is Fixed. And Other Most Rare and Curious Things*, trans. D. Sashott [London: John Streater, 1658], 34.139–40). Another resolution is posited by Henry More (1614–1687), whose *Divine Dialogues* (1668) reasoned that God could save extraterrestrials by revealing to them what he had done on the earth—just as God had saved Americans by revealing the good news from overseas (*Divine Dialogues Containing Disquisitions Concerning the Attributes and Providence of God*, 3 vols. [Glasgow: Robert Foulis, 1743], 3:420–21). Thomas Paine (1737–1809), by contrast, accepted Melanchthon's argument in order to turn it against him:

> [A]re we to suppose that every world in the boundless creation, had an Eve, an apple, a serpent, and a Redeemer? In this case, the person who is irreverently called the Son of God, and sometimes God himself, would have nothing else to do, than to travel from world to world, in an endless succession of death, with scarcely a momentary interval of life. . . . Such is the strange construction of the Christian system of faith, that every evidence the heavens afford to man, either directly contradicts it or renders it absurd. (*The Age of Reason: Being an Investiga-*

tion of True and Fabulous Theology, electronic reproduction [Farmington Hills, Mich.: Cengage Gale, 2009], 48)

27. This question finds a parodic formulation in Gottfried Leibniz's *New Essays on Human Understanding* (written 1704, published 1765), cited in Crowe, *Extraterrestrial Life Debate*, 29.

28. John Donne, "An Anatomie of the World," in *The Poems of John Donne: Edited from the Old Editions and Numerous Manuscripts with Introductions and Commentary*, ed. Herbert J. C. Grierson, 2 vols. (Oxford: Clarendon Press, 1912), 1:237.

29. Blaise Pascal, *Pensées*, trans. A. J. Krailsheimer (New York: Penguin, 1966), nos. 202 and 42, pp. 95, 38. For the original French, see Blaise Pascal, *Pensées* (Paris: Bookking International, 1995), nos. 206–7, p. 85.

30. Although *Somnium* was written in 1609, it was not published until 1634, four years after Kepler's death. A recent English translation can be found in Johannes Kepler, *Kepler's Somnium: The Dream, or Posthumous Works on Lunar Astronomy*, trans. Edward Rosen, with commentary (Mineola, N.Y.: Dover, 2003).

31. Johannes Kepler, *Kepler's Conversation with Galileo's Sidereal Messenger*, trans. Edward Rosen, with introduction and notes (New York: Johnson Reprint, 1965), 11, 37, and *De stella nova* (1606), quoted in Koyré, *From the Closed World to the Infinite Universe*, 46–47.

32. Kepler, *De stella nova*, quoted in Koyré, *From the Closed World to the Infinite Universe*, 47.

33. Ibid.

34. As Kepler writes in *Conversation with the Sidereal Messenger* (1610), "[T]his system of planets, on one of which we humans dwell, is located in the very bosom of the world, around the heart of the universe, that is, the sun. . . . [Therefore,] we humans live on the globe which by right belongs to the primary rational creature, the noblest of the (corporeal) creatures" (*Kepler's Conversation with Galileo's Sidereal Messenger*, 43).

35. Ibid., 35–36.

36. Galileo had similarly said that even if there were an "infinite space superior to the fixed stars," it would be "imperceptible to us" and therefore not worth wondering about (*Dialogue*, quoted in Koyré, *From the Closed World to the Infinite Universe*, 72). In a number of letters, Galileo acknowledged that he had no idea whether the universe was finite or infinite, preferring to "defer" in such matters to "the higher disciplines" ("Letter to Ingoli," quoted in Kevin Decker, "The Open System and Its Enemies: Bruno, the Idea of Infinity, and Speculation in Early Modern Philosophy of Science," *Catholic Philosophical Quarterly* 74, no. 4 [2000]: 616).

37. Quoted in Miguel A. Granada, "Kepler and Bruno on the Infinity of the Universe and of Solar Systems," *Journal for the History of Astronomy* 39 (2008): 479.

38. Kepler, *Kepler's Conversation with Galileo's Sidereal Messenger*, 36–37 (emphasis added).

39. Guthrie, *Last Frontier*, 99.

40. Crowe, *Extraterrestrial Life Debate*, 12.

41. Johannes Kepler, *Mysterium Cosmographicum: The Secret of the Universe*, trans. A. M. Duncan (New York: Abaris Books, 1981), 93.

42. Steven J. Dick, "Plurality of Worlds," in *Cosmology: Historical, Literary, Philosophical, Religious, and Scientific Perspectives*, ed. Norris S. Hetherington (New York: Garland, 1993), 522.

43. Borel makes this point repeatedly throughout his treatise, arguing, for example, that "it seems requisite, that the object be the measure of the power; but this world not being infinite as God is, there must needs be an infinity of them" (*New Treatise Proving a Multiplicity of Worlds*, 39.163). Apart from Walter Charleton, the most notable exception of this era was Gerhard de Vries (1648–1705), who advocated a retrenchment of medieval cosmology and theology in the face of such rampant pluralism. See Dick, *Plurality of Worlds*, 120–21.

44. Walter Charleton, *Physiologia Epicuro-Gassendo-Charltoniana: Or a Fabrick of Science Natural Upon the Hypothesis of Atoms* (London: Tho. Newcomb for Thomas Heath, 1654), 14.

45. Ibid., 13, 12.

46. Ibid., 13.

47. Ibid., 9, 15.

48. Kepler, *Kepler's Conversation with Galileo's Sidereal Messenger*, 44.

49. Henry More, Pierre Borel, and Bernard le Bovier de Fontenelle are the most commonly cited names in this regard, but intellectual historians have uncovered scores of others. See Peter Harrison, "The Influence of Cartesian Cosmology in England," in *Descartes' Natural Philosophy*, ed. Stephen Gaukroger, John Schuster, and John Sutton, Routledge Studies in Seventeenth-Century Philosophy (New York: Routledge, 2000), 168–92; and Dick, *Plurality of Worlds*, 106–41.

50. The popularity of this book, which Guthrie calls the "astronomical best-seller of the Age of Enlightenment," seems to have been undiminished by its having been placed on the Catholic Church's Index of Prohibited Books: "By the time of the author's death in 1757, thirty-three French editions had already appeared; several English and German translations went through a number of editions" (*Last Frontier*, 227–28).

51. Fontenelle, *Conversations on the Plurality of Worlds*, 4.

52. On the relationship between Fontenelle's and Huygens's books, see Guthrie, *Last Frontier*, 239–44.

53. Huygens, *Cosmotheoros*, 115.

54. Ibid., 119.

55. In response to Pierre Chanut, who wrote on behalf of Queen Christina of Sweden to express Her Majesty's concern that Cartesianism might involve believing in other, more intelligent beings, Descartes stated that "I always leave undecided questions of this kind rather than denying or affirming anything" (quoted in Crowe, *Extraterrestrial Life Debate*, 16). That said, he assured the queen that if there were "men" on other planets, nothing would prevent God from redeeming even an infinite number of kinds of them.

56. "I dare not call [the world] infinite as I perceive that God is greater than the world, not in respect to His extension . . . but in regard to his position" (René Descartes, "Letter to Henry More," quoted in Koyré, *From the Closed World to the Infinite Universe*, 90).

57. René Descartes, *Principles of Philosophy*, in *The Philosophical Writings of Descartes*, ed. John Cottingham, Robert Stoothoff, and Dugald Murdoch (Cambridge: Cambridge University Press, 1984), 2.22.

58. René Descartes to Marin Mersenne, October 8, 1629, quoted in Michael Sean Mahoney, introduction to René Descartes, *Le monde, ou Traité de la lumière*, trans. Michael Sean Mahoney, Janus Library (New York: Abaris Books, 1979), viii. See also chap. 3, sec. "End Without End."

59. Steven J. Dick, *Life on Other Worlds: The 20th-Century Extraterrestrial Life Debate* (Cambridge: Cambridge University Press, 1998), 12.

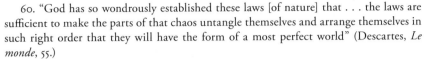

60. "God has so wondrously established these laws [of nature] that . . . the laws are sufficient to make the parts of that chaos untangle themselves and arrange themselves in such right order that they will have the form of a most perfect world" (Descartes, *Le monde*, 55.)

61. Descartes to Mersenne, December 18, 1629, quoted in Mahoney, introduction to ibid., x.

62. Descartes to Mersenne, November 1633, quoted in Mahoney, introduction to ibid., xii.

63. Ibid.

64. Descartes, *Principles of Philosophy*, 1.26.

65. Koyré, *From the Closed World to the Infinite Universe*, 85–88.

66. Descartes, *Principles of Philosophy*, 1.27.

67. René Descartes to Henry More, February 5, 1649, quoted in Koyré, *From the Closed World to the Infinite Universe*, 86.

68. Descartes, *Le monde*, 49 (subsequent references are cited in the text).

69. Paul J. Steinhardt and Neil Turok, *Endless Universe: Beyond the Big Bang—Rewriting Cosmic History* (New York: Broadway Books, 2008), 226.

70. Descartes, *Principles of Philosophy*, 2.10.

71. Ibid., 2.16.

72. Ibid., 2.20.

73. Ibid.

74. Ibid., 1.26.

75. Ibid., 2.22.

76. In his published work, Descartes attempts to temper his heliocentrism by saying that "the earth is at rest in its own heaven, but nonetheless it is carried along by it" (ibid., 3.26.)

77. Descartes's ideas were partially revived and reconceived in the nineteenth-century "vortex theory of matter." With the exception of Peter Guthrie Tait and Balfourt Stewart's *The Unseen Universe* (1875), however, this theory did not offer a cosmology, much less a cosmogony. See Helge Kragh, *Higher Speculations: Grand Theories and Failed Revolutions in Physics and Cosmology* (Oxford: Oxford University Press, 2011), 35–57.

78. In this equation, F is the force of gravity, g is the gravitational constant (9.9 m/s^2), M is the mass of one body, m is the mass of the other, and r is the distance between them.

79. Richard S. Westfall, "Newtonian Cosmology," in *Cosmology*, ed. Hetherington, 273.

80. In the *Principia*, Newton writes, "I do not define time, space, place, and motion, as are well known to all," before going on to do so anyway. "Absolute, true, and mathematical time, of itself, and from its own nature flows equibly without regard to anything external," he explains and similarly says, "Absolute space, in its own nature, without regard to anything external, remains always similar and unmovable" (*The Principia*, trans. Andrew Motte, Great Minds Series [Amherst, N.Y.: Prometheus Books, 1995], 13). An early articulation of the difference between space and matter can be found in the "De gravitatione," an essay most likely written at some point between 1664 and 1668. See Isaac Newton, "De gravitatione et aequipondio fluidorum," in *Unpublished Scientific Papers of Isaac Newton*, ed. A. Rupert Hall and Marie Boas Hall (Cambridge: Cambridge University Press, 1962), 89–156, esp. 142–43, 148.

81. Gottfried Leibniz, "Third Letter" (February 25, 1716), in G. W. Leibniz and Samuel Clarke, *Correspondence*, ed. Roger Ariew (Indianapolis: Hackett, 2000), 14.

82. "Revolution in Science: New Theory of the Universe, Newtonian Ideas Overthrown," *Times* (London), November 17, 1919.

83. Newtonian physics works well at the level of bouncing balls and even solar systems, but it is far less reliable at the level of large-scale structures such as galaxies, which require the use of general relativity.

84. At the very end of the *Principia*, Newton confesses to this difficulty:

> Hitherto we have explained the phaenomena of the heavens and of our sea by the power of gravity, but have not yet assigned the cause of this power. This is certain, that it must proceed from a cause that penetrates to the very centres of the sun and planets, without suffering the least diminution of its force. . . . But hitherto I have not been able to discover the cause of those properties of gravity from phaenomena, and I frame no hypotheses; for whatever is not deduced from the phaenomena is to be called an hypothesis; and hypotheses, whether metaphysical or physical, whether of occult qualities or mechanical, have no place in experimental philosophy. (442–43)

This qualification notwithstanding, both Leibniz and Huygens accused Newton of introducing an "occult force" into the universe: a force through which objects can operate on other objects without coming into contact with them. For a colorful treatment of this debate, see Edward Dolnick, *The Clockwork Universe: Isaac Newton, the Royal Society, and the Birth of the Modern World* (New York: Harper Perennial, 2012), 305. Even years later, in the Bentley correspondence, Newton is careful to correct the misattribution to him of what Einstein would call "spooky action at a distance": "You sometimes speak of Gravity as essential and inherent to matter," he writes toward the end of his second letter. "Pray do not ascribe that notion to me; for the cause of gravity is what I do not pretend to know, and therefore would take more time to consider of it" (Isaac Newton, *Four Letters from Sir Isaac Newton to Doctor Bentley, Containing Some Arguments in Proof of a Deity* [London: Dodsley, 1756], 20).

85. Edward R. Harrison, "Newton and the Infinite Universe," *Physics Today* 39, no. 2 (1986): 27.

86. Newton, *Four Letters*, 1 (capitalization regularized in this and all subsequent quotations from the letters). In the later *Opticks* (1704), Newton will integrate a Gassendi-style atomism into his theory of light, writing that "it seems probable to me that God in the beginning formed matter in solid, massy, hard, impenetrable, movable particles" (quoted in Marcelo Gleiser, *The Dancing Universe: From Creation Myths to the Big Bang* [Hanover, N.H.: University Press of New England/Dartmouth College Press, 2005], 143).

87. Newton, *Four Letters*, 3–4.

88. Ibid., 20.

89. Ibid., 2.

90. This was not always Newton's view; in his early, anti-Cartesian "De gravitatione," he affirms a finite cosmos suspended in an infinite void. See Newton, "De gravitatione et aequipondio fluidorum"; and Harrison, "Newton and the Infinite Universe."

91. Newton, *Four Letters*, 2–3.

92. Quoted in Harrison, "Newton and the Infinite Universe," 28.

93. Newton, *Four Letters*, 15.

94. Leibniz, "First Letter" (December 1715), in Leibniz and Clarke, *Correspondence*, 4.

95. Newton, *Four Letters*, 24.

96. Ibid., 25.

97. Richard Bentley, "The Folly of Atheism and (What Is Now Called) Deism, Even with Respect to the Present Life," in *The Works of Richard Bentley*, ed. Alexander Dyce (London: Francis McPherson, 1838), 3:1–26; Thomas Wright, *An Original Theory or New Hypothesis of the Universe (1750): A Fascimile Reprint Together with the First Publication of "A Theory of the Universe" (1734)*, introduction and transcription by Michael A. Hoskin (New York: American Elsevier, 1971), in which Wright glosses the phrase "a finite view of infinity" (174); Johann Lambert, *Cosmologische Briefe über die Einrichtung des Weltbaues* (Augsburg: Kletts Wittib, 1761), translated as *Cosmological Letters on the Arrangement of the World-Edifice*, trans. Stanley L. Jaki (New York: Science History, 1976). The possibility of many worlds was also briefly entertained in Ruggero Giuseppe Boscovich, *Theoria philosophiae naturalis* (1758), translated as *A Theory of Natural Philosophy, Put Forward by Roger Joseph Boscovich* (Cambridge, Mass.: MIT Press, 1966).

98. Immanuel Kant, *Allgemeine Naturgeschichte und Theorie des Himmels* (Berlin: Aufbau, 1995), frontispiece, translation (altered slightly) from Immanuel Kant, *Universal Natural History and Theory of the Heavens*, trans. W. Hastie (Ann Arbor: University of Michigan Press, 1969).

99. Milton Munitz, "Introduction," in Kant, *Universal Natural History*, trans. Hastie, xiii.

100. Crowe, *Extraterrestrial Life Debate*, 58. On the similarities and differences between Kant and Lambert, see Stanley L. Jaki, introduction to Lambert, *Cosmological Letters*, 17–18.

101. Munitz, "Introduction," vii.

102. The first description is from Michel Serres, *The Birth of Physics*, trans. Jack Hawkes (Manchester, Eng.: Clinamen Press, 2000), 36; the second, from Peter Coles, "Key Themes and Major Figures," in *The Routledge Companion to the New Cosmology*, ed. Peter Coles (New York: Routledge, 2001), 225.

103. For a roundup of the little secondary material there is on the *Natural History*, see—in addition to other sources cited in this chapter, especially Serres, *Birth of Physics*; and Peter D. Fenves, *A Peculiar Fate: Metaphysics and World History in Kant* (Ithaca, N.Y.: Cornell University Press, 1991)—Martin Schönfeld, "The Phoenix of Nature: Kant and the Big Bounce," *Collapse* 5 (2009): 361–76.

104. Stanley Jaki, introduction to Immanuel Kant, *Universal Natural History and Theory of the Heavens*, trans. Stanley L. Jaki (Edinburgh: Scottish Academic Press, 1981), 8, 5, 12, 8. Although there is nothing objectionable about Jaki's effort to correct a perhaps facile philosophical reception of this work, it is a bit jarring to find this argument laid out in the introduction to his own translation of it. Jaki's dismissive, often angry framing does not dispose the reader to think highly of what follows, which is perhaps part of the reason the book is out of print. In fact, there remains no standard English translation of the *Universal Natural History*: Hastie's translation is incomplete; Jaki's is out of print; Johnson's is printed by a very small publishing house; and Cambridge University Press has not yet delivered the edition it has long promised.

105. Schönfeld, "Phoenix of Nature," 364–65; Fenves, *Peculiar Fate*, 21; Crowe, *Extraterrestrial Life Debate*, 53.

106. Kant, *Universal Natural History*, trans. Jaki, 81; "Solche Einsichten scheinen sehr weit die Kräfte der menschlichen Vernunft zu überschreiten" (*Allgemeine Naturgeschichte*,

39). Subsequent references are cited in the text, with the page numbers of the Jaki translation followed by the page numbers of the German text. I have also consulted Hastie's translation and provide notes when using his language.

107. As Serres puts it, the *Natural History* "opens with the principles of mechanics in the manner of the atomists, it cites Epicurus, while apologizing for this reference to an atheist.... Kant never abandoned Lucretius, he thought, perhaps, of leaving him behind in the introduction, but he remains . . . when all is said and done, an Epicurean" (*Birth of Physics*, 36). Or, in the words of Fenves, Kant is "compulsively drawn" to Lucretius (*Peculiar Fate*, 28).

108. Lucretius, *De rerum natura*, trans. W. H. D. Rouse, rev. Martin F. Smith, Loeb Classical Library 181 (Cambridge, Mass.: Harvard University Press, 1975), 2.216–17; Serres, *Birth of Physics*, 5.

109. Lucretius, *De rerum natura*, 2.220. See also chap. 2, sec. "Accident and Infinity."

110. Newton, *Four Letters*, 34.

111. Kant, *Natural History*, trans. Hastie, 73; *Allgemeine Naturgeschichte*, 86.

112. Lucretius, *De rerum natura*, 1.63–79, 1.73–74.

113. Ibid., 5.8–12.

114. Although Kant's most famous refutations of the classical arguments for the existence of God can be found in *Critique of Pure Reason* (1787), his first refutation of the teleological and cosmological proofs can be found in his essay "Der einzig möglich Beweisgrund zu einer Demonstration des Daseyns Gottes" (1763). For a translation, see Immanuel Kant, "The Only Possible Argument in Support of a Demonstration of the Existence of God," in *Theoretical Philosophy, 1755–1770*, ed. David Walford (Cambridge: Cambridge University Press, 1992), 107–201.

115. "Es ist ein Gott eben deswegen, weil die Natur auch selbst in Chaos nicht anders als regelmäßig und ordentlich verfahren kann" (emphasis removed).

116. Lucretius, *De rerum natura*, 5.193–95.

117. Kant, *Universal Natural History*, trans. Hastie, 139; *Allgemeine Naturgeschichte*, 145.

118. Hugyens states, "I must be of the same opinion with all the philosophers of our age, that the sun is of the same nature with the fixed stars" (*Cosmotheoros*, 113–14).

119. Wright had, however, speculated that there might be other systems of stars beyond the Milky Way, even an "endless immensity" of them—a prospect that seems to have caught Kant's imagination (quoted in John D. Barrow, *The Book of Universes: Exploring the Limits of the Cosmos* [New York: Norton, 2011], 27).

120. According to Steven Weinberg, the "apparent position" of the relatively fast-moving Barnard's star "shifts in one year by an angle of 0.0029 degrees" (*The First Three Minutes* [New York: Basic Books, 1977], 12).

121. Lambert writes that "the satellites belong to the planets, these to the sun, the sun to its system, and this to the system of the whole Milky Way. Farther our eyes do not reach and I leave it undecided whether the Milky Way visible to us still belongs to uncounted other and forms with these a whole system" (*Cosmological Letters*, 111; see also 88). G. J. Whitrow finds one passing reference to the possibility that nebulae might be "Milky Ways" in a 1657 lecture by Christopher Wren, in "Kant and the Extragalactic Nebulae," *Quarterly Journal of the Royal Astronomical Society* 8 (1967): 55.

122. Quoted in Agnes M. Clerke, *A Popular History of Astronomy During the Nineteenth Century* (London: Adam and Charles Black, 1908), 23.

123. Whitrow, "Kant and the Extragalactic Nebulae," 53.

124. Quoted in ibid. According to Weinberg, the "hazy patch" in Andromeda was first mentioned in writing by Abdurrahman al-Sufi in 964 C.E., who "described it as 'a little cloud'" (*First Three Minutes*, 17).

125. It would take astronomers two and a half centuries to agree. Edwin Hubble's discoveries notwithstanding, the vastness of Kant's vision would not be matched among scientists until 1995, when the Hubble Deep Field Telescope was trained on an "empty" patch of sky for ten days. Astonishingly, each of the smudges of light in the resulting composite photograph is itself a galaxy, which astronomers now tell us can contain anywhere between 10 million and 100 trillion stars.

126. "Finds Spiral Nebulae Are Stellar Systems," *New York Times*, November 23, 1924.

127. Kant's cosmic hierarchy was restored to infinity in Edmund Fournier d'Albe's *Two New Worlds* (1907), which imagined the universe as an endless series of "clusters of clusters of clusters . . . ad infinitum." This idea was quickly refined and expanded by Carl Charlier and then by Franz Selety. See Barrow, *Book of Universes*, 87–89.

128. This strange conclusion, impossible for a Descartes or a Leibniz (or an Einstein), is the result of Kant's Newtonian notion of absolute space.

129. Kant, *Universal Natural History*, trans. Hastie, 148; *Allgemeine Naturgeschichte*, 153.

5. Bangs, Bubbles, and Branes

1. Pierre Simon Laplace, *Exposition du système du monde* (Paris: Bachelier, 1835), 28–33, translated as *The System of the World*, trans. J. Pond, 2 vols. (London: Phillips, 1809). Although Kant's and Laplace's cosmogonies are often mentioned in the same breath (along with Emanuel Swedenborg's), the nineteenth century saw them as markedly different. According to Agnes M. Clerke,

Laplace's primitive nebula was a coherent mass. It rotated as a whole; it divided only under considerable strain; its separated parts had individual unity. . . . Kant's elemental matter, on the contrary, was a loose aggregate of independent particles, each pursuing its way, disturbed, indeed, by its neighbours, but essentially isolated from them. They were, in short, genuine Lucretian atoms, intended to stand for the irreducible minima of Nature. (*Modern Cosmogonies* [London: Adam and Charles Black, 1905], 28)

2. Laplace does say that solar systems (not galaxies) are formed out of these nebular rotations, but he also writes that "it is . . . probable, that those nebulae, without distinct stars, are groups of stars seen from a distance, and which, if approached, would present appearances similar to the milky way [*sic*]." He then goes on to indulge in a brief Kantian reflection on what seems to be an "infinite number of nebulae," confessing that "the imagination, lost in the immensity of the universe, will have difficulty to conceive its bounds" (Laplace, *System of the World*, 2:369–70).

3. Agnes M. Clerke, *The System of the Stars*, 2nd ed. (London: Adam and Charles Black, 1905), 281. On the state and categories of nebular astronomy at the end of the nineteenth century, see 242–81.

4. Ibid., 349. When one is reading turn-of-the century accounts of the nebulae, it almost seems as though the authors do not *want* to know that there are other galaxies, even though they have a great deal of the evidence of such galaxies before them. For example, in the same chapter of *System of the Stars* that declares it ridiculous to consider the nebulae to be akin to galaxies, Clerke writes that "just as the Milky Way might be described as a great compound cluster made up of innumerable subordinate clusters, so the greater Magellanic Cloud seems to be a gigantic nebula combining into some kind of systematic unity multitudes of separate nebula" (351).

5. Edgar Allan Poe's very strange, multiply prescient *Eureka* (1848) argues that "we have no reason to suppose the Milky Way *really* more extensive than the least of these 'nebulae'" (*Eureka*, ed. Stuart Levine and Susan F. Levine [Champaign: University of Illinois Press, 2004], 73). The two other significant exceptions are Edmund Fornier d'Albe, whose *Two New Worlds* (1907) sets forth a kind of unbound Kantianism, suggesting that the universe might have a hierarchical structure that stretches infinitely outward; and Franz Selety, who argues for an infinite, centerless universe. See John D. Barrow, *The Book of Universes: Exploring the Limits of the Cosmos* (New York: Norton, 2011), 87, 90.

6. "Finds Spiral Nebulae Are Stellar Systems," *New York Times*, November 23, 1924.

7. Ibid.

8. "Ten years ago," Annie Dillard wrote in 1999,

> I read that there were two galaxies for everyone alive. Lately, since we loosed the Hubble space telescope, we have revised our figures. There are maybe nine galaxies for each of us—eighty billion galaxies. Each galaxy harbors at least one hundred billion suns. In our galaxy, the Milky Way, there are four hundred billion suns— give or take 50 percent—or sixty-nine suns for each person alive. The Hubble shows, said an early report, that the stars are "not 12 but 13 billion years old." Two galaxies, nine galaxies . . . one hundred billion suns, four hundred billion suns . . . twelve billion years, thirteen billion years. . . . These astronomers are nickel-and-diming us to death. (*For the Time Being* [New York: Knopf, 1999], 71)

9. Luce Irigaray, *This Sex Which Is Not One*, trans. Catherine Porter and Carolyn Burke (Ithaca, N.Y.: Cornell University Press, 1985).

10. The resulting "Hubble's Law" ($V = H_0 l$) states that the velocity of receding galaxies is proportional to their distance from the point of observation; that is, the farther galaxies are, the faster they are flying away from us. See Alan H. Guth, "The Inflationary Universe," in *Cosmology: Historical, Literary, Philosophical, Religious, and Scientific Perspectives*, ed. Norris S. Hetherington (New York: Garland, 1993), 413.

11. Woody Allen, dir., *Annie Hall* (United Artists, 1977).

12. The story that broke the news in London after observational confirmation of general relativity bore the headline "Revolution in Science: New Theory of the Universe, Newtonian Ideas Overthrown," *Times* (London), November 17, 1919. The American counterpart read, "LIGHTS ALL ASKEW IN THE HEAVENS; Men of Science More or Less Agog over Results of Eclipse Observations. EINSTEIN THEORY TRIUMPHS. Stars Not Where They Seemed or Were Calculated to Be, but Nobody Need Worry," *New York Times*, November 10, 1919.

13. For Leibniz's debate with Newton (through Samuel Clark), see chap. 4, sec. "An Immeasurable Abyss."

14. Albert Einstein, "Cosmological Considerations on the General Theory of Relativity (1917)," in *Cosmological Constants: Papers in Modern Cosmology*, ed. Jeremy Bernstein and Gerald Feinberg (New York: Columbia University Press, 1986), 16–26. Einstein was influenced in this regard by the astronomers Hugo Seeliger (1849–1924) and Carl von Neumann (1832–1925), who had argued at the end of the nineteenth century that Newton's equations required a (nontheological) repulsive force in order to function cosmologically. See John D. Norton, "The Cosmological Woes of Newtonian Gravitation Theory," in *The Expanding Worlds of General Relativity*, ed. Hubert Goenner, Jürgen Renn, Jim Ritter, and Tilman Sauer (Boston: Center for Einstein Studies, 1999), 271–322.

15. Quoted in Dan Hooper, *Dark Cosmos: In Search of Our Universe's Missing Mass and Energy* (New York: HarperCollins/Smithsonian Books, 2006), 143. For the next sixty years, the cosmological constant became, in the words of Robert P. Kirshner, "theoretical poison ivy." It was just assumed that lambda had a value of 0. It was not until the late 1990s that two independent research teams began detecting a lambda value that was not 0; in other words, Einstein may have thrown out the cosmological constant too soon. Dark energy might be that constant. See Robert P. Kirshner, *The Extravagant Universe: Exploding Stars, Dark Energy, and the Accelerating Cosmos* (Princeton, N.J.: Princeton University Press, 2002), xi, 215–21. See also introduction, sec. "The Rise of the Dark Lord," and chap. 5, sec. "And the Darkness Has Overcome It." Of course, some theorists did think that Einstein had abandoned lambda too soon; in particular, both Georges Lemaître and Arthur Eddington continued to figure the term in their calculations, and although Einstein doubted that such an endeavor would be successful, he reportedly said to Lemaître, "[T]out de même . . . si vous parveniez à démonstrer que la constante cosmologique n'est pas nulle, ce serait important" (Nevertheless . . . if you were able to show that the cosmological constant isn't zero, that would be important) (quoted in Georges Lemaître, "Rencontre avec A. Einstein," *Revue des questions scientifiques* 129 [1958]: 131). In the late 1980s and the 1990s, a few brave physicists—most notably Steven Weinberg, George Efstathiou, Michael Turner, and Lawrence Krauss—suggested reintroducing the term. See Steven Weinberg, "Anthropic Bound on the Cosmological Constant," *Physical Review Letters* 59 (1987): 2607–10; Brian Greene, *The Hidden Reality: Parallel Universes and the Deep Laws of the Cosmos* (New York: Knopf, 2011), 92, 146; and Lawrence Krauss, *A Universe from Nothing: Why There Is Something Rather Than Nothing* (New York: Free Press, 2012), 56.

16. Quoted in André Deprit, "Monsignor Georges Lemaître," in *The Big Bang and Georges Lemaître*, ed. André L. Berger (New York: Springer, 1984), 370. The way Lemaître himself phrases Einstein's comment to him is: "[A]près quelques remarkques techniques favorables, il conclut en disant que du point de vue physique cela lui paraissait tout à fait abominable" ("Rencontre avec A. Einstein," 129). Six years later, at a conference in California, Einstein "declared Lemaître's theory to be 'the most beautiful and satisfactory explanation of creation to which I have ever listened'" (quoted in Greene, *Hidden Reality*, 12). Lemaître's findings were initially published as Georges Lemaître, "Un univers homogène de masse constante et de rayon croissant rendant compte de la vitesse radiale des nébuleuses extragalactique," *Annales de la Société scientifique de Bruxelles* 47 (1927): 49–59.

17. Alexander Friedmann, "Über die Krümmung des Raumes," *Zeitschrift für Physik* 10 (1922): 377–86.

18. Georges Lemaître, "The Expanding Universe," *Monthly Notices of the Royal Astronomical Society* 91 (1931): 490–501. Both Lemaître's and Friedmann's calculations formed

the basis for George Gamow and Ralph Alpher's "alpha-beta-gamma" theory of 1948, the first articulation of what would become over the next few decades the "standard model" of big bang cosmology. See Ralph Alpher, Hans Bethe, and George Gamow, "The Origin of Chemical Elements," *Physical Review* 73 (1948): 803–4. Bethe actually had nothing to do with the writing of this paper; Gamow added his name so that the list of authors sounded like the first three letters of the Greek alphabet. Gamow apparently also tried to convince Robert Herman, who had collaborated with him and Alpher for a number of years, to change his name to Delter. See Helge Kragh, "Big Bang Cosmology," in *Cosmology*, ed. Hetherington, 378.

19. Clerke, *System of the Stars*, 349.

20. Stephen Weinberg, *The First Three Minutes* (New York: Basic Books, 1977), 4.

21. According to Paul Davies, "As we pass back to progressively earlier epochs of the fireball phase . . . we reach a moment when all distances in the universe have shrunk to zero. The density of matter, and the curvature of spacetime, become infinite. At this point the theory of relativity, and indeed the whole of physics, ceases to apply. Such a pathological circumstance is called a *singularity*" ("Spacetime Singularities in Cosmology," in *The Study of Time*, ed. J. T. Fraser [New York: Springer, 1978], 78).

22. Although physicists clamored throughout the mid-twentieth century to find alternatives, Stephen Hawking and Roger Penrose demonstrated in the 1970s that the singularity is inescapable ("The Singularities of Gravitational Collapse and Cosmology," *Proceedings of the Royal Society of London A* 314, no. 1519 [1970]: 529–48), and as Hawking explains it, "nowadays nearly everyone assumes the universe started with a big bang singularity" (*A Brief History of Time: The Updated and Expanded Tenth Anniversary Edition* [New York: Bantam, 1998], 53). Hawking then changed his mind, rounding out the edge of space and time by appealing to quantum mechanics rather than general relativity to describe the earliest state of the universe. See Stephen Hawking and George F. R. Ellis, "Space-Time Singularities," in *The Large-Scale Structure of Space-Time* (Cambridge: Cambridge University Press, 1973), 256–98; and James Hartle and Stephen Hawking, "Wave Function of the Universe," *Physical Review D* 28 (1983): 2960–75. But even now he claims that the universe was born "out of nothing" (Stephen Hawking and Leonard Mlodinow, *The Grand Design* [New York: Bantam, 2010], 8).

23. See, in particular, Ted Peters, "Cosmos as Creation," in *Cosmos as Creation: Theology and Science in Consonance*, ed. Ted Peters (Nashville: Abingdon Press, 1989), 45–113; Paul Copan and William Lane Craig, *Creation out of Nothing: A Biblical, Philosophical, and Scientific Exploration* (Grand Rapids, Mich.: Baker Academic, 2004); John Polkinghorne, *The Faith of a Physicist: Reflections of a Bottom-Up Thinker* (Minneapolis: Fortress, 1996); and William Lane Craig, "Philosophical and Scientific Pointers to Creation *ex Nihilo*," in *Contemporary Perspectives on Religious Epistemology*, ed. R. Douglas Geivett and Brendan Sweetman (New York: Oxford University Press, 1992), 185–200; "The Ultimate Question of Origins: God and the Beginning of the Universe," *Astrophysics and Space Science* 269–70 (1999): 723–40; and *The Kalam Cosmological Argument* (New York: Harper & Row, 1979).

24. Pius XII, "Modern Science and the Existence of God," *Catholic Mind* 49 (1952): 190 (emphasis added).

25. On the connections between Christian creation theology and big bang cosmology, see Mary-Jane Rubenstein, "Cosmic Singularities: On the Nothing and the Sovereign," *Journal of the American Association of Religion* 80, no. 2 (2012): 485–517.

26. Gerhard May, *Creatio ex nihilo: The Doctrine of "Creation out of Nothing" in Early Christian Thought*, trans. A. S. Worrall (Edinburgh: Clark, 1994); Catherine Keller, *Face of the Deep: A Theology of Becoming* (New York: Routledge, 2003), chaps. 1 and 2; Thomas Oord and Richard Livingston, eds., *Creation Options: Rethinking Initial Creation* (New York: Routledge, forthcoming); David D. Burrell, Carlo Cogliati, Janet M. Soskice, and William R. Stoeger, eds., *Creation and the God of Abraham* (Cambridge: Cambridge University Press, 2010).

27. Robert Jastrow, *God and the Astronomers* (New York: Norton, 1978), 116; Peters, "Cosmos as Creation," 79. See also Fred Hoyle, *The Nature of the Universe* (New York: Mentor, 1950).

28. Fred Hoyle, *Astronomy Today* (London: Heinemann, 1975), 165.

29. Fred Hoyle, "A New Model for the Expanding Universe," *Monthly Notices of the Royal Astronomical Society* 108 (1948): 372–82; Herman Bondi and Thomas Gold, "The Steady State Theory of the Expanding Universe," *Monthly Notices of the Royal Astronomical Society* 108 (1948): 252–70. For a full treatment of the rise and fall of this model, see Helge Kragh, "Steady State Theory," in *Cosmology*, ed. Hetherington, 391–403.

30. Arno Penzias and Robert Wilson received the Nobel Prize in Physics for this discovery in 1978. For an account of the CMB's effect on the discipline of cosmology, see "Unendings," sec. "On the Genealogy of Cosmology." A highly technical summary of these astronomers' findings can be found in A. A. Penzias and R. W. Wilson, "A Measurement of Excess Antenna Temperature at 4080 Mc/S," *Astrophysical Journal* 142 (1965): 419–21. For more narrative and less equation-heavy descriptions of the rather dramatic circumstances surrounding this discovery, see Kirshner, *Extravagant Universe*, 248–49; and George Gamow, *The Creation of the Universe*, rev. ed. (New York: Mentor, 1957), 44–76.

31. This temperature would have been approximately 3,000 Kelvin. See Gamow, *Creation of the Universe*, 64.

32. Jastrow, *God and the Astronomers*, 116.

33. This idea is technically not "biblical" at all; rather, it is the product of centuries of (primarily Christian) theological development. For a lengthier treatment of the relationship between big bang cosmology and the doctrine of *creatio ex nihilo*, see Rubenstein, "Cosmic Singularities," and "Myth and Modern Physics: On the Power of Nothing," in *Creation Options*, ed. Oord and Livingston.

34. In the twentieth-century West at least, the idea of an oscillating universe traces back to Friedmann's 1923 solutions of Einstein's gravitational equations. If there is enough matter in the universe, gravity will pull the cosmos back into a "big crunch." Friedman did consider the idea that a collapsed universe might then cycle back into life. See Barrow, *Book of Universes*, 63. For the revival of this idea in the 1960s and 1970s, see R. H. Dicke, P. J. E. Peebles, P. G. Roll, and D. T. Wilkinson, "Cosmic Black-Body Radiation," *Physical Review Letters* 98 (1965): 414–19; P. J. E. Peebles and David T. Wilkinson, "The Primeval Fireball," *Scientific American*, June 1967, 28–37; Charles Misner, Kip Thorne, and John Archibald Wheeler, "Beyond the End of Time," in *Gravitation*, ed. Charles Misner, Kip Thorne, and John Archibald Wheeler (San Francisco: Freeman, 1970), 1196–217; and C. M. Patton and John Archibald Wheeler, "Is Physics Legislated by Cosmogony?" in *Quantum Gravity: An Oxford Symposium*, ed. C. J. Isham, Roger Penrose, and Dennis W. Sciama (Oxford: Clarendon Press, 1975), 538–605.

35. John Gribbin, "Oscillating Universe Bounces Back," *Nature*, January 1, 1976, 15. Considering the persistent pairing of the ex nihilo and absolute singularity, it is perhaps

no surprise that the oscillating model's efforts to avoid the former also unsettled the latter, producing a "singularity" that is not one, but rather one of many, each of which creates a world.

36. Quoted in Barrow, *Book of Universes*, 63.

37. Perhaps the most enthusiastic of these theorists was Carl Sagan, who traveled all the way to India to explain the oscillating universe on his television show *Cosmos*. Footage of women braiding flowers into their hair, men bathing cows, and lotus flowers floating peacefully on water—all in 1979—purportedly illustrates the wisdom of "the Ancient Hindus," who knew all along that the cosmos was cyclical. Walking through temples he does not name, Sagan explains, "The Hindu religion is the only one of the world's great faiths dedicated to the idea that the cosmos itself undergoes an immense, indeed infinite, number of deaths and rebirths. It is the only religion in which the time scales correspond . . . to those of modern scientific cosmology. Of course," Sagan points out, this correspondence is "no doubt by accident" ("The Edge of Forever," *Cosmos: A Personal Voyage*, episode 10 [Cosmos Studios, 1980]).

38. Weinberg, *First Three Minutes*, 153. For his more technical refutation of the oscillating model, see Steven Weinberg, *Gravitation and Cosmology* (New York: Wiley, 1972).

39. Richard Tolman offered this critique in the same set of papers that explored the cyclical idea: *Relativity, Thermodynamics, and Cosmology* (Oxford: Clarendon Press, 1934). It was then more definitively set out by Herman Zanstra in "On the Pulsating or Expanding Universe and Its Thermodynamical Aspect," *Proceedings of the Royal Dutch Academy of Sciences, Series B* 60 (1957): 286–307.

40. Brian Greene concludes, "[E]ven in Tolman's cyclical framework, the universe would have a beginning" (*The Fabric of the Cosmos: Space, Time, and the Texture of Reality* [New York: Vintage, 2005], 406).

41. According to Weinberg, the temperature at this first instant after the bang ($t = 0.0108$ second) would have been 100,000 million degrees Kelvin, with the universe in "nearly perfect thermal equilibrium" (*First Three Minutes*, 5, 102). It is interesting to note that thermal equilibrium is equivalent to maximum entropy, or disorder, which physicists also call "chaos." See John Gribbin, *In Search of the Multiverse* (London: Allen Lane, 2009), 90; and Alex Vilenkin, *Many Worlds in One: The Search for Other Universes* (New York: Hill and Wang, 2006), 25.

42. This would mean that the dark energy's equation of state would be $w=-1$, and dark energy would be the "cosmological constant." For the most recent data on the energy density of lambda, see Miao Li, Xiao-Dong Li, and Xin Zhang, "Comparison of Dark Energy Models: A Perspective from the Latest Observational Data," *Science China: Physics, Mechanics, & Astronomy* 53, no. 9 (2010): 1631–45.

43. Einstein's law of special relativity (1905) states that nothing can travel faster than light *through space*. But there is no upper limit on the speed of space itself.

44. Kirshner, *Extravagant Universe*, 258.

45. Sean M. Carroll, *From Eternity to Here: The Quest for the Ultimate Theory of Time* (New York: Dutton, 2010), 62.

46. "*This is the way the world ends/This is the way the world ends/This is the way the world ends/Not with a bang, but a whimper*" (T. S. Eliot, "The Hollow Men," in *Collected Poems: 1909–1962* [New York: Harcourt, Brace, 1991], 82, emphasis in original).

47. Robert Caldwell and Paul J. Steinhardt, "Quintessence," *Physics World*, November 1, 2000, 31–38; Paul J. Steinhardt and Neil Turok, *Endless Universe: Beyond the Big Bang—Rewriting Cosmic History* (New York: Broadway Books, 2008), 45.

48. Robert Caldwell, Marc Kamionkowski, and Nevin N. Weinberg, "Phantom Energy and Cosmic Doomsday," *Physical Review Letters* 91 (2003): 1–4, arXiv:astro-ph/0302506v1; Ron Cowen, "Embracing the Dark Side: Looking Back on a Decade of Cosmic Acceleration," *Science News*, February 2, 2008.

49. This new consideration was due in large part to Weinberg's paper "Anthropic Bound on the Cosmological Constant."

50. Dark matter was first discovered in the early 1950s by Vera Rubin, who realized that galaxies seemed to be spinning too quickly to stay in orbit. She suggested that galaxies must therefore be more massive than we think they are; there must be some invisible substance weighing them down and preventing them from spinning into oblivion. Although dark matter does not reflect light or interact with electricity or magnetism, it can be indirectly detected by the way its gravitational pull *bends* light from distant galaxies. In fact, a "filament" of dark energy seems to have been discovered recently using the Subaru Telescope in Mauna Kea, Hawai'i. See Jörg P. Dietrich, Norbert Werner, Douglas Clowe, Alexis Finoguenov, Tom Kitching, Lance Miller, and Aurora Simionescu, "A Filament of Dark Matter Between Two Clusters of Galaxies," *Nature*, July 4, 2012, http://www.nature.com/nature/journal/vaop/ncurrent/full/nature11224.html (accessed June 6, 2013). For an introduction to dark matter geared to nonspecialists, see Lawrence Krauss, *Quintessence: The Mystery of Missing Mass in the Universe*, rev. ed. (New York: Basic Books, 2000); and Hooper, *Dark Cosmos*.

51. For an explanation of these two types of mass–energy, see National Aeronautics Space Administration (NASA), "Dark Energy, Dark Matter," *NASA Science: Astrophysics*, National Aeronautics and Space Administration, http://science.nasa.gov/astrophysics/focus-areas/what-is-dark-energy/ (accessed April 13, 2013).

52. As Rüdiger Vaas rather colorfully puts it, "If you imagine the universe as a cosmic cappuccino, the coffee stands for dark energy, the milk for dark matter, both of which we know almost nothing about; only the powdered chocolate would be what we are familiar with, namely ordinary matter made of protons, neutrons, electrons, et. cetera" ("Dark Energy and Life's Ultimate Future," in *The Future of Life and the Future of Our Civilization*, ed. Vladimir Burdyuzha [Dordrecht: Springer, 2006], 233, http://philsci-archive.pitt.edu/archive/00003271/).

53. Quoted in Richard Panek, "Out There," *New York Times*, March 11, 2007.

54. Marcelo Gleiser, *A Tear at the Edge of Creation: A Radical New Vision for Life in an Imperfect Universe* (New York: Free Press, 2010), 128.

55. Quoted in Panek, "Out There."

56. Quoted in Dennis Overbye, "Dark, Perhaps Forever," *New York Times*, June 3, 2008.

57. Lawrence Krauss and Robert J. Scherrer, "The End of Cosmology?" *Scientific American*, March 2008, 51; compare Krauss, *Universe from Nothing*, 109–18.

58. Their conclusion is not that, by analogy, there might be some major cosmic truth that is inaccessible to *us*, but that "we may be living in the only epoch in the history of the universe when scientists can achieve an accurate understanding of the true nature of the universe" (Krauss and Scherrer, "End of Cosmology?" 47).

59. Michael D. Lemonick, "The End," *Time*, June 25, 2001.

60. Kirshner, *Extravagant Universe*, 258.

61. Greene, *Fabric of the Cosmos*, 301.

62. In "Runaway Universe," *NOVA*, PBS, November 11, 2000.

63. Seth Shostak, "The Lugubrious Universe," *Huffington Post*, November 26, 2010 (emphasis in original).

64. Vilenkin, *Many Worlds in One*, 93. Andrei Linde similarly states: "One can draw some optimism from knowing that even if our civilization dies, there will be other places in the universe where life will emerge again and again, in all its possible forms" ("The Self-Reproducing Inflationary Universe," *Scientific American*, November 1994, 55).

65. Alan H. Guth, "Inflationary Universe: A Possible Solution to the Horizon and Flatness Problems," *Physical Review D* 23 (1981): 347–56. For an introduction friendly to nonspecialists, see Guth, "Inflationary Universe."

66. Greene, *Fabric of the Cosmos*, 272.

67. Weinberg, *First Three Minutes*, 8.

68. Steinhardt and Turok, *Endless Universe*, 52.

69. Greene, *Hidden Reality*, 24.

70. The density parameter of the universe is expressed as $\Omega_{tot}=\rho_{tot}/\rho_c$, "where ρ_{tot} is the average total mass density of the universe and $\rho_c=3H^2/8\pi G$ is the critical density, the density that would make the universe spatially flat." A broadly cited measurement is that "the present value of Ω_{tot} is equal to one within a few percent ($\Omega_{tot}=1.012\pm^{0.018}_{0.022}$)" (Alan H. Guth, "Eternal Inflation and Its Implications," *Journal of Physics A: Mathematical and Theoretical* 40 [2007]: 6813, arXiv:hep-th/0702178). See also Max Tegmark, Michael A. Strauss, Michael R. Blanton, Kevork Abazajian, Scott Dodelson, Havard Sandvik, Xiaomin Wang, et al., "Cosmological Parameters from SDSS and WMAP," *Physical Review D* 69 (2004): 1–26.

71. Greene, *Fabric of the Cosmos*, 291; Martin A. Bucher and David N. Spergel, "Inflation in a Low-Density Universe," *Scientific American*, January 1999, 64.

72. Andrei Linde, "The Inflationary Multiverse," in *Universe or Multiverse?* ed. Bernard Carr (Cambridge: Cambridge University Press, 2007), 127.

73. Ibid.

74. In a balletic ceremony marking a medical school graduation, the doctor-to-be intones, "Mihi a docto doctore / Domandatur causem et rationem quare / Opium facit dormire / A quoi respondeo / Quia est in eo / Virtus dormitiva / Cujus est natura / Sensus assoupire" (roughly translated: "The learned doctor asks me for the cause and the reason that opium puts people to sleep, to which I respond, 'It is because of a dormitive property, whose nature is to make the senses drowsy' "). The chorus then replies, "Bene, bene, bene, bene respondere!" (Molière, *Le malade imaginaire*, in *Théatre Complet de Molière*, ed. Robert Jouanny [Paris: Éditions Garnier Frères, 1956], 1636–730).

75. Bucher and Spergel, "Inflation in a Low-Density Universe," 64.

76. Inflation also addresses and resolves the "monopole problem." See Guth, "Eternal Inflation and Its Implications," 6813; and Barrow, *Book of Universes*, 197.

77. Guth, "Inflationary Universe: A Possible Solution."

78. Hawking and Mlodinow, *Grand Design*, 129. Some theorists call inflationary energy a kind of super dark energy, as do Carroll, *From Eternity to Here*; and Gribbin, *In Search of the Multiverse*. Others are a bit more conservative; when asked about the relationship between inflationary and dark energies, Michael Turner told a geophysicist-

journalist that thanks to their vastly different strengths, "it seems unlikely that they're related," but then he added, "which is a good reason to pursue that idea" (quoted in Eric Hand, "The Test of Inflation," *Nature*, April 16, 2009, 821).

79. Hawking and Mlodinow, *Grand Design*, 129.

80. Sounding remarkably like Kant, Greene explains that

[a]t the close of inflation in our bubble universe, regions with slightly more energy . . . exerted a slightly stronger gravitational pull, attracting more particles from their surroundings and thus growing larger. The larger aggregate, in turn, exerted an even stronger gravitational pull, thus attracting yet more matter and growing larger still. In time, this snowball effect resulted in the formation of clumps of matter and energy that, over billions of years, evolved into galaxies and the stars within them. (*Hidden Reality*, 60–61)

81. Max Tegmark, "Parallel Universes," *Scientific American*, May 2003, 41–51, and "The Multiverse Hierarchy," in *Universe or Multiverse?* ed. Carr, 99–125.

82. For efforts to demonstrate universal finitude, see M. Lachieze-ray and Jean-Pierre Luminet, "Cosmic Topology," *Physics Reports* 254 (1995): 135–214, arXiv:gr-qc/ 9605010 (updated version); Jean-Pierre Luminet, Glenn D. Starkman, and Jeffrey R. Weeks, "Is Space Finite?" *Scientific American*, April 1999, 90–97; and John D. Barrow and Janna Levin, "The Copernican Principle in Compact Spacetimes," *Monthly Notices of the Royal Astronomical Society* 346 (2003): 615–18, arXiv:gr-qc/0304038v1. For popular introductions to the problem of cosmic infinity, see John D. Barrow, *The Infinite Book: A Short Guide to the Boundless, Timeless, and Endless* (New York: Vintage, 2005); and Janna Levin, *How the Universe Got Its Spots: Diary of a Finite Time in a Finite Space* (New York: Anchor, 2003). This work notwithstanding, Greene has written that "although observations leave the finite-versus-infinite issue undecided, I've found that when pressed, physicists and cosmologists tend to favor the proposition that the universe is infinite" (*Hidden Reality*, 26).

83. "Cosmologists assume that our universe, with an almost uniform distribution of matter and initial density fluctuations of one part in 100,000, is a fairly typical one (at least among those that contain observers)" (Tegmark, "Parallel Universes," 42).

84. Greene, *Hidden Reality*, 27.

85. See, for example, ibid., 10; and Tegmark, "Parallel Universes," 41.

86. For this mashup, see KimKierkegaardashian, https://twitter.com/#!/KimKierkeg aard (accessed June 6, 2013).

87. Foundational papers include Paul J. Steinhardt, "Natural Inflation," in *The Very Early Universe*, ed. Gary W. Gibbons, Stephen W. Hawking, and S. T. C. Siklos (Cambridge: Cambridge University Press, 1983), 251–66; Alexander Vilenkin, "The Birth of Inflationary Universes," *Physical Review D* 27 (1983): 2848–55; and Andrei Linde, "Eternally Existing Self-Reproducing Chaotic Inflationary Universe," *Physics Letters B* 175 (1986): 395–400.

88. See a very helpful mapping in Guth, "Eternal Inflation and Its Implications."

89. Vilenkin, *Many Worlds in One*, 180. See also Alexander Vilenkin, "Creation of Universes from Nothing," *Physics Letters B* 117 (1982): 25–28.

90. Rubenstein, "Cosmic Singularities," 505–7.

91. Vilenkin, *Many Worlds in One*, 179.

92. Ibid., 183. Edward P. Tryon's resulting paper was published as "Is the Universe a Vacuum Fluctuation?" *Nature*, December 14, 1973, 396–97.

93. The question of whether inflationary cosmology begins with a singularity or not is a fairly vexed one. Linde says that "it is not necessary to assume that the universe as a whole was created at some initial moment $t=0$. . . . [T]he whole process can be considered as an infinite chain reaction which has no end and which may have no beginning" ("Eternally Existing," 398; compare Linde, "Self-Reproducing Inflationary Universe," 54). In 2003, however, Arvind Borde, Alan Guth, and Alexander Vilenkin published a highly influential paper arguing that inflationary models cannot be "past eternal"; that is, they must always produce a singularity at the beginning. This is what is known as the "BVG Theorem" ("Inflationary Spacetimes Are Incomplete in Past Directions," *Physical Review Letters* 90 [2003]: 1–4). As far as Guth in particular is concerned, this means that "some new physics . . . would be needed to describe the past boundary of the inflating region" ("Eternal Inflation and Its Implications," 6823). But not everybody agrees that the singularity is inescapable; for a counterargument, see Anthony Aguirre and Steven Gratton, "Steady State Eternal Inflation," *Physical Review D* 65 (2002): 1–6.

94. Vilenkin, *Many Worlds in One*, 82. See also Greene, *Hidden Reality*, 277.

95. As Linde has put it, "[F]rom this perspective, inflation is not part of the big bang theory . . . the big bang is a part of the inflationary model" ("Self-Reproducing Inflationary Universe," 55).

96. This host includes Paul Steinhardt, Andreas Albrecht, Alex Vilenkin, and Andrei Linde, among others. There have been efforts to secure noneternal versions of inflation, but very few physicists seem to find them convincing. See Paul J. Steinhardt, "The Inflation Debate," *Scientific American*, April 2011, 42.

97. For an explanation and diagram of this process, see Guth, "Eternal Inflation and Its Implications," 6815–16.

98. Linde, "Self-Reproducing Inflationary Universe," 54.

99. Guth, "Eternal Inflation and Its Implications," 6816.

100. Joshua Knobe, Ken D. Olum, and Alexander Vilenkin, "Philosophical Implications of Inflationary Cosmology," *British Journal for the Philosophy of Science* 57, no. 1 (2006): 50; Tegmark, "Parallel Universes," 44; Greene, *Hidden Reality*, 57.

101. Gribbin, *In Search of the Multiverse*, 132.

102. Specifically, it is dark energy that will decay as the universe ages, so that gravity will eventually collapse the bubble universe in a "big crunch" (Vilenkin, *Many Worlds in One*, 198).

103. Kate Becker, "When Worlds Collide," *fq(x) News*, August 1, 2008.

104. Stephen Feeney, Matthew C. Johnson, Daniel J. Mortlock, and Hiranya V. Peiris, "First Observational Tests of Eternal Inflation: Analysis Methods and WMAP 7-Year Results," *Physical Review D* 84 (2011): 1–36; Jason Palmer, " 'Multiverse' Theory Suggested by Microwave Background," BBC News: Science and Environment, August 3, 2011, http://www.bbc.co.uk/news/science-environment-14372387 (accessed June 6, 2013); Lisa Zyga, "Scientists Find First Evidence That Many Universes Exist," December 17, 2010, Phys.org, http://phys.org/news/2010-12-scientists-evidence-universes.html (accessed June 6, 2013).

105. Brian Greene presents a remarkably cogent explanation of special relativity in *The Elegant Universe: Superstrings, Hidden Dimensions, and the Quest for the Ultimate Theory* (New York: Norton, 1999), 23–52.

106. Greene, *Hidden Reality*, 68.

107. Ibid., 58.

108. Vilenkin, *Many Worlds in One*, 91.

109. Hawking, *Brief History of Time*, 9; Hawking and Mlodinow, *Grand Design*, 51. For a more technical critique of the "instability" of pre–big bang scenarios, including inflationary and cyclic cosmologies, see Stephen Hawking, "Cosmology from the Top Down," in *Universe or Multiverse?* ed. Carr, 91–98.

110. Jaume Garriga and Alexander Vilenkin, "Many Worlds in One," *Physical Review D* 64 (2001): 1–5; Gribbin, *In Search of the Multiverse*, 135; Laura Mersini-Houghton, "Thoughts on Defining the Multiverse," April 27, 2008, available only through arXiv /0804.4280; Greene, *Hidden Reality*, 7.

111. Lucretius, *De rerum natura*, trans. W. H. D. Rouse, rev. Martin Ferguson Smith, Loeb Classical Library 181 (Cambridge, Mass.: Harvard University Press, 1975), 2.1090–91.

112. Andrei Linde, "Inflation in Supergravity and String Theory: Brief History of the Multiverse," March 21, 2012, text at http://www.ctc.cam.ac.uk/stephen70/talks/swh70 _linde.pdf; video at http://sms.cam.ac.uk/media/1228717 (accessed June 6, 2013). Linde refers primarily to a paper by Raphael Bousso and Joseph Polchinski, "Quantization of Four-Form Fluxes and Dynamical Neutralization of the Cosmological Constant," *Journal of High Energy Physics*, no. 6 (2000): 1–25.

113. William Connolly, "The Evangelical–Capitalist Resonance Machine," *Political Theology* 33, no. 6 (2005): 869–86.

114. For a basic introduction to string theory, see Greene, *Elegant Universe*, 135–230. For a historical account of its emergence, see Helge Kragh, *Higher Speculations: Grand Theories and Failed Revolutions in Physics and Cosmology* (Oxford: Oxford University Press, 2011), 291–315.

115. Kragh, *Higher Speculations*, 266–67.

116. Some important early papers include Bousso and Polchinski, "Quantization of Four-Form Fluxes"; Raphael Bousso and Joseph Polchinski, "The String Theory Landscape," *Scientific American*, September 2004, 78–87; and Steven B. Giddings, Shamit Kachru, and Joseph Polchinski, "Hierarchies from Fluxes in String Compactifications," *Physical Review D* 66 (2002): 1–6, arXiv:hep-th/0105097. As Bousso and Polchinksi explain vacuum energy, it is calculated by adding up the energy of "fluxes, branes and the curvature itself of the curled-up dimensions" ("String Theory Landscape," 81).

117. The first papers that produced these numbers are Bousso and Polchinski, "Quantization of Four-Form Fluxes"; and Sujay K. Ashok and Michael R. Douglas, "Counting String Vacua," *Journal of High Energy Physics*, no. 1 (2004): 1–35.

118. Some string theorists argue that if "nongeometric" compactifications are taken into account, the number of possible vacua becomes infinite. This view, of course, poses a problem for any claim that universes must necessarily repeat. See Jessie Shelton, Washington Taylor, and Brian Wecht, "Generalized Flux Vacua," *Journal of High Energy Physics*, no. 2 (2007): 1–27, arXiv:hep-th/0607015v2; and Brian Wecht, "Lectures on Nongeometric Flux Compactifications," August 29, 2007, available only through arXiv :hep-th/0708.3984v1.

119. Guth, "Eternal Inflation and Its Implications," 6819.

120. Leonard Susskind, "The Anthropic Landscape of String Theory," in *Universe or Multiverse?* ed. Carr, 247–66, arXiv:hep-th/0302219v1.

121. Leonard Susskind, *The Cosmic Landscape: String Theory and the Illusion of Intelligent Design* (Boston: Back Bay Books, 2006), 109.

122. Steven Weinberg, "Living in the Multiverse," in *Universe or Multiverse?* ed. Carr, 37–38; Stephen Hawking and Thomas Hertog, "Populating the Landscape: A Top-Down Approach," *Physical Review D* 73 (2006): 1–9.

123. According to Linde,

> At present it seems absolutely improbable that all domains contained in our expo-nentially large universe are of the same type. On the contrary, all types of mini-universes in which inflation is possible should be produced during the expansion of the universe, and it is unreasonable to expect that our domain is the only pos-sible one or the best one. From this point of view, an enormously large number of possible types of compactification which exist e.g. in the theories of superstrings should be considered not as a difficulty but as a virtue of these theories, since it increases the probability of the existence of mini-universes in which life of our type may appear. ("Eternally Existing," 399)

124. Guth, "Eternal Inflation and Its Implications," 6819; Bousso and Polchinski, "String Theory Landscape," 46; Linde, "Inflation in Supergravity and String Theory"; Susskind, "Anthropic Landscape of String Theory," 257.

125. Vilenkin, *Many Worlds in One*, 91.

126. Brian Wecht has dramatized this problem with unparalleled humor in "Contro-versies in Modern Theoretical Particle Physics" (paper presented at the Northeast Con-ference on Science and Skepticism, New York City, April 22, 2012, slide 10).

127. Greene, *Hidden Reality*, 313.

128. Ibid.

129. Susskind, *Cosmic Landscape*, 21.

130. Steinhardt, "Inflation Debate," 40.

131. Ibid. For the most recent data from the Planck datellite, see P. A. R. Ade, N. Aghanim, C. Armitage-Caplan, M. Arnaud, M. Ashdown, F. Atrio-Barandela, J. Au-mont, et al., "Planck 2013 Results: Overview of Products and Scientific Results," March 20, 2013, available only through arXiv:1303.5062.

132. Quoted in Nathan Schneider, "The Multiverse Problem," *Seed Magazine*, April 14, 2009; Paul J. Steinhardt and Neil Turok, "The Cyclic Model Simplified," *New As-tronomy Reviews* 49, no. 206 (2005): 44.

133. Steinhardt, "Inflation Debate," 42.

134. Steinhardt and Turok, *Endless Universe*, 222.

135. Ibid., 138.

136. Ibid., 135.

137. Having learned of this connection from a colleague in classics, Steinhardt and Turok do not cite many ancient sources or secondary material in relation to *ekpyrosis*. The one exception is Cicero, whose *De natura deorum* they provide as evidence of the doctrine (ibid., 171), but without recognizing (or at least without acknowledging) the passage's half-hearted and ironic tone (see chap. 2, sec. "Fire and the Phoenix").

138. Justin Khoury, Burt A. Ovrut, Paul J. Steinhardt, and Neil Turok, "Ekpyrotic Universe: Colliding Branes and the Origin of the Hot Big Bang," *Physical Review D* 64 (2001): 1–24; Paul J. Steinhardt and Neil Turok, "Cosmic Evolution in a Cyclic Uni-verse," *Physical Review D* 65 (2002): 1–20; Steinhardt and Turok, "Cyclic Model Simpli-

fied"; Evgeny I. Buchbinder, Justin Khoury, and Burt A. Ovrut, "New Ekpyrotic Cosmology," *Physical Review D* 76 (2007): 1–18.

139. Jean-Luc Lehners and Paul J. Steinhardt, "Dark Energy and the Return of the Phoenix Universe," *Physical Review D* 79 (2009): 1–5, arXiv:hep-th/0812.3388, and "Dynamical Selection of the Primordial Density Fluctuation Amplitude," *Physical Review Letters* 106 (2011) : 1–4, arXiv:hep-th/1008.4567v1.

140. Steinhardt and Turok, *Endless Universe*, 139.

141. Ibid., 8. For a video illustration of this cycle, see http://wwwphy.princeton.edu/~steinh/cycliccosmology.html (accessed June 6, 2013).

142. Steinhardt and Turok, *Endless Universe*, 169–70.

143. Greene, *Hidden Reality*, 123.

144. Steinhardt and Turok, *Endless Universe*, 165; compare Lehners and Steinhardt, "Dark Energy and the Return of the Phoenix Universe," which imagines "the sudden quantum creation from nothing of a positive- and negative-tension orbifold plane pair with random, but smooth, initial conditions" (4).

145. Steinhardt and Turok, *Endless Universe*, 223.

146. Ibid., 68, 241.

147. Ibid., 67 (emphasis added).

148. Ibid., 166.

149. Ibid.

150. Greene, *Hidden Reality*, 207–8.

151. Gribbin, *In Search of the Multiverse*, 160.

152. Ibid., 165. Although Lisa Randall and Raman Sundrum are suspicious of the term *multiverse*, they have explored the possibility of numerous branes in an extradimensional "bulk." See Lisa Randall, *Warped Passages: Unraveling the Mysteries of the Universe's Hidden Dimensions* (New York: HarperCollins, 2005), 60–62. The groundbreaking and highly technical paper on which this model relies is Lisa Randall and Raman Sundrum, "Large Mass Hierarchy from a Small Extra Dimension," *Physical Review Letters* 83 (1999): 3370–73.

153. Quoted in Hand, "Test of Inflation," 822. See also Renata Kallosh, Jin U. Kang, Andrei Linde, and Viatscheslav Mukhanov, "The New Ekpyrotic Ghost," *Journal of Cosmology and Astroparticle Physics* 2008 (2008): 1–23, arXiv:0712.2040.

154. Quoted in Michio Kaku, *Parallel Worlds: A Journey Through Creation, Higher Dimensions, and the Future of the Cosmos* (New York: Doubleday, 2004), 224. Kaku's book offers a chatty account of the exchange between Linde and Steinhardt. Hawking issues a similar critique of all pre–big bang scenarios in "Cosmology from the Top Down."

155. Lehners and Steinhardt, "Dark Energy and the Return of the Phoenix Universe."

156. Neil Turok, "The Cyclic Universe: A Talk with Neil Turok," *Edge* 210, May 17, 2007, http://www.edge.org/documents/archive/edge210.html#turok2 (accessed June 6, 2013).

157. Lehners and Steinhardt, "Dark Energy and the Return of the Phoenix Universe."

158. Kragh, *Higher Speculations*, 207 (emphasis added).

159. Lauris Baum and Paul Frampton, "Turnaround in Cyclic Cosmology," *Physical Review Letters* 98 (2007): 1–4.

160. Its equation of state, in other words, would be less than −1. See Caldwell, Kamionkowski, and Weinberg, "Phantom Energy and Cosmic Doomsday." See also Tolman,

Relativity, Thermodynamics, and Cosmology; Zanstra, "On the Pulsating or Expanding Universe"; Greene, *Fabric of the Cosmos*, 406; Weinberg, *First Three Minutes*, 5, 102; Gribbin, *In Search of the Multiverse*, 90; Vilenkin, *Many Worlds in One*, 25; Miao et al., "Comparison of Dark Energy Models"; Kirshner, *Extravagant Universe*, 258; and Carroll, *From Eternity to Here*, 62

161. Paul Frampton, *Did Time Begin? Will Time End? Maybe the Big Bang Never Occurred* (Singapore: World Scientific, 2009), 96.

162. Ibid.

163. "No Big Bang? Endless Universe Made Possible by New Model," January 30, 2007, Phys.org, http://phys.org/news89399974.html (emphasis added; accessed June 6, 2013).

164. Lauris Baum, Paul Frampton, and Shinya Matsuzaki, "Constraints on Deflation from the Equation of State of Dark Energy," *Journal of Cosmology and Astroparticle Physics* 2008 (2008): 7, arXiv/0801.4420v2.

165. Baum and Frampton, "Turnaround in Cyclic Cosmology."

166. Paul H. Frampton, "Cyclic Universe and Infinite Past," *Modern Physics Letters A* 22 (2007): 2587–92, arXiv/0705.2730v2.

167. Baum and Frampton, "Turnaround in Cyclic Cosmology."

168. For an "insider's" perspective on loop quantum gravity, see Lee Smolin, *Three Roads to Quantum Gravity* (New York: Basic Books, 2001); and Martin Bojowald, *Once Before Time: A Whole Story of the Universe* (New York: Vintage, 2010). For an "outsider's" perspective, see Kragh, *Higher Speculations*, 316–20.

169. Bojowald, *Once Before Time*, 84–89, 11. See also chap. 1, sec. "So Let Us Begin Again . . ."

170. Bojowald, *Once Before Time*, 114, 24, 49.

171. Kragh, *Higher Speculations*, 210.

172. Bojowald, *Once Before Time*, 111.

173. Cristiano Germani, Nicolás Grandi, and Alex Kehagias, "The Cosmological Slingshot Scenario: A Stringy Proposal for the Early Time Cosmology," *AIP Conference Proceedings* 1031 (2007): 19–21, arXiv.org/abs/0805.2073. For an introduction geared to nonspecialists, see Brian Clegg, *Before the Big Bang: The Prehistory of Our Universe* (New York: St. Martin's Press, 2009), 218–25.

174. Germani, Grandi, and Kehagias, "Cosmological Slingshot Scenario."

175. Roger Penrose, *Cycles of Time: An Extraordinary New View of the Universe* (New York: Knopf, 2011), 66, 172–73.

176. Maurizio Gasperini and Gabriele Veneziano, "Pre–Big Bang in String Cosmology," *Astroparticle Physics* 1 (1993): 317–39, and "The Pre–Big Bang Scenario in String Cosmology," *Physics Reports* 373 (2003): 1–212, arXiv/0207130v1; Gabriele Veneziano, "The Myth of the Beginning of Time," *Scientific American*, April 2004, 54–59, 62–65; Penrose, *Cycles of Time*, 172.

177. Penrose, *Cycles of Time*, 147.

178. Jason Palmer, "Cosmos May Show Echoes of Events Before Big Bang," November 27, 2010, BBC News: Science and Environment, http://www.bbc.co.uk/news/science-environment-11837869 (accessed June 6, 2013).

179. V. G. Gurzadyan and Roger Penrose, "Concentric Circles in WMAP Data May Provide Evidence of Violent Pre–Big Bang Activity," November 16, 2010, available only through arXiv/1011.3706v1.

180. For such arguments, see I. K. Wehus and H. K. Eriksen, "A Search for Concentric Circles in the 7-Year WMAP Temperature Sky Maps," *Astrophysical Journal Letters* 733 (2011): 1–6, arXiv:astro-ph/1012.1268v1; and Adam Moss, Douglas Scott, and James P. Zibin, "No Evidence for Anomalously Low Variance Circles on the Sky," *Journal of Cosmology and Astroparticle Physics* 2011 (2011): 1–7, arXiv:astro-ph/1012.1305v3.

181. Quoted in Dennis Overbye, "Rings in Sky Leave Alternate Visions of Universes," *New York Times*, December 13, 2010.

6. Ascending to the Ultimate Multiverse

1. Hugh Everett, "'Relative State' Formulation of Quantum Mechanics," *Review of Modern Physics* 29 (1957): 454–62.

2. Brian Greene, *The Hidden Reality: Parallel Universes and the Deep Laws of the Cosmos* (New York: Knopf, 2011), 31.

3. This is Niels Bohr's understanding of the uncertainty principle, to which Werner Heisenberg eventually acceded in a postscript to his original paper: "The Physical Content of Quantum Kinematics and Mechanics," *Zeitschrift für Physik* 43 (1927): 172–98. According to Bohr, the indeterminacy of a quantum particle is not an epistemological failure; that is, the problem is not that we lack sufficiently precise instruments with which to measure it or that shining any light on a particle necessarily moves it from where it "was." The problem, for Bohr, is *ontological*; until the moment a particle is measured, it simply does not *have* a determinate position. On this distinction and on Heisenberg's eventual deference to Bohr, see Karen Barad, *Meeting the Universe Halfway: Quantum Physics and the Entanglement of Matter and Meaning* (Durham, N.C.: Duke University Press, 2007), esp. 7–19, 115–18.

4. John Lasseter, dir., *Toy Story* (Walt Disney Pictures, Pixar Studio, 1995); Eric Till, dir., *The Christmas Toy* (Henson Associates and Sony Pictures, 1986).

5. Quoted in John Gribbin, *In Search of the Multiverse* (London: Allen Lane, 2009), 25.

6. In David Deutsch's version of the MWI, the universe does not split. Rather, there are a vast number of universes from the beginning, all of which start out the same and remain identical until specific quantum choices set them on different courses. See David Deutsch, *The Fabric of Reality* (New York: Penguin, 1997), and "The Structure of the Multiverse," *Proceedings of the Royal Society of London A* 458 (2002): 2911–23.

7. Max Tegmark, "Parallel Universes," *Scientific American*, May 2003, 48.

8. Bryce DeWitt, "Quantum Mechanics and Reality," *Physics Today* 23, no. 9 (1970): 161.

9. Deutsch, *Fabric of Reality*.

10. This is a very loose summary of part of the famed "double-slit experiment," which is addressed at greater length in the final chapter. For a popular introduction to this experiment, see Brian Greene, *The Elegant Universe: Superstrings, Hidden Dimensions, and the Quest for the Ultimate Theory* (New York: Norton, 1999), 85–116.

11. For such an argument, see Barad, *Meeting the Universe Halfway*, 85.

12. Ibid., 33, 118–31.

13. Colin Bruce, *Schrödinger's Rabbits: The Many-Worlds of Quantum* (Washington, D.C.: Joseph Henry, 2004), 69. For a serious treatment of the possibility that the universe

did not "exist" until there were conscious beings to observe it, see John Wheeler, "Genesis and Observership," in *Foundational Porblems in the Special Sciences*, ed. Robert E. Butts and Jaakko Hintikka (Dordrecht: Reidel, 1977), 3–33.

14. Bruce, *Schrödinger's Rabbits*, 154.

15. Tegmark, "Parallel Universes," 48.

16. Sean M. Carroll, "Does This Ontological Commitment Make Me Look Fat?" *Discover*, June 4, 2012, http://blogs.discovermagazine.com/cosmicvariance/2012/06/04/does-this-ontological-commitment-make-me-look-fat/ (accessed May 29, 2013).

17. See, for example, Leonard Susskind, *The Cosmic Landscape: String Theory and the Illusion of Intelligent Design* (Boston: Back Bay Books, 2006), 316; and Barad, *Meeting the Universe Halfway*, 27.

18. Martin Gardner, *Are Universes Thicker Than Blackberries? Discourses on Gödel, Magic Hexagrams, Little Red Riding Hood, and Other Mathematical and Pseudoscientific Topics* (New York: Norton, 2003), 5.

19. Stephen Hawking, "Cosmology from the Top Down," in *Universe or Multiverse?* ed. Bernard Carr (Cambridge: Cambridge University Press, 2007), 91.

20. Stephen Hawking and Thomas Hertog, "Populating the Landscape: A Top-Down Approach," *Physical Review D* 73 (2006): 3.

21. Hawking, "Cosmology from the Top Down," 93.

22. Michel Foucault, "Nietzsche, Genealogy, History," trans. Donald F. Bouchard and Sherry Simon, in *The Foucault Reader*, ed. Paul Rabinow (New York: Pantheon, 1984), 76–100.

23. Hawking and Hertog, "Populating the Landscape," 2.

24. James Hartle and Stephen Hawking, "Wave Function of the Universe," *Physical Review D* 28 (1983): 2960–75. For a popular introduction to this model, see Stephen Hawking, *A Brief History of Time: The Updated and Expanded Tenth Anniversary Edition* (New York: Bantam, 1998). From the beginning, Hawking and Hartle have deployed the rhetorical strategy of claiming that the universe is therefore created "from nothing." On the ironic persistence of this Christian trope, especially among physicists who are seeking to unsettle Christian explanations of creation, see Mary-Jane Rubenstein, "Cosmic Singularities: On the Nothing and the Sovereign," *Journal of the American Association of Religion* 80, no. 2 (2012): 485–517, and "Myth and Modern Physics: On the Power of Nothing," in *Creation Options: Rethinking Initial Creation*, ed. Thomas Oord and Richard Livingston (New York: Routledge, forthcoming).

25. Hawking, *Brief History of Time*, 141.

26. Hawking and Hertog, "Populating the Landscape," 3.

27. Ibid., 8.

28. Ibid., 2.

29. Hawking, "Cosmology from the Top Down," 95.

30. Ibid., 95, 97 (emphasis added). Hawking's fairly squirm-inducing analogy is that "it would be like asking for the amplitude that I am Chinese. I know I am British, even though there are more Chinese [people on the earth]" (96).

31. For a treatment of the realist/antirealist debate in the history of philosophy of science, see Ian Hacking, *The Social Construction of What?* (Cambridge, Mass.: Harvard University Press, 1999).

32. Stephen Hawking and Leonard Mlodinow, *The Grand Design* (New York: Bantam, 2010), 42.

33. Ibid., 43, 59 (emphasis in original).

34. Hawking, "Cosmology from the Top Down."

35. Susskind, *Cosmic Landscape*, 316.

36. As we may recall from the earliest cosmogonists, "chaos" is a state of undifferentiation, whereas order is a state of distinction. Chaos is Thales's water, Anaximenes's air, Timaeus's "traces," the biblical *tehom*. Order is the distinction between land and sea, day and night, cold and heat, solid and liquid.

37. Sean M. Carroll, "The Cosmic Origins of Time's Arrow," *Scientific American*, June 2008, 50.

38. Ibid., 52, 50.

39. Laura Mersini-Houghton, "Birth of the Universe from the Multiverse," September 22, 2008, 1, available only through arXiv/0809.3623.

40. Laura Mersini-Houghton, "Thoughts on Defining the Multiverse," April 27, 2008, 3, available only through arXiv/0804.4280.

41. Mersini-Houghton, "Birth of the Universe from the Multiverse," 1–2.

42. Laura Mersini-Houghton, "Notes on Time's Enigma," in *The Arrows of Time: A Debate in Cosmology*, ed. Laura Mersini-Houghton and Rüdiger Vaas, Fundamental Theories of Physics 172 (Berlin: Springer, 2012), 157–59, arXiv/0909.2330.

43. Laura Mersini-Houghton and Richard Holman, "'Tilting' the Universe with the Landscape Multiverse: The 'Dark' Flow," *Journal of Cosmology and Astroparticle Physics* 2009 (2009): 1, arXiv/0810.5388.

44. Mersini-Houghton, "Birth of the Universe from the Multiverse," 7.

45. Annie Dillard, *For the Time Being* (New York: Knopf, 1999), 8.

46. Mersini-Houghton, "Notes on Time's Enigma," 4.

47. Mersini-Houghton, "Thoughts on Defining the Multiverse," 7.

48. Mersini-Houghton, "Birth of the Universe from the Multiverse," 9.

49. "Scientists Believe That They Have Discovered Another Universe," December 15, 2009, Current.com.

50. Mersini-Houghton and Holman, "'Tilting' the Universe," 1.

51. A. Kashlinsky, F. Atrio-Barandela, H. Ebeling, A. Edge, and D. Kocevski, "A New Measurement of the Bulk Flow of X-Ray Luminous Clusters of Galaxies," *Astrophysical Journal Letters* 71 (2010): L81–L85, arXiv:astro-ph/0910.4958.

52. These two predictions are the quadrupole–octopole alignment and power suppression at low multipoles. See P. A. R. Ade, N. Aghanim, C. Armitage-Caplan, M. Arnaud, M. Ashdown, F. Atrio-Barandela, J. Aumont, et al., "Planck 2013 Results: Overview of Products and Scientific Results," March 20, 2013, available only through arXiv:1303.5062.

53. Mersini-Houghton and Holman, "'Tilting' the Universe," 2, 1.

54. On the quantum theory of entanglement, see Barad, *Meeting the Universe Halfway*, 71–94; and Brian Greene, *The Fabric of the Cosmos: Space, Time, and the Texture of Reality* (New York: Vintage, 2005), 77–123. For a historical introduction, see Louisa Gilder, *The Age of Entanglement: When Quantum Physics Was Reborn* (New York: Vintage, 2009). For differing theological engagements of entanglement, see Catherine Keller, "The Cloud of the Impossible" (manuscript); and Kirk Wegter-McNelly, *The Entangled God: Divine Relationality and Quantum Physics* (New York: Routledge, 2011).

55. Mersini-Houghton, "Thoughts on Defining the Multiverse," 7.

56. Ibid., 5, 4.

57. Mersini-Houghton and Holman, "'Tilting' the Universe," 1.

58. On the dramatic development of this principle, see Leonard Susskind, *The Black Hole War: My Battle with Stephen Hawking to Make the World Safe for Quantum Mechanics* (New York: Little, Brown, 2008).

59. Mersini-Houghton, "Birth of the Universe from the Multiverse," 8, 9.

60. Mersini-Houghton, "Thoughts on Defining the Multiverse," 7.

61. Nicholas of Cusa, *On Learned Ignorance*, in *Nicholas of Cusa: Selected Spiritual Writings*, trans. H. Lawrence Bond (New York: Paulist Press, 1997), 2.2.100. See also chap. 3, sec. "End Without End."

62. Mersini-Houghton, "Notes on Time's Enigma," 2.

63. Mersini-Houghton, "Thoughts on Defining the Multiverse," 7.

64. Giordano Bruno, *On the Infinite Universe and Worlds*, in *Giordano Bruno: His Life and Thought with Annotated Translation of His Work "On the Infinite Universe and Worlds,"* ed. Dorothea Singer (New York: Schuman, 1950), 225–378. See also chap. 3, sec. "Infinity Unbound." William James, *A Pluralistic Universe* (Lincoln: University of Nebraska Press, 1996), 325, 27. See also introduction, sec. "The One and the Many."

65. Mersini-Houghton, "Thoughts on Defining the Multiverse," 7.

66. James, *Pluralistic Universe*, 325.

67. On the femininity of liquid metaphors, see Luce Irigaray, *Marine Lover of Friedrich Nietzsche*, trans. Gillian C. Gill (New York: Columbia University Press, 1991). On the relationship of this feminine fluidity to creation narratives, see Catherine Keller, *Face of the Deep: A Theology of Becoming* (New York: Routledge, 2003).

68. Karl Schwarzschild presented this theory in "Uber das Gravitationsfeld eineses Massenpunktes nach der Einsteinschen Theorie," translated as "On the Gravitational Field of a Point-Mass, According to Einstein's Theory," trans. Larissa Borissova and Dmitri Rabounski, *Abraham Zelmanov Journal* 1 (2008): 10–19. See also Wheeler, "Genesis and Observership."

69. On Hawking's discovery of black-hole radiation, see John D. Barrow, *The Infinite Book: A Short Guide to the Boundless, Timeless, and Endless* (New York: Vintage, 2005), 108–9.

70. Lee Smolin, *The Life of the Cosmos* (New York: Oxford University Press, 1997), 87, 88 (subsequent references are cited in the text).

71. Lee Smolin, "Scientific Alternatives to the Anthropic Principle," in *Universe or Multiverse?* ed. Carr, 335.

72. David Hume, *Dialogues Concerning Natural Religion*, ed. Richard H. Popkin, 2nd ed. (Indianapolis: Hackett, 1998), 36. See also introduction, sec. "Whence the Modern Multiverse?"

73. Smolin, "Scientific Alternatives to the Anthropic Principle."

74. Gribbin, *In Search of the Multiverse*, 185 (emphasis in original).

75. Lee Smolin, *Three Roads to Quantum Gravity* (New York: Basic Books, 2001), 17. Jacques Derrida's infamous utterance is that "there is nothing outside the text" (*Of Grammatology*, trans. Gayatri Chakravorty Spivak [Baltimore: Johns Hopkins University Press, 1976], 158).

76. Smolin, "Scientific Alternatives to the Anthropic Principle," 338.

77. Ibid.

78. Having told Abraham in Genesis 15 that he would have as many descendants as there were stars in the sky, God reaffirms his promise after Abraham's near sacrifice of

Isaac, saying, "I will indeed bless you, and I will make your offspring as numerous as the stars of heaven and as the sand that is on the seashore" (Genesis 22:17 [NRSV]).

79. J. Richard Gott and Li-Xin Li, "Can the Universe Create Itself?" *Physical Review D* 58 (1998): 37, 2.

80. Paraphrased in Greene, *Hidden Reality*, 227.

81. Dennis Overbye, "Gauging a Collider's Odds of Creating a Black Hole," *New York Times*, April 15, 2008.

82. Steven K. Blau, E. I. Guendelman, and Alan H. Guth, "Dynamics of False-Vacuum Bubbles," *Physical Review D* 35 (1987): 1747–66.

83. Edward R. Harrison, "The Natural Selection of Universes Containing Intelligent Life," *Quarterly Journal of the Royal Astronomical Society* 36, no. 3 (1995): 198.

84. Greene explains that, according to Alan Guth and Edward Farhi, you would need

a strong kick-start to get the inflationary expansion off and running. So strong that there's only one entity that can provide it: a white hole. A white hole, the opposite of a black hole, is a hypothetical object that spews matter out rather than drawing it in. This requires conditions so extreme that known mathematical methods break down (much as is the case at the center of a black hole); suffice it to say, no one anticipates generating white holes in the laboratory. Ever. (*Hidden Reality*, 279)

85. Ibid., 278.

86. Harrison, paraphrased in Barrow, *Infinite Book*, 201.

87. Heinz Pagels has also explored this idea, considering the baffling explicability of the universe to be a function of its having been designed by beings like us. The laws of physics, in other words, might be something like a secret code inscribed on the universe by its creator. This would mean that "scientists in discovering this code are deciphering the Demiurge's hidden message, the tricks he used in creating the universe. No human mind could have arranged for any message so flawlessly coherent, so strangely imaginative, and sometimes downright bizarre. It must be the work of an Alien Intelligence" (*The Dreams of Reason* [New York: Bantam, 1989], 156). Even Andrei Linde has considered this possibility, in "Hard Art of the Universe Creation," October 15, 1991, esp. 23–24, available only through arXiv:hep-th/9110037v1.

88. Harrison, "Natural Selection of Universes," 199.

89. Gribbin, *In Search of the Multiverse*, 199.

90. Harrison, "Natural Selection of Universes," 200.

91. Hume, *Dialogues Concerning Natural Religion*, 15.

92. Harrison, "Natural Selection of Universes," 199.

93. Greene, *Hidden Reality*, 280 (subsequent references are in the text).

94. For these games, see *Sim City*, http://www.simcity.com/en_US; *Second Life*, http://secondlife.com/; *McCurdy's Armor*, Forge FX Simulators, http://www.forgefx.com /casestudies/mccurdys/military-training-simulator.htm; and *Sindome*, http://www.sin dome.org/. For similar gaming communities, see *Phantasy World*, http://pworld.dyndns .org/index.php?page=home; and The Artemis Project, Artemis Society International, http://www.asi.org/adb/09/08/04/moo.html (all accessed June 7, 2013).

95. Hans Moravec, *Robot: Mere Machine to Transcendent Mind* (New York: Oxford University Press, 2000); Nick Bostrom, "A Short History of Transhumanist Thought," in

Man into Superman: The Startling Potential of Human Evolution—and How to Be a Part of It, ed. Robert C. W. Ettinger (Palo Alto, Calif.: Ria University Press, 2005), 315–49; Ray Kurzweil, *The Singularity Is Near: When Humans Transcend Biology* (New York: Penguin, 2006).

96. Nick Bostrom, "Are We Living in a Computer Simulation?" *Philosophical Quarterly* 53, no. 211 (2003): 255.

97. Hume, *Dialogues Concerning Natural Religion*, 37.

98. Nick Bostrom, "The Simulation Argument: Some Explanations," *Analysis* 69, no. 3 (2009): 458 (emphasis added).

99. Or, as Bostrom phrases the negative argument, "if we do not think that we are currently living in a computer simulation, we are not entitled to believe that we shall have descendants who will run lots of simulations on their forebears" ("Are We Living in a Computer Simulation?" 243).

100. Martin Rees, "In the Matrix," *Edge*, May 19, 2003.

101. In addition to Greene, *Hidden Reality*, 274–306, consult the interviews with Greene and reflections on his hypotheses in Jad Abumrad and Robert Krulwich, "DIY Universe," *Radiolab*, WNYC, March 25, 2009; and Geek's Guide to the Galaxy, "Theoretical Physicist Brian Greene Thinks You Might Be a Hologram" [interview with Brian Greene], *Wired*, May 16, 2012, http://www.wired.com/underwire/2012/05/geeks-guide-brian-greene/ (accessed June 7, 2013).

102. René Descartes, *Meditations on First Philosophy*, trans. Donald A. Cress, 3rd ed. (Indianapolis: Hackett, 1993) (subsequent references are cited in the text).

103. Paul C. W. Davies, "A Brief History of the Multiverse," *New York Times*, April 12, 2003, quoted in Davies, *Cosmic Jackpot: Why Our Universe Is Just Right for Life* (New York: Houghton Mifflin, 2007), 188, 187.

104. Davies, *Cosmic Jackpot*, 188.

105. Robin Hanson, "How to Live in a Simulation," *Journal of Evolution and Technology* 7, no. 1 (2001), http://www.jetpress.org/volume7/simulation.htm (accessed June 7, 2013).

106. "Let us then examine this point, and let us say: 'Either God is or he is not.' But to which view shall we be inclined? Reason cannot decide this question. Infinite chaos separates us. . . . Yes, but you must wager. . . . Which will you choose then? . . . Let us weigh up the gain and the loss involved in calling heads that God exists. Let us assess the two cases: if you win[,] you win everything, if you lose[,] you lose nothing. Do not hesitate then; wager that he does exist" (Blaise Pascal, *Pensées*, trans. A. J. Krailsheimer [New York: Penguin, 1966], no. 418, pp. 149–53; for the original French, see Blaise Pascal, *Pensées* [Paris: Bookking International, 1995], nos. 233–418, pp. 89–93).

107. Hanson, "How to Live in a Simulation."

108. Ibid.

109. Greene often capitalizes the term *simulators* in *Hidden Reality*.

110. Barrow, *Infinite Book*, 210. See also Davies, "Brief History of the Multiverse."

111. For example, see Gribbin, *In Search of the Multiverse*; Greene, *Hidden Reality*; Rees, "In the Matrix."

112. Barrow, *Infinite Book*, 205. Similar arguments can be found in Davies, "Brief History of the Multiverse," and "Universes Galore: Where Will It All End?" in *Universe or Multiverse?* ed. Carr, 487–505.

113. Davies, *Cosmic Jackpot*, 186.

114. Paraphrased in Gribbin, *In Search of the Multiverse*, 198.

115. Rees, "In the Matrix," emphasis added.

116. Plato, *Republic*, trans. G. M. A. Grube, ed. C. D. C. Reeve (Indianapolis: Hackett, 1992), book VII; Plato, *Timaeus*, in *Timaeus and Critias*, trans. Desmond Lee (New York: Penguin Books, 1977), 30a, 35a. See also chap. 1, sec. "So Let Us Begin Again . . ."

117. Tegmark, "Parallel Universes," 49.

118. Davies makes a similar argument, saying that "most physicists working on fundamental topics" think of mathematics as prescriptive of physical laws rather than descriptive of them (*Cosmic Jackpot*, 12). Whether this is the case or not, most physicists do not take Tegmark's step of ascribing physical reality to mathematical forms.

119. Tegmark, "Parallel Universes," 50.

120. Plato, *Timaeus*, 30a, 31b.

121. Ibid., 31a–33b.

122. Tegmark, "Parallel Universes," 50.

123. Max Tegmark, "The Multiverse Hierarchy," in *Universe or Multiverse?* ed. Carr, 118.

124. David Lewis, *On the Plurality of Worlds* (Oxford: Oxford University Press, 1986).

125. Robert Nozick, *Philosophical Explanations* (Cambridge, Mass.: Harvard University Press, 1981). See also chap. 2, sec. "Accident and Infinity."

126. The Neoplatonist and sixth-century Christian author Pseudo-Dionysius coined the term *hierarchy*, by which he meant "a sacred order, a state of understanding, and an activity approximating as closely as possible to the divine" (*The Celestial Hierarchy*, in *The Complete Works*, ed. Colm Luibheid [New York: Paulist Press, 1987], 3.1.164d). See Mary-Jane Rubenstein, "Dionysius, Derrida, and the Problem of Ontotheology," *Modern Theology* 24, no. 2 (2008): 733.

127. Tegmark, "Multiverse Hierarchy."

128. Helge Kragh, *Higher Speculations: Grand Theories and Failed Revolutions in Physics and Cosmology* (Oxford: Oxford University Press, 2011), 273.

129. Tegmark, "Parallel Universes," 50. Of course, this claim leaves out "inconsistent" structures; as Kragh puts it, "one can easily imagine a flat-space universe where the circumference of a circle differs from 2πr (at least I can), but no such mathematical universe exists" (*Higher Speculations*, 273).

130. Greene, *Hidden Reality*, 314, 338–39; Davies, *Cosmic Jackpot*, 210.

131. This is Gribbin's parody of the ekpyrotic scenario, in *In Search of the Multiverse*, 165. See also chap. 5, sec. "Other Answers."

132. Greene, *Hidden Reality*, 297.

133. Tegmark, "Parallel Universes," 49.

Unendings

1. Neil A. Manson, "Introduction," in *God and Design: The Teleological Argument and Modern Science*, ed. Neil A. Manson (New York: Routledge, 2003), 18. With this phrase, Manson is reporting on this position, not necessarily adopting it. One of the scholars who best embodies this position is the evangelical theologian William Lane Craig, who argues that "the very fact that detractors of design have to resort to such a remarkable hypothesis underlies the point that cosmic fine-tuning is not explicable in terms of a physical necessity alone or in terms of sheer chance in the absence of a World Ensemble.

The Many-Worlds Hypothesis is [therefore] a sort of backhanded compliment to the design hypothesis" ("Design and the Anthropic Fine-tuning of the Universe," in ibid., 171).

2. Christof Schönborn, "Finding Design in Nature," *New York Times*, July 7, 2005.

3. Ibid.

4. Karl Popper, *The Logic of Scientific Discovery* (New York: Basic Books, 1959).

5. Most prominent and subtle among these critics are George Ellis, a cosmologist, Quaker, and self-professed "Platonist"; William Stoeger, an astronomer and staff scientist to the Vatican; Paul Davies, a physicist and chair of the Search for Extra-Terrestrial Intelligence Institute; and John Barrow, a cosmologist and self-identified "deist." See their critiques of various multiverse hypotheses in G. F. R. Ellis, U. Kirchner, and W. R. Stoeger, "Multiverses and Physical Cosmology," *Monthly Notices of the Royal Astronomical Society* 347 (2004): 921–36, arXiv:astro-ph/0305292v3; William R. Stoeger, G. F. R. Ellis, and U. Kirchner, "Multiverses and Cosmology: Philosophical Issues," January 19, 2006, available only through arXiv/0407329v2; George Ellis, "Does the Multiverse Really Exist?" *Scientific American*, August 2011, 38–43; Paul C. W. Davies, "A Brief History of the Multiverse," *New York Times*, April 12, 2003, and *Cosmic Jackpot: Why Our Universe Is Just Right for Life* (New York: Houghton Mifflin, 2007); and John D. Barrow, *The Book of Universes: Exploring the Limits of the Cosmos* (New York: Norton, 2011).

6. For resonant but *very* different accounts of the complementary relationship between reason and revelation stemming from Roman Catholic, Anglican, and process positions, respectively, see Denys Turner, *Faith, Reason, and the Existence of God* (Cambridge: Cambridge University Press, 2004); John Milbank, *Theology and Social Theory: Beyond Secular Reason* (Oxford: Blackwell, 1993); and John B. Cobb and David Ray Griffin, *Process Theology: An Introductory Exposition* (Philadelphia: Westminster John Knox Press, 1996).

7. This is a *very* loose paraphrase of Thomas Aquinas, *Summa theologiae*, trans. Fathers of the English Dominican Province, 5 vols. (Allen, Tex.: Christian Classics, 1981), 1.2.3, repl. obj. 2.

8. In addition to Jeffrey Zweernick, *Who's Afraid of the Multiverse?* (Pasadena, Calif.: Reasons to Believe, 2008), works by Christian (or "Neoplatonic") physicists and philosophers who affirm of the multiverse include John Leslie, *Universes* (New York: Routledge, 1996); Don N. Page, "Does God So Love the Multiverse?" in *Science and Religion in Dialogue*, ed. Melville Y. Stewart (Malden, Mass.: Wiley-Blackwell, 2010), 1:380–95, arXiv/0801.0246, and "Predictions and Tests of Multiverse Theories," in *Universe or Multiverse?* ed. Bernard Carr (Cambridge: Cambridge University Press, 2007), 411–29; Stephen Barr, *Modern Physics and Ancient Faith* (Notre Dame, Ind.: Notre Dame University Press, 2003); and Robin Collins, "The Multiverse Hypothesis: A Theistic Perspective," in *Universe or Multiverse?* ed. Carr, 459–80.

9. Zweernick, *Who's Afraid of the Multiverse?* 53. Craig sees this point as an argument *against* the multiverse and in favor of the "simpler" single universe, in "Design and the Anthropic Fine-tuning of the Universe," 172. Compare Davies, *Cosmic Jackpot*, 204.

10. This line of thinking seems to rely on the work of Collins, who argues that

the multiverse-generator itself, whether of the inflationary variety or some other type, seems to need to be "well designed" in order to produce life-sustaining universes. After all, even a mundane item like a bread machine, which only produces loaves of bread instead of universes, must be well designed as an appliance and must have the right ingredients (flour, water and yeast) to produce decent loaves of bread. If this is right, then invoking some sort of multiverse-generator as an explanation of the fine-tuning serves to kick the issue of design up one level, to the question of who designed the multiverse-generator. ("Multiverse Hypothesis," 464)

It should be said that this is not the most apt of analogies; unless something goes wrong, a bread machine makes a "good" loaf of bread every time it runs. Inflation, by contrast, produces mostly empty space and failed universes. The reason it can spit out a functioning universe from time to time is that it has an infinite amount of materials with which to work. If a bread machine were able to rely on an infinite supply of flour, water, and yeast, it might be terribly designed or just haphazardly thrown together and still produce a decent loaf of bread once in a while anyway.

11. See introduction, sec. "Whence the Modern Multiverse?" For an explanation of the difference between modern and traditional arguments from design, see Manson, "Introduction," 5–8. Although Manson says that the form of these arguments delivers them from Philo's critiques, he does not engage Hume's text in order to demonstrate as much. I tend to be of the opinion that design arguments in their modern form are little more than, as Robert O'Connor has argued, "old wine in new wineskins" ("The Design Inference: Old Wine in New Wineskins," in *God and Design*, ed. Manson, 66–87). Several of Philo's critiques (especially his battery of questions beginning "Have worlds ever formed before your eyes") are reframed and leveled against modern arguments from design in Timothy McGrew, Lydia McGrew, and Eric Vestrup, "Probabilities and the Fine-tuning Argument: A Skeptical View," in ibid., 200–208.

12. Blaise Pascal, *Pensées*, trans. A. J. Krailsheimer (New York: Penguin, 1966), no. 913, p. 309.

13. Dietrich Bonhoeffer, *Letters and Papers from Prison*, trans. Isabel Best, Lisa E. Dahill, Reinhard Krauss, Nancy Lukens, Barbara Rumscheidt, and Martin Rumscheidt, vol. 8 of *Dietrich Bonhoeffer Works*, ed. Eberhard Bethge, Ernst Feil, Christian Gremmels, Wolfgang Huber, Hans Pfeifer, Albrecht Schönherr, Heinz Eduard Tödt, and Ilse Tödt (Minneapolis: Augsburg Fortress Press, 2010), letter 137 (April 30, 1944), 366.

14. Ibid.

15. Ibid., 367.

16. Bonhoeffer, "After Ten Years," in ibid., 42.

17. It is doubtless his overexposure to Oxbridgian teleological arguments that prompts Martin Rees to confess in a recent lecture that "as far as the practice of religion is concerned, I appreciate and participate in it . . . but I doubt . . . that theological insights can help me with my physics" ("In the Matrix," *Edge*, May 19, 2003). Rees concludes this same lecture with a series of reflections on global warming, nuclear proliferation, and biological warfare—ethical considerations to which he claims cosmologists are particularly sensitive because they are aware of how rare our "pale blue dot" is—even if it is situated in an infinite multiverse. I would suggest that such ethical crises, which occupy the center of a number of theologies with which Rees is probably far less familiar, would be a much more productive space for the "dialogue" between cosmology and theology than the question of whether an eternal Being got the thing going in the first place. For an introduction to modern ecologically minded theologies, see Laurel Kearns and Catherine Keller, eds., *Ecospirit: Religions and Philosophies for the Earth* (New York: Fordham University Press, 2007).

18. George Ellis makes this connection in "Physics Ain't What It Used to Be: Review of *The Cosmic Landscape*, by Leonard Susskind," *Nature*, December 8, 2005, 739–40.

19. Brian Greene calls this model the "Pac-Man model" of the universe, in *The Hidden Reality: Parallel Universes and the Deep Laws of the Cosmos* (New York: Knopf, 2011), 22.

20. Jean-Pierre Luminet, Glenn D. Starkman, and Jeffrey R. Weeks, "Is Space Finite?" *Scientific American*, October 2002, 63. See also M. Lachieze-ray and Jean-Pierre Luminet, "Cosmic Topology," *Physics Reports* 254 (1995): 135–214, arXiv:gr-qc/9605010 (updated version).

21. According to Aristotle, "That which is finite moves in a circle" (*On the Heavens* [*De caelo*], trans. J. L. Stocks, in *The Complete Works of Aristotle: The Revised Oxford Translation*, ed. Jonathan Barnes [Princeton, N.J.: Princeton University Press, 1971], 1:271b26). See also chap. 1, sec. "Reflecting Singularity," and chap. 3, sec. "Infinity Unbound."

22. John D. Barrow, *The Infinite Book: A Short Guide to the Boundless, Timeless, and Endless* (New York: Vintage, 2005), 149.

23. Ellis, "Does the Multiverse Really Exist?" For a response to this charge and a defense of well-grounded extrapolation in inflationary scenarios, see Joshua Knobe, Ken D. Olum, and Alexander Vilenkin, "Philosophical Implications of Inflationary Cosmology," *British Journal for the Philosophy of Science* 57, no. 1 (2006): 55.

24. For the evidence gathered so far in favor of inflation, see N. Jarosik and C. L. Bennett, "Seven-Year Wilkinson Microwave Anistrophy Probe (WMAP) Observations: Sky Maps, Systematic Errors, and Basic Results," *Astrophysical Journal Supplement Series* 192 (2011): 1–47, arXiv/1001.4744.

25. "We are assuming quantum field theory remains valid far beyond the domain where it has been tested. . . . [W]e have faith in that extreme extrapolation despite all the unsolved problems at the foundation of quantum theory, the divergences of quantum field theory, and the failure of that theory to provide a satisfactory resolution of the cosmological constant problem" (Stoeger, Ellis, and Kirchner, "Multiverses and Cosmology," 30).

26. Even a theorist sympathetic to string theory such as Greene writes that "remarkable as string theory may be, rich as its mathematical structure may have become, the dearth of testable predictions, and the concomitant absence of contact with observations or experiments, relegates [*sic*] it to the realm of scientific speculation" (*Hidden Reality*, 171). For a polemical articulation of this critique, see Lee Smolin, *The Trouble with Physics: The Rise of String Theory, the Fall of a Science, and What Comes Next* (New York: First Mariner, 2007).

27. Davies, "Brief History of the Multiverse."

28. Alan P. Lightman, "The Accidental Universe: Science's Crisis of Faith," *Harper's Magazine*, December 22, 2011.

29. Paul J. Steinhardt, "The Inflation Debate," *Scientific American*, April 2011, 36–43. See also chap. 5, sec. "Other Answers."

30. D. H. Mellor argues that it makes no sense to talk about the parameters being "improbable" because this universe is the only one we have. The solution, then, is to stop asking why the universe is the way it is and instead "accept" it ("Too Many Universes," in *God and Design*, ed. Manson, 227). Davies similarly suggests that the very question "How come existence?" is a remnant of an age we ought to outgrow. Once information technology expands our understanding of the universe, "those age-old questions of exis-

tence may evaporate away, exposed as nothing more than the befuddled musings of biological beings trapped in a mental straitjacket" (*Cosmic Jackpot*, 259).

31. Martin Gardner, *Are Universes Thicker Than Blackberries? Discourses on Gödel, Magic Hexagrams, Little Red Riding Hood, and Other Mathematical and Pseudoscientific Topics* (New York: Norton, 2003), 9.

32. For example, Craig, "Design and the Anthropic Fine-Tuning of the Universe," 171; Richard Swinburne, *The Existence of God* (New York: Oxford University Press, 1979).

33. Richard Dawkins, "The Improbability of God," *Free Inquiry* 18, no. 3 (1998): 6, quoted in Davies, *Cosmic Jackpot*, 219.

34. For a preliminary exploration of the theological implications of *ekpyrosis*, see Mary-Jane Rubenstein, "The Fire Each Time: Dark Energy and the Breath of Creation," in *Cosmology, Ecology, and the Energy of God*, ed. Donna Bowman and Clayton Crockett (New York: Fordham University Press, 2011), 26–41. See also chap. 5, sec. "Other Answers."

35. Lee Smolin, "Scientific Alternatives to the Anthropic Principle," in *Universe or Multiverse?* ed. Carr, 323–65, and *Life of the Cosmos* (New York: Oxford University Press, 1997). See also chap. 6, sec. "Black Holes and Baby Universes."

36. William Dembski, "The Chance of the Gaps," in *God and Design*, ed. Manson, 260.

37. Sean M. Carroll, "Does This Ontological Commitment Make Me Look Fat?" *Discover*, June 4, 2012, http://blogs.discovermagazine.com/cosmicvariance/2012/06/04/does-this-ontological-commitment-make-me-look-fat/ (accessed May 29, 2013); Colin Bruce, *Schrödinger's Rabbits: The Many-Worlds of Quantum* (Washington, D.C.: Joseph Henry, 2004).

38. Max Tegmark, "Parallel Universes," *Scientific American*, May 2003, 48.

39. Greene, *Hidden Reality*, 212.

40. Paraphrased in Karen Barad, *Meeting the Universe Halfway: Quantum Physics and the Entanglement of Matter and Meaning* (Durham, N.C.: Duke University Press, 2007), 287.

41. Evelyn Fox Keller, "Cognitive Repression in Contemporary Physics," in *Reflections on Gender and Science* (New Haven, Conn.: Yale University Press, 1985), 139–49.

42. Barad, *Meeting the Universe Halfway*, 46.

43. Bruce, *Schrödinger's Rabbits*, 154.

44. John Wheeler, "Genesis and Observership," in *Foundational Porblems in the Special Sciences*, ed. Robert E. Butts and Jaakko Hintikka (Dordrecht: Reidel, 1977), 8. See also Tim Folger, "Does the Universe Exist If We're Not Looking?" *Discover*, June 1, 2002.

45. For an explanation of this configuration and its implications, see Davies, *Cosmic Jackpot*, 242–49.

46. Folger, "Does the Universe Exist If We're Not Looking?" (emphasis in original).

47. Wheeler, "Genesis and Observership," 27.

48. Departing somewhat from Wheeler and the Copenhagen Interpretation, and routing his thought experiment through David Deutsch's vision of infinite computation, Davies imagines the universe eventually becoming "saturated by mind," so that "the *entire universe* would be brought within the scope of observer-participancy." The idea, he admits, remains rather half-baked, but he finds it an attractive alternative to all other explanations of the fine-tuning of the universe: (1) dumb luck, (2) an intelligent designer, and (3) infinite worlds (*Cosmic Jackpot*, 249–50, 67). As I suggest in this chapter, however,

it would seem to me important to think of consciousness *itself* as a product of the intra-actions that compose the universe.

49. A "senior adviser" to George W. Bush, quoted in Ron Suskind, "Faith, Certainty, and the Presidency of George W. Bush," *New York Times Magazine*, October 17, 2004, http://www.nytimes.com/2004/10/17/magazine/17BUSH.html?_r=0 (accessed June 8, 2013).

50. Barad, *Meeting the Universe Halfway*, 337, 341 (emphasis added).

51. Differently from Barad, Wheeler writes that the observer has an "indispensable role in genesis" and that "the architecture of existence [is] such that only through 'observer-ship' does the universe have a way to come into being" ("Genesis and Observership," 7). In the spirit of a more nuanced, less anthropocentric reading of the Copenhagen Interpretation, I have borrowed Barad's term *intra-action*. If "interaction" presumes the self-constitution of entities that subsequently enter into relation, "intra-action" signals that the relation constitutes the entities as such. See Barad, *Meeting the Universe Halfway*, 46.

52. Tegmark, "Parallel Universes," 50.

53. Ellis, "Does the Multiverse Really Exist?" 42; Marcelo Gleiser, *A Tear at the Edge of Creation: A Radical New Vision for Life in an Imperfect Universe* (New York: Free Press, 2010), 220; Stoeger, Ellis, and Kirchner, "Multiverses and Cosmology," 26.

54. Greene, *Hidden Reality*, 314.

55. Ibid., 300.

56. Ibid., 302.

57. Walter Chalmers Smith, "Immortal, Invisible, God Only Wise," in *The English Hymnal (Second Edition), with Tunes*, ed. J. H. Arnold (London: Oxford University Press, 1933), http://www.oremus.org/hymnal/i/i225.html (accessed April 15, 2013).

58. Helge Kragh, *Higher Speculations: Grand Theories and Failed Revolutions in Physics and Cosmology* (Oxford: Oxford University Press, 2011), 2.

59. Bruno Latour, "Thou Shalt Not Freeze-Frame; or, How Not to Misunderstand the Science and Religion Debate," in *Science, Religion, and the Human Experience*, ed. James D. Proctor (Oxford: Oxford University Press, 2005), 36.

60. Max Tegmark, "The Multiverse Hierarchy," in *Universe or Multiverse?* ed. Carr, 100; compare Steven Weinberg, "Living in the Multiverse," in ibid., 39.

61. Quoted in Tim Folger, "Science's Alternative to an Intelligent Creator: The Multiverse Theory" [interview with Andrei Linde], *Discover*, November 10, 2008.

62. Tegmark, "Multiverse Hierarchy," 122; compare Tegmark, "Parallel Universes," 51.

63. Paul C. W. Davies, "Universes Galore: Where Will It All End?" in *Universe or Multiverse?* ed. Carr, 497; Gross, quoted in Weinberg, "Living in the Multiverse," 39; Steinhardt, quoted in Nathan Schneider, "The Multiverse Problem," *Seed Magazine*, April 14, 2009. One of the sections in Davies's book *Cosmic Jackpot* is simply titled "Many Scientists Hate the Multiverse Idea" (170).

64. On the Augustinian "God forbid," see chap. 2, sec. "Once More, with Feeling." Giordano Bruno's speaker is Burchio, who expresses this contrary opinion in *On the Infinite Universe and Worlds*, in *Giordano Bruno: His Life and Thought with Annotated Translation of His Work "On the Infinite Universe and Worlds*," ed. Dorothea Singer (New York: Schuman, 1950), 250.

65. Quoted in Folger, "Science's Alternative." A similar position can be found in Rees, "In the Matrix"; and Tegmark, "Multiverse Hierarchy."

66. Bernard Carr and George Ellis, "Universe or Multiverse?" *Astronomy and Geophysics* 49 (2008): 2.33.

67. Davies, "Universes Galore," 494.

68. Andrei Linde, "The Inflationary Multiverse," in *Universe or Multiverse?* ed. Carr, 145.

69. Leonard Susskind, *The Cosmic Landscape: String Theory and the Illusion of Intelligent Design* (Boston: Back Bay Books, 2006), 196. Susskind is fond of calling those scientists who *do* adhere to this philosopher's dictum "the Popperazi."

70. Alan H. Guth, "Eternal Inflation and Its Implications," *Journal of Physics A: Mathematical and Theoretical* 40 (2007): 6811–26, arXiv:hep-th/0702178; Weinberg, "Living in the Multiverse," 40. Weinberg tells the story of finding a popular article on the multiverse in which "Martin Rees said that he was sufficiently confident about the multiverse to bet his dog's life on it, while Andrei Linde said he would bet his own life. As for me," Weinberg writes, "I have just enough confidence about the multiverse to bet the lives of both Andrei Linde *and* Martin Rees's dog" (40).

71. Guth, "Eternal Inflation and Its Implications," 6819; Steven Weinberg, "The Cosmological Constant Problems" (lecture given at Dark Matter 2000 Conference, Marina del Ray, Calif., February 22–24, 2000), 4, arXiv:astro-ph/0005265v1.

72. Weinberg, "Cosmological Constant Problems," 4.

73. Quoted in Lightman, "Accidental Universe."

74. Kepler had offered one such deduction, imagining, as Davies explains it, the planets as "attached to spheres inside perfect polyhedra nested inside one another. Following the mystical tradition of the Pythagoreans" (*Cosmic Jackpot*, 153).

75. Rees, "In the Matrix."

76. Quoted in Carr and Ellis, "Universe or Multiverse?" 2.33.

77. Kragh, *Higher Speculations*, 279.

78. Greene, *Hidden Reality*, 9.

79. "Vocatus atque nonvocatus, Deus aderit." Jung had this sentence inscribed over the front doorway of his house, having found it in the works of Desiderius Erasmus. It is also inscribed on the border of his gravestone in Zurich. See Deirdre Bair, *Jung: A Biography* (New York: Little, Brown, 2003), 126.

80. Weinberg, "Living in the Multiverse," 33. See also Anthony Aguirre, "Making Predictions in a Multiverse: Conundrums, Dangers, Coincidences," in *Universe or Multiverse?* ed. Carr, 367–86. Similar efforts are under way with respect to the probabilities of eternally inflating multiverses. See Jaume Garriga and Alexander Vilenkin, "Testable Anthropic Predictions for Dark Energy," *Physical Review D* 67 (2003): 1–11; and Andrei D. Linde, Dmitri Linde, and Arthur Mezhlumian, "Do We Live in the Center of the World?" *Physics Letters B* 345 (1995): 203–10.

81. Weinberg, "Living in the Multiverse," 33.

82. Quoted in Greene, *Hidden Reality*, 280.

83. Bernard Carr, "The Anthropic Principle Revisited," in *Universe or Multiverse?* ed. Carr, 84.

84. Quoted in Greene, *Hidden Reality*, 280.

85. Stephen Feeney, Matthew C. Johnson, Daniel J. Mortlock, and Hiranya V. Peiris, "First Observational Tests of Eternal Inflation: Analysis Methods and WMAP 7-Year Results," *Physical Review D* 84 (2011): 1–36.

86. Lisa Zyga, "Scientists Find First Evidence That Many Universes Exist," December 17, 2010, Phys.org, http://phys.org/news/2010-12-scientists-evidence-universes.html (accessed June 6, 2013).

87. V. G. Gurzadyan and Roger Penrose, "Concentric Circles in WMAP Data May Provide Evidence of Violent Pre–Big Bang Activity," November 16, 2010, arXiv/1011.3706v1. However, physicists who are not partial to Penrose's model interpret the same circles as quantum fluctuations within our own universe at early stages of its development (see chap. 5, sec. "Other Answers").

88. Laura Mersini-Houghton, "Birth of the Universe from the Multiverse," September 22, 2008, available only through arXiv/0809.3623; Laura Mersini-Houghton and Richard Holman, "'Tilting' the Universe with the Landscape Multiverse: The 'Dark' Flow," *Journal of Cosmology and Astroparticle Physics* 2009 (2009): 1–12, arXiv/0810.5388; A. Kashlinsky, F. Atrio-Barandela, H. Ebeling, A. Edge, and D. Kocevski, "A New Measurement of the Bulk Flow of X-Ray Luminous Clusters of Galaxies," *Astrophysical Journal Letters* 71 (2010): L81–L85, arXiv:astro-ph/0910.4958.

89. Mersini-Houghton and Holman, "'Tilting' the Universe."

90. Tegmark, "Parallel Universes," 41.

91. Martin Heidegger, *Introduction to Metaphysics*, trans. Gregory Fried and Richard Polt (New Haven, Conn.: Yale University Press, 2000). See also Mary-Jane Rubenstein, *Strange Wonder: The Closure of Metaphysics and the Opening of Awe* (New York: Columbia University Press, 2009), 17–19, 25–33.

92. Plato, *Timaeus*, in *Timaeus and Critias*, trans. Desmond Lee (New York: Penguin, 1977), 31a.

93. Michel Serres, *Genesis*, trans. Genevieve James and James Nielson (Ann Arbor: University of Michigan Press, 1995), 112. See also chap. 1, sec. "So Let Us Begin Again . . ."

94. Aristotle, *The Metaphysics*, trans. Hugh Lawson-Tancred (New York: Penguin, 1998), 1074a. See also chap. 1, sec. "Reflecting Singularity."

95. Thomas Aquinas, *Summa theologiae*, 1.47.3; René Descartes, *Principles of Philosophy*, in *The Philosophical Writings of Descartes*, ed. John Cottingham, Robert Stoothoff, and Dugald Murdoch (Cambridge: Cambridge University Press, 1984), 2.22. See also chap. 3, sec. "Ending the Endless," and chap. 4, sec. "From Infinity to Pluralism."

96. Bruno, *On the Infinite Universe*, 229 (translation altered slightly). See also chap. 3, sec. "Infinity Unbound."

97. Serres, *Genesis*, 111.

98. William James, *A Pluralistic Universe* (Lincoln: University of Nebraska Press, 1996), 325.

99. Friedrich Nietzsche, *Thus Spoke Zarathustra*, trans. Walter Kaufman (New York: Penguin, 1978), 323.

100. Friedrich Nietzsche, *On the Genealogy of Morals*, trans. Walter Kaufmann (New York: Vintage, 1989), third essay (subsequent references are cited in the text).

101. David Hume, *An Enquiry Concerning the Principles of Morals*, ed. J. B. Schneewind (Indianapolis: Hackett, 1983), 9.1, p. 73.

102. Nietzsche hinges this allegation almost entirely on the first few verses of Jesus's Sermon on the Mount:

Blessed are the poor in spirit, for theirs is the kingdom of heaven. Blessed are those who mourn, for they will be comforted. Blessed are the meek, for they will inherit the earth. Blessed are those who hunger and thirst for righteousness, for they will be filled. Blessed are the merciful, for they will receive mercy. Blessed are the pure in heart, for they will see God. Blessed are the peacemakers, for they will be called children of God. Blessed are those who are persecuted for righteousness' sake, for theirs is the kingdom of heaven. Blessed are you when people revile you and persecute you and utter all kinds of evil against you falsely on my account. Rejoice and be glad, for your reward is great in heaven, for in the same way they persecuted the prophets who were before you. (Matthew 5:3–12 [NRSV])

103. Nietzsche names priests *after* philosophers in the third essay, but he has already spent the first two essays lambasting them, saying, for example, that "the truly great haters in world history have always been priests" (*On the Genealogy of Morals*, 1.7), so his audience most likely has them in mind by the time we reach the third essay.

104. Gleiser, *Tear at the Edge of Creation*, 7.

105. Davies mentions that

[a]ccording to Cambridge University folklore, the nuclear physicist Ernest Rutherford is said to have issued an edict to his subordinates against fanciful and grandiose speculation. "Don't let me catch anyone talking about the universe in my department!" he warned. That was in the 1930s, and, to be fair to Rutherford, cosmology didn't then exist as a proper science. Even when I was a student in London in the 1960s, cynics quipped that there is speculation, speculation squared—and [then] cosmology. (*Cosmic Jackpot*, 18)

106. David Hume, *Dialogues Concerning Natural Religion*, ed. Richard H. Popkin, 2nd ed. (Indianapolis: Hackett, 1998), 22.

107. For a detailed account of this story, see Robert P. Kirshner, *The Extravagant Universe: Exploding Stars, Dark Energy, and the Accelerating Cosmos* (Princeton, N.J.: Princeton University Press, 2002), chap. 7, "A Hot Day in Holmdel."

108. For the new image, see European Space Agency, http://www.esa.int/Our_Activities/Space_Science/Planck (accessed June 8, 2013).

109. Davies, *Cosmic Jackpot*, 18.

110. "Comme je lui parlais de mes idées sur l'origine des rayons cosmiques, il réagissait vivement . . . mais lorsque je lui parlais de l'atome primitif, il m'arrêtait; 'Non, pas cela, cela suggère trop la création'" (Georges Lemaître, "Rencontre avec A. Einstein," *Revue des questions scientifiques* 129 [1958]: 130, my translation, emphasis added).

111. I have borrowed the last pair from Barad, *Meeting the Universe Halfway*, whose subtitle is *Quantum Physics and the Entanglement of Matter and Meaning*.

112. Søren Kierkegaard [Johannes Climacus, pseud.], *Philosophical Fragments, or a Fragment of Philosophy*, trans. Howard V. Hong and Edna H. Hong (Princeton, N.J.: Princeton University Press, 1985), 7.

113. For a scathing, very funny critique of the category of "belief" by means of a parodic appeal to hunches, see Russell McCutcheon, "I Have a Hunch," *Method and Theory in the Study of Religion* 24, no. 1 (2012): 81–92.

BIBLIOGRAPHY

Abumrad, Jad, and Robert Krulwich. "DIY Universe." *Radiolab*, WNYC, March 25, 2009.

———. "The (Multi) Universe(S)." *Radiolab*, WNYC, August 12, 2008.

Adams, Edward. "Graeco-Roman and Ancient Jewish Cosmology." In *Cosmology and New Testament Theology*, edited by Jonathan T. Pennington and Sean M. McDonald, 5–27. New York: Clark, 2008.

———. *The Stars Will Fall from Heaven: Cosmic Catastrophes in the New Testament and Its World.* New York: Clark, 2007.

Ade, P. A. R., N. Aghanim, C. Armitage-Caplan, M. Arnaud, M. Ashdown, F. Atrio-Barandela, J. Aumont, et al. "Planck 2013 Results: Overview of Products and Scientific Results." March 20, 2013. Available only through arXiv:1303.5062.

Aguirre, Anthony. "Making Predictions in a Multiverse: Conundrums, Dangers, Coincidences." In *Universe or Multiverse?* edited by Bernard Carr, 367–86. Cambridge: Cambridge University Press, 2007.

Aguirre, Anthony, and Steven Gratton. "Steady State Eternal Inflation." *Physical Review D* 65 (2002): 1–6.

Albrecht, Andreas, and Paul J. Steinhardt. "Cosmology for Grand Unified Theories with Radiatively Induced Symmetry Breaking." *Physical Review Letters* 48 (1982): 1220–23.

Allen, Woody, dir. *Annie Hall*. United Artists, 1977.

Alpher, Ralph, Hans Bethe, and George Gamow. "The Origin of Chemical Elements." *Physical Review* 73 (1948): 803–4.

Anderson, William S. "Discontinuity in Lucretian Symbolism." *Transactions and Proceedings of the American Philological Association* 91 (1960): 1–29.

Anim, Hans Frederich August von, ed. *Chryssipi fragmenta logica et physica*. Vol. 2 of *Stoicorum veterum fragmenta*. Leipzig: Teubneri, 1903.

————, ed. *Zeno et Zenonis discipuli: Exemplar nastatice iteratum*. Vol. 1 of *Stoicorum veterum fragmenta*. Leipzig: Teubneri, 1921.

Antoninus, Marcus Aurelius. *Meditations*. Translated by Maxwell Staniforth. New York: Penguin, 1964.

————. *Thoughts of Marcus Aurelius Antoninus*. 2005. http://www.felix.org/node/34464.

Aristotle. *The Metaphysics*. Translated by Hugh Lawson-Tancred. New York: Penguin, 1998.

————. *Metaphysics*. Translated by Hugh Tredennick. Loeb Classical Library 271. Cambridge, Mass.: Harvard University Press, 1961.

————. *Metaphysics*. In *The Complete Works of Aristotle: The Revised Oxford Translation*, edited by Jonathan Barnes, 2:1552–728. Princeton, N.J.: Princeton University Press, 1971.

————. *On the Heavens (De caelo)*. Translated by J. L. Stocks. In *The Complete Works of Aristotle: The Revised Oxford Translation*, edited by Jonathan Barnes, 1:447–511. Princeton, N.J.: Princeton University Press, 1971.

————. *Physics*. Translated by R. P. Hardie and R. K. Gaye. In *The Complete Works of Aristotle*, edited by Jonathan Barnes, 1:315–446. Princeton, N.J.: Princeton University Press, 1984.

Arnold, J. H., ed. *The English Hymnal (Second Edition), with Tunes*. London: Oxford University Press, 1933.

Ashok, Sujay K., and Michael R. Douglas. "Counting String Vacua." *Journal of High Energy Physics*, no. 1 (2004): 1–35.

Asmis, Elizabeth. "Lucretius' Venus and Stoic Zeus." *Hermes* 110 (1982): 458–70.

Augustine of Hippo. *Arianism and Other Heresies*. Edited by Roland J. Teske. Vol. 1 of *The Works of St. Augustine: A Translation for the 21st Century*, edited by John E. Rotelle. Hyde Park, N.Y.: New City Press, 1995.

————. *The City of God*. Vol. 4. Translated by Philip Levine. Loeb Classical Library 414. Cambridge, Mass.: Harvard University Press, 1966.

————. *Concerning the City of God Against the Pagans*. Translated by Henry Bettenson. New York: Penguin, 2003.

————. *Confessions*. Translated by Owen Chadwick. Oxford: Oxford University Press, 1991.

Bair, Deirdre. *Jung: A Biography*. New York: Little, Brown, 2003.

Barad, Karen. *Meeting the Universe Halfway: Quantum Physics and the Entanglement of Matter and Meaning*. Durham, N.C.: Duke University Press, 2007.

Baretti, Guiseppe Marco Antonio. *The Italian Library*. Electronic reproduction. Farmington Hills, Mich.: Thomson Gale, 2003.

Barr, Stephen. *Modern Physics and Ancient Faith*. Notre Dame, Ind.: Notre Dame University Press, 2003.

Barrow, John D. *The Book of Universes: Exploring the Limits of the Cosmos*. New York: Norton, 2011.

————. *The Constants of Nature*. New York: Pantheon, 2002.

————. *The Infinite Book: A Short Guide to the Boundless, Timeless, and Endless*. New York: Vintage, 2005.

Barrow, John D., and Janna Levin. "The Copernican Principle in Compact Spacetimes." *Monthly Notices of the Royal Astronomical Society* 346 (2003): 615–18. arXiv:gr-qc/0304038v1.

Barrow, John D., and Frank Tipler. *The Anthropic Cosmological Principle*. Oxford: Clarendon Press, 1986.

Baum, Lauris, and Paul Frampton. "Turnaround in Cyclic Cosmology." *Physical Review Letters* 98 (2007): 1–4.

Baum, Lauris, Paul Frampton, and Shinya Matsuzaki. "Constraints on Deflation from the Equation of State of Dark Energy." *Journal of Cosmology and Astroparticle Physics* 2008 (2008): 1–18. arXiv/0801.4420v2.

Bauman, Whitney. "*Creatio ex Nihilo, Terra Nullius*, and the Erasure of Presence." In *Ecospirit: Religions and Philosophies for the Earth*, edited by Laurel Kearns and Catherine Keller, 353–72. New York: Fordham University Press, 2007.

Becker, Kate. "When Worlds Collide." August 1, 2008. fq$^{(x)}$ News, Foundational Questions Institute. http://fqxi.org/data/articles/Garriga_Guth_Vilenkin.pdf.

Bentley, Richard. "The Folly of Atheism and (What Is Now Called) Deism, Even with Respect to the Present Life." In *The Works of Richard Bentley*, edited by Alexander Dyce, 3:1–26. London: Francis McPherson, 1838.

Blau, Steven K., E. I. Guendelman, and Alan H. Guth. "Dynamics of False-Vacuum Bubbles." *Physical Review D* 35 (1987): 1747–66.

Blumenberg, Hans. *The Legitimacy of the Modern Age*. Translated by Robert M. Wallace. Cambridge, Mass.: MIT Press, 1983.

Bojowald, Martin. *Once Before Time: A Whole Story of the Universe*. New York: Vintage, 2010.

Bondi, Herman, and Thomas Gold. "The Steady State Theory of the Expanding Universe." *Monthly Notices of the Royal Astronomical Society* 108 (1948): 252–70.

Bonhoeffer, Dietrich. *Letters and Papers from Prison*. Translated by Isabel Best, Lisa E. Dahill, Reinhard Krauss, Nancy Lukens, Barbara Rumscheidt, and Martin Rumscheidt. Vol. 8 of *Dietrich Bonhoeffer Works*, edited by Eberhard Bethge, Ernst Feil, Christian Gremmels, Wolfgang Huber, Hans Pfeifer, Albrecht Schönherr, Heinz Eduard Tödt, and Ilse Tödt. Minneapolis: Augsburg Fortress Press, 2010.

Borde, Arvind, Alan H. Guth, and Alexander Vilenkin. "Inflationary Spacetimes Are Incomplete in Past Directions." *Physical Review Letters* 90 (2003): 1–4.

Borel, Pierre. *A New Treatise Proving a Multiplicity of Worlds: That the Planets Are Regions Inhabited, and the Earth a Star, and That It Is out of the Center of the World in the Third Heaven, and Turns Round Before the Sun Which Is Fixed. And Other Most Rare and Curious Things*. Translated by D. Sashott. London: John Streater, 1658.

Borges, Jorge Luis. "The Aleph." In *Collected Fictions*, 274–86. Translated by Andrew Hurley. New York: Penguin, 1998.

———. "The Writing of the God." In *Collected Fictions*, 250–54. Translated by Andrew Hurley. New York: Penguim, 1998.

Boscovich, Ruggero Giuseppe. *A Theory of Natural Philosophy, Put Forward and Explained by Roger Joseph Boscovich*. Cambridge, Mass.: MIT Press, 1966.

Bostrom, Nick. *Anthropic Bias: Observation Selection Effects in Science and Philosophy*. New York: Routledge, 2002.

———. "Are We Living in a Computer Simulation?" *Philosophical Quarterly* 53, no. 211 (2003): 243–55.

———. "A Short History of Transhumanist Thought." In *Man into Superman: The Startling Potential of Human Evolution—and How to Be a Part of It*, edited by Robert C. W. Ettinger, 315–49. Palo Alto, Calif.: Ria University Press, 2005.

———. "The Simulation Argument: Some Explanations." *Analysis* 69, no. 3 (2009): 458–61.

Bousso, Raphael, and Joseph Polchinski. "Quantization of Four-Form Fluxes and Dynamical Neutralization of the Cosmological Constant." *Journal of High Energy Physics*, no. 6 (2000): 1–25.

———. "The String Theory Landscape." *Scientific American*, September 2004, 78–87.

Brient, Elizabeth. "Transitions to a Modern Cosmology: Meister Eckhart and Nicholas of Cusa on the Intensive Infinite." *Journal of the History of Philosophy* 37, no. 4 (1999): 575–600.

Bruce, Colin. *Schrödinger's Rabbits: The Many-Worlds of Quantum*. Washington, D.C.: Joseph Henry, 2004.

Brumfiel, George. "Outrageous Fortune." *Nature*, January 5, 2006, 10–12.

Bruno, Giordano. *The Ash Wednesday Supper*. Translated by Edward A. Gosselin and Lawrence S. Lerner. Renaissance Society of America Reprint Texts 4. Toronto: University of Toronto Press, 1995.

———. *Cause, Principle, and Unity*. Translated by Richard J. Blackwell. In *Cause, Principle, and Unity and Essays on Magic*, edited by Richard J. Blackwell and Robert de Lucca, 1–102. Cambridge Texts in the History of Philosophy. Cambridge: Cambridge University Press, 1998.

———. *On the Infinite Universe and Worlds*. In *Giordano Bruno: His Life and Thought with Annotated Translation of His Work "On the Infinite Universe and Worlds,"* edited by Dorothea Singer, 225–378. New York: Schuman, 1950.

Buchanan, Mark. "Many Worlds: See Me Here, See Me There." *Nature*, July 5, 2007, 15–17.

Buchbinder, Evgeny I., Justin Khoury, and Burt A. Ovrut. "New Ekpyrotic Cosmology." *Physical Review D* 76 (2007): 1–18.

Bucher, Martin A., and David N. Spergel. "Inflation in a Low-Density Universe." *Scientific American*, January 1999, 63–69.

Burrell, David D., Carlo Cogliati, Janet M. Soskice, and William R. Stoeger, eds. *Creation and the God of Abraham*. Cambridge: Cambridge University Press, 2010.

Butler, Judith. *Parting Ways: Jewishness and the Critique of Zionism*. New York: Columbia University Press, 2012.

———. *Precarious Life: The Powers of Mourning and Violence*. New York: Verso, 2004.

———. *Undoing Gender*. New York: Routledge, 2004.

Calcagno, Antonio. *Giordano Bruno and the Logic of Coincidence: Unity and Multiplicity in the Philosophical Thought of Giordano Bruno*. Renaissance and Baroque Studies and Texts 23. New York: Peter Lang, 1998.

Caldwell, Robert, Marc Kamionkowski, and Nevin N. Weinberg. "Phantom Energy and Cosmic Doomsday." *Physical Review Letters* 91 (2003): 1–4. arXiv:astro-ph/0302506v1.

Caldwell, Robert, and Paul J. Steinhardt. "Quintessence." *Physics World*, November 1, 2000, 31–38.

Carr, Bernard. "The Anthropic Principle Revisited." In *Universe or Multiverse?* edited by Bernard Carr, 77–89. Cambridge: Cambridge University Press, 2007.

———. "Introduction and Overview." In *Universe or Multiverse?* edited by Bernard Carr, 3–28. Cambridge: Cambridge University Press, 2007.

———, ed. *Universe or Multiverse?* Cambridge: Cambridge University Press, 2007.

Carr, Bernard, and George Ellis. "Universe or Multiverse?" *Astronomy and Geophysics* 49 (2008): 2.29–2.37.

Carr, Bernard, and Martin Rees. "The Antrophic Principle and the Structure of the Physical World." *Nature*, April 12, 1979, 605–12.

Carroll, Sean M. "The Cosmic Origins of Time's Arrow." *Scientific American*, June 2008, 48–57.

———. "Does This Ontological Commitment Make Me Look Fat?" *Discover*, June 4, 2012. http://blogs.discovermagazine.com/cosmicvariance/2012/06/04/does-this-onto-logical-commitment-make-me-look-fat/.

———. *From Eternity to Here: The Quest for the Ultimate Theory of Time.* New York: Dutton, 2010.

Carter, Brandon. "Anthropic Principle in Cosmology." In *Current Issues in Cosmology*, edited by Jean-Claude Pecker and Jayant Narlikar, 173–80. Cambridge: Cambridge University Press, 2011. arXiv:gr-qc/0606117v1.

———. "Large Number Coincidences and the Anthropic Principle in Cosmology." In *Physical Cosmology and Philosophy*, edited by John Leslie and Paul Edwards, 125–33. Cambridge: Cambridge University Press, 1990.

Cassirer, Ernst. *The Individual and the Cosmos in Renaissance Philosophy.* Translated by Mario Domandi. New York: Harper Torchbooks, 1964.

Catchpole, Heather. "Weird Data Suggests Something Big Beyond the Edge of the Universe." *Cosmos*, November 24, 2009.

Catto, Bonnie A. "Venus and Natura in Lucretius' *De rerum natura* 1.1–23 and 2.167–74." *Classical Journal* 84 (1989): 97–104.

Charleton, Walter. *Physiologia Epicuro-Gassendo-Charltoniana: Or a Fabrick of Science Natural Upon the Hypothesis of Atoms.* London: Tho. Newcomb for Thomas Heath, 1654.

Cicero. *De natura deorum. Academica.* Translated by H. Rackham. Loeb Classical Library 268. Cambridge, Mass.: Harvard University Press, 2005.

———. *On the Nature of the Gods.* Translated by Horace C. P. McGregor. New York: Penguin, 1972.

Clegg, Brian. *Before the Big Bang: The Prehistory of Our Universe.* New York: St. Martin's Press, 2009.

Clerke, Agnes M. *Modern Cosmogonies.* London: Adam and Charles Black, 1905.

———. *A Popular History of Astronomy During the Nineteenth Century.* London: Adam and Charles Black, 1908.

———. *The System of the Stars.* London: Adam and Charles Black, 1905.

Clifford, Richard. *Creation Accounts in the Ancient Near East and in the Bible.* Washington, D.C.: Catholic Biblical Association of America, 1994.

Cobb, John B., and David Ray Griffin. *Process Theology: An Introductory Exposition.* Philadelphia: Westminster John Knox Press, 1996.

Coles, Peter, ed. *The Routledge Companion to the New Cosmology.* New York: Routledge, 2001.

Collins, Robin. "God, Design, and Fine-tuning." In *God Matters: Readings in the Philosophy of Religion*, edited by Raymond Martin and Christopher Bernard, 119–35. New York: Longman, Pearson, 2002.

———. "The Multiverse Hypothesis: A Theistic Perspective." In *Universe or Multiverse?* edited by Bernard Carr, 459–80. Cambridge: Cambridge University Press, 2007.

Connolly, William. "The Evangelical–Capitalist Resonance Machine." *Political Theology* 33, no. 6 (2005): 869–86.

Copan, Paul, and William Lane Craig. *Creation out of Nothing: A Biblical, Philosophical, and Scientific Exploration.* Grand Rapids, Mich.: Baker Academic, 2004.

Copernicus, Nicolaus. *On the Revolutions of the Heavenly Spheres.* Translated by A. M. Duncan. New York: Barnes and Noble Books, 1976.

Cornford, Francis MacDonald. *Plato's Cosmology: The "Timaeus" of Plato.* Translated by Francis MacDonald Cornford. Library of Liberal Arts. New York: Bobbs-Merrill, 1957.

Cowen, Ron. "Embracing the Dark Side: Looking Back on a Decade of Cosmic Acceleration." *Science News*, February 2, 2008.

Craig, William Lane. "Barrow and Tipler on the Anthropic Principle vs. Divine Design." *British Journal for the Philosophy of Science* 38 (1988): 389–95.

———. "Design and the Anthropic Fine-tuning of the Universe." In *God and Design: The Teleological Argument and Modern Science,* edited by Neil A. Manson, 155–77. New York: Routledge, 2003.

———. *The Kalam Cosmological Argument.* New York: Harper & Row, 1979.

———. "Philosophical and Scientific Pointers to Creation *ex Nihilo.*" In *Contemporary Perspectives on Religious Epistemology,* edited by R. Douglas Geivett and Brendan Sweetman, 185–200. New York: Oxford University Press, 1992.

———. "The Ultimate Question of Origins: God and the Beginning of the Universe." *Astrophysics and Space Science* 269–70 (1999): 723–40.

Crowe, Michael J. *The Extraterrestrial Life Debate, 1750–1900: The Idea of a Plurality of Worlds from Kant to Lowell.* Cambridge: Cambridge University Press, 1986.

Danielson, Dennis Richard, ed. *The Book of the Cosmos: Imagining the Universe from Heraclitus to Hawking.* Cambridge, Mass.: Perseus, 2000.

Daston, Lorraine. "The Coming into Being of Scientific Objects." In *Biographies of Scientific Objects,* edited by Lorraine Daston, 1–14. Chicago: University of Chicago Press, 2000.

Davies, Paul C. W. *The Accidental Universe.* Cambridge: Cambridge University Press, 1982.

———. "A Brief History of the Multiverse." *New York Times,* April 12, 2003.

———. *Cosmic Jackpot: Why Our Universe Is Just Right for Life.* New York: Houghton Mifflin Harcourt, 2007.

———. "Spacetime Singularities in Cosmology." In *The Study of Time,* edited by J. T. Fraser, 78–79. New York: Springer, 1978.

———. "Universes Galore: Where Will It All End?" In *Universe or Multiverse?* edited by Bernard Carr, 487–505. Cambridge: Cambridge University Press, 2007.

Dawkins, Richard. "The Improbability of God." *Free Inquiry* 18, no. 3 (1998). Academic OneFile, April 20, 2013.

Decker, Kevin. "The Open System and Its Enemies: Bruno, the Idea of Infinity, and Speculation in Early Modern Philosophy of Science." *Catholic Philosophical Quarterly* 74, no. 4 (2000): 599–620.

Deleuze, Gilles. *Nietzsche and Philosophy.* Translated by Hugh Tomlinson. New York: Columbia University Press, 1983.

Dembski, William. "The Chance of the Gaps." In *God and Design: The Teleological Argument and Modern Science,* edited by Neil A. Manson, 251–74. New York: Routledge, 2003.

Denton, Michael J. *Nature's Destiny*. New York: Free Press, 1998.

Deprit, André. "Monsignor Georges Lemaître." In *The Big Bang and Georges Lemaître*, edited by André L. Berger, 363–92. New York: Springer, 1984.

Derrida, Jacques. "Khôra." In *On the Name*, edited by Thomas Dutoit, 89–130. Stanford, Calif.: Stanford University Press, 1995.

———. *Of Grammatology*. Translated by Gayatri Chakravorty Spivak. Baltimore: Johns Hopkins University Press, 1976.

Descartes, René. *Meditations on First Philosophy*. Translated by Donald A. Cress. 3rd ed. Indianapolis: Hackett, 1993.

———. *Le monde, ou Traité de la lumière*. Translated by Michael Sean Mahoney. Janus Library. New York: Abaris Books, 1979.

———. *Principles of Philosophy*. In *The Philosophical Writings of Descartes*, edited by John Cottingham, Robert Stoothoff, and Dugald Murdoch, 177–291. Cambridge: Cambridge University Press, 1984.

Deutsch, David. *The Fabric of Reality*. New York: Penguin, 1997.

———. "The Structure of the Multiverse." *Proceedings of the Royal Society of London A* 458 (2002): 2911–23.

DeWitt, Bryce. "Quantum Mechanics and Reality." *Physics Today* 23, no. 9 (1970): 155–65.

DeWitt, Bryce, and Neill Graham, eds. *The Many-Worlds Interpretation of Quantum Mechanics*. Princeton, N.J.: Princeton University Press, 1973.

Dick, Steven J. *Life on Other Worlds: The 20th-Century Extraterrestrial Life Debate*. Cambridge: Cambridge University Press, 1998.

———. "Plurality of Worlds." In *Cosmology: Historical, Literary, Philosophical, Religious, and Scientific Perspectives*, edited by Norris S. Hetherington, 515–32. New York: Garland, 1993.

———. *Plurality of Worlds: The Origins of the Extraterrestrial Life Debate from Democritus to Kant*. Cambridge: Cambridge University Press, 1982.

Dicke, R. H., P. J. E. Peebles, P. G. Roll, and D. T. Wilkinson. "Cosmic Black-Body Radiation." *Physical Review Letters* 98 (1965): 414–19.

Dietrich, Jörg P., Norbert Werner, Douglas Clowe, Alexis Finoguenov, Tom Kitching, Lance Miller, and Aurora Simionescu. "A Filament of Dark Matter Between Two Clusters of Galaxies." *Nature*, July 4, 2012. http://www.nature.com/nature/journal/vaop/ncurrent/full/nature11224.html.

Dillard, Annie. *For the Time Being*. New York: Knopf, 1999.

Diogenes Laertius. *Lives of Eminent Philosophers*. Translated by R. D. Hicks. 2 vols. Loeb Classical Library 184 and 185. Cambridge, Mass.: Harvard University Press, 1942.

Dolnick, Edward. *The Clockwork Universe: Isaac Newton, the Royal Society, and the Birth of the Modern World*. New York: Harper Perennial, 2012.

Donne, John. *The Poems of John Donne: Edited from the Old Editions and Numerous Manuscripts with Introductions and Commentary*. Edited by Herbert J. C. Grierson. 2 vols. Oxford: Clarendon Press, 1912.

Duhem, Pierre. *Medieval Cosmology: Theories of Infinity, Place, Time, Void, and the Plurality of Worlds*. Translated by Roger Ariew. Chicago: University of Chicago Press, 1985.

Eddington, Arthur. "The Arrow of Time, Entropy, and the Expansion of the Universe." In *The Concepts of Space and Time*, edited by Milic Capec, 463–70. Dordrecht: Reidel, 1976.

———. *The Nature of the Physical World*. Cambridge: Cambridge University Press, 1935.

Einstein, Albert. "Cosmological Considerations on the General Theory of Relativity (1917)." In *Cosmological Constants: Papers in Modern Cosmology*, edited by Jeremy Bernstein and Gerald Feinberg, 16–26. New York: Columbia University Press, 1986.

———. "On the Electrodynamics of Moving Bodies." 1905. Fourmilab Switzerland. http://www.fourmilab.ch/etexts/einstein/specrel/www/.

Eliade, Mircea. *The Myth of the Eternal Return: Cosmos and History*. Translated by Willard R. Trask. Princeton, N.J.: Princeton University Press, 2005.

Eliot, T. S. *Collected Poems: 1909–1962*. New York: Harcourt, Brace, 1991.

Ellis, George. "Does the Multiverse Really Exist?" *Scientific American*, August 2011, 38–43.

———. "Physics Ain't What It Used to Be: Review of *The Cosmic Landscape*, by Leonard Susskind." *Nature*, December 8, 2005, 739–40.

Ellis, G. F. R., U. Kirchner, and W. R. Stoeger. "Multiverses and Physical Cosmology." *Monthly Notices of the Royal Astronomical Society* 347 (2004): 921–36. arXiv: astro-ph/0305292v3.

Everett, Hugh. "'Relative State' Formulation of Quantum Mechanics." *Review of Modern Physics* 29 (1957): 454–62.

"The Fabric of the Cosmos: Universe or Multiverse?" *NOVA*, PBS, November 23, 2011.

Feeney, Stephen, Matthew C. Johnson, Daniel J. Mortlock, and Hiranya V. Peiris. "First Observational Tests of Eternal Inflation: Analysis Methods and WMAP 7-Year Results." *Physical Review D* 84 (2011): 1–36.

Fenves, Peter D. *A Peculiar Fate: Metaphysics and World History in Kant*. Ithaca, N.Y.: Cornell University Press, 1991.

"Finds Spiral Nebulae Are Stellar Systems." *New York Times*, November 23, 1924.

Finocchiaro, Maurice A. "Philosophy Versus Religion and Science Versus Religion: The Trials of Bruno and Galileo." In *Giordano Bruno: Philosopher of the Renaissance*, edited by Hilary Gatti, 51–96. Burlington, Vt.: Ashgate, 2002.

Folger, Tim. "Does the Universe Exist If We're Not Looking?" *Discover*, June 1, 2002.

———. "Science's Alternative to an Intelligent Creator: The Multiverse Theory." [Interview with Andrei Linde.] *Discover*, November 10, 2008.

Fontenelle, Bernard le Bovier de. "But One Little Family of the Universe." Translated by Aphra Behn. In *The Book of the Cosmos: Imagining the Universe from Heraclitus to Hawking*, edited by Dennis Richard Danielson, 206–18. Cambridge, Mass.: Perseus, 2000

———. *Conversations on the Plurality of Worlds*. Translated by H. A. Hargreaves. Berkeley: University of California Press, 1990.

Foucault, Michel. "Nietzsche, Genealogy, History." Translated by Donald F. Bouchard and Sherry Simon. In *The Foucault Reader*, edited by Paul Rabinow, 76–100. New York: Pantheon, 1984.

Frampton, Paul. "Cyclic Universe and Infinite Past." *Modern Physics Letters A* 22 (2007): 2587–92. arXiv/0705.2730v2.

———. *Did Time Begin? Will Time End? Maybe the Big Bang Never Occurred*. Singapore: World Scientific, 2009.

Friedmann, Alexander. "Über die Krümmung des Raumes." *Zeitschrift für Physik* 10 (1922): 377–86.

Furley, David J. *Cosmic Problems: Essays on Greek and Roman Philosophy of Nature*. Cambridge: Cambridge University Press, 1989.

————. *The Greek Cosmologists: The Formation of the Atomic Theory and Its Earliest Critics.* Cambridge: Cambridge University Press, 1987.

Galileo Galilei. *Dialogue Concerning the Two Chief World Systems: Ptolemaic and Copernican.* Translated by Stillman Drake. Foreword by Albert Einstein. Berkeley: University of California Press, 1967.

————. *Sidereus Nuncius, or, A Sidereal Message.* Translated by William R. Shea. Sagamore Beach, Mass.: Science History, 2009.

Gamow, George. *The Creation of the Universe.* Rev. ed. New York: Mentor, 1957.

Gardner, Martin. *Are Universes Thicker Than Blackberries? Discourses on Gödel, Magic Hexagrams, Little Red Riding Hood, and Other Mathematical and Pseudoscientific Topics.* New York: Norton, 2003.

————. "WAP, SAP, PAP, & FAP." *New York Review of Books*, May 8, 1986, 22–25.

Garriga, Jaume, and Alexander Vilenkin. "Many Worlds in One." *Physical Review D* 64 (2001): 1–5.

————. "Testable Anthropic Predictions for Dark Energy." *Physical Review D* 67 (2003): 1–11.

Gasperini, Maurizio, and Gabriele Veneziano. "Pre–Big Bang in String Cosmology." *Astroparticle Physics* 1 (1993): 317–39.

————. "The Pre–Big Bang Scenario in String Cosmology." *Physics Reports* 373 (2003): 1–212. arXiv/0207130v1.

Gassendi, Pierre. *The Selected Works of Pierre Gassendi.* Translated by Craig B. Brush. New York: Johnson Reprint, 1972.

Gatti, Hilary. "Giordano Bruno's Copernican Diagrams." *Filozofski Vestnik* 25, no. 2 (2004): 25–50.

Geek's Guide to the Galaxy. "Theoretical Physicist Brian Greene Thinks You Might Be a Hologram." [Interview with Brian Greene.] *Wired*, May 16, 2012. http://www.wired.com/underwire/2012/05/geeks-guide-brian-greene/.

Germani, Cristiano, Nicolás Grandi, and Alex Kehagias. "The Cosmological Slingshot Scenario: A Stringy Proposal for the Early Time Cosmology." *AIP Conference Proceedings* 1031 (2007): 19–21. http://arxiv.org/abs/0805.2073.

Giddings, Steven B., Shamit Kachru, and Joseph Polchinski. "Hierarchies from Fluxes in String Compactifications." *Physical Review D* 66 (2002): 1–16. arXiv:hep-th/0105097.

Gilder, Louisa. *The Age of Entanglement: When Quantum Physics Was Reborn.* New York: Vintage, 2009.

Gleiser, Marcelo. *The Dancing Universe: From Creation Myths to the Big Bang.* Hanover, N.H.: University Press of New England/Dartmouth College Press, 2005.

————. *A Tear at the Edge of Creation: A Radical New Vision for Life in an Imperfect Universe.* New York: Free Press, 2010.

Gonzalez, Guillermo, and Jay W. Richards. *The Privileged Planet.* Washington, D.C.: Regnery, 2004.

Gott, J. Richard, and Li-Xin Li. "Can the Universe Create Itself?" *Physical Review D* 58 (1998): 1–43.

Gourinat, Jean-Baptiste. "The Stoics on Matter and Prime Matter: 'Coporealism' and the Imprint of Plato's *Timaeus.*" In *God and Cosmos in Stoicism*, edited by Ricardo Salles, 46–70. Oxford: Oxford University Press, 2009.

Granada, Miguel A. "Aristotle, Copernicus, Bruno: Centrality, the Principle of Movement, and the Extension of the Universe." *Studies in History and Philosophy of Science* 35 (2004): 91–114.

———. "L'infinité de l'univers et la conception du Système Solaire chez Giordano Bruno." *Revue des sciences philosophiques et théologiques* 82 (1998): 243–75.

———. "Kepler and Bruno on the Infinity of the Universe and of Solar Systems." *Journal for the History of Astronomy* 39 (2008): 469–95.

———. "Mersenne's Critique of Giordano Bruno's Conception of the Relation Between God and the Universe: A Reappraisal." *Perspectives on Science* 18, no. 1 (2010): 26–49.

Grant, Edward. *A History of Natural Philosophy: From the Ancient World to the Nineteenth Century*. Cambridge: Cambridge University Press, 2007.

Greenblatt, Stephen. *The Swerve: How the World Became Modern*. New York: Norton, 2011.

Greene, Brian. *The Elegant Universe: Superstrings, Hidden Dimensions, and the Quest for the Ultimate Theory*. New York: Norton, 1999.

———. *The Fabric of the Cosmos: Space, Time, and the Texture of Reality*. New York: Vintage, 2005.

———. *The Hidden Reality: Parallel Universes and the Deep Laws of the Cosmos*. New York: Knopf, 2011.

Gribbin, John. *In Search of the Multiverse*. London: Allen Lane, 2009.

———. "Oscillating Universe Bounces Back." *Nature*, January 1, 1976, 15–16.

Gurzadyan, V. G., and Roger Penrose. "Concentric Circles in WMAP Data May Provide Evidence of Violent Pre–Big Bang Activity." November 16, 2010. Available only through arXiv/1011.3706v1.

Guth, Alan H. "Eternal Inflation and Its Implications." *Journal of Physics A: Mathematical and Theoretical* 40 (2007): 6811–26. arXiv:hep-th/0702178.

———. "The Inflationary Universe." In *Cosmology: Historical, Literary, Philosophical, Religious, and Scientific Perspectives*, edited by Norris S. Hetherington, 411–45. New York: Garland, 1993.

———. "Inflationary Universe: A Possible Solution to the Horizon and Flatness Problems." *Physical Review D* 23 (1981): 347–56.

Guth, Alan H., and Paul J. Steinhardt. "The Inflationary Universe." *Scientific American*, May 1984, 116–28.

Guthrie, Karl S. *The Last Frontier: Imagining Other Worlds, from the Copernican Revolution to Modern Science Fiction*. Translated by Helen Atkins. Ithaca, N.Y.: Cornell University Press, 1990.

Hacking, Ian. *The Social Construction of What?* Cambridge, Mass.: Harvard University Press, 1999.

Hahm, David E. *The Origins of Stoic Cosmology*. Columbus: Ohio State University Press, 1977.

Hand, Eric. "The Test of Inflation." *Nature*, April 16, 2009, 820–25.

Hanson, Robin. "How to Live in a Simulation." *Journal of Evolution and Technology* 7, no. 1 (2001). http://www.jetpress.org/volume7/simulation.htm.

Harries, Karsten. "The Infinite Sphere: Comments on the History of a Metaphor." *Journal of the History of Philosophy* 13, no. 1 (1975): 5–15.

Harrison, Edward R. "The Natural Selection of Universes Containing Intelligent Life." *Quarterly Journal of the Royal Astronomical Society* 36, no. 3 (1995): 193–203.

———. "Newton and the Infinite Universe." *Physics Today* 39, no. 2 (1986): 24–32.

Harrison, Peter. "The Influence of Cartesian Cosmology in England." In *Descartes' Natural Philosophy*, edited by Stephen Gaukroger, John Schuster, and John Sutton, 168–92. Routledge Studies in Seventeenth-Century Philosophy. New York: Routledge, 2000.

Hartle, James, and Stephen Hawking. "Wave Function of the Universe." *Physical Review D* 28 (1983): 2960–75.

Haskings, Charles Homer. *The Renaissance of the Twelfth Century.* Cambridge, Mass.: Harvard University Press, 1971.

Hawking, Stephen. *A Brief History of Time: The Updated and Expanded Tenth Anniversary Edition.* New York: Bantam, 1998.

———. "Cosmology from the Top Down." In *Universe or Multiverse?* edited by Bernard Carr, 91–98. Cambridge: Cambridge University Press, 2007.

Hawking, Stephen, and George F. R. Ellis. "Space-Time Singularities." In *The Large-Scale Structure of Space-Time*, 256–98. Cambridge: Cambridge University Press, 1973.

Hawking, Stephen, and Thomas Hertog. "Populating the Landscape: A Top-Down Approach." *Physical Review D* 73 (2006): 1–9.

Hawking, Stephen, and Leonard Mlodinow. *The Grand Design.* New York: Bantam, 2010.

Hawking, Stephen, and Roger Penrose. "The Singularities of Gravitational Collapse and Cosmology." *Proceedings of the Royal Society of London A* 314, no. 1519 (1970): 529–48.

Heidegger, Martin. *Introduction to Metaphysics.* Translated by Gregory Fried and Richard Polt. New Haven, Conn.: Yale University Press, 2000.

Heisenberg, Werner. "The Physical Content of Quantum Kinematics and Mechanics." *Zeitschrift für Physik* 43 (1927): 172–98.

Hesiod. *Theognis.* Translated by Dorothea Wender. In *Theogony, Works and Days, Theognis, Elegies*, 87–148. New York: Penguin, 1973.

Hetherington, Norris S., ed. *Cosmology: Historical, Literary, Philosophical, Religious, and Scientific Perspectives.* New York: Garland, 1993.

———. "Introduction: A New Physics and a New Cosmology." In *Cosmology: Historical, Literary, Philosophical, Religious, and Scientific Perspectives*, edited by Norris S. Hetherington, 227–34. New York: Garland, 1993.

High Energy Physics Advisory Panel. "Quantum Universe: The Revolution in 21st Century Particle Physics." October 22, 2003. InterActions. http://www.interactions.org/pdf/Quantum_Universe.pdf.

Hinch, Jim. "Why Stephen Greenblatt Is Wrong—and Why It Matters." *Los Angeles Review of Books*, December 1, 2012.

Homer. *The Iliad.* Translated by Richmond Lattimore. Chicago: University of Chicago Press, 1961.

———. *The Odyssey.* Translated by Richmond Lattimore. New York: Harper, 2007.

Hooper, Dan. *Dark Cosmos: In Search of Our Universe's Missing Mass and Energy.* New York: HarperCollins/Smithsonian Books, 2006.

Hopkins, Jasper. *Nicholas of Cusa's Debate with John Wenck: A Translation and an Appraisal of "De ignota litteratura" and "Apologia doctae ignorantiae."* Minneapolis: Banning, 1981.

Hoyle, Fred. *Astronomy Today.* London: Heinemann, 1975.

———. *The Nature of the Universe.* New York: Mentor, 1950.

———. "A New Model for the Expanding Universe." *Monthly Notices of the Royal Astronomical Society* 108 (1948): 372–82.

Hume, David. *Dialogues Concerning Natural Religion.* Edited by Richard H. Popkin. 2nd ed. Indianapolis: Hackett, 1998.

———. *An Enquiry Concerning Human Understanding.* Edited by Eric Steinberg. 2nd ed. Indianapolis: Hackett, 1993.

———. *An Enquiry Concerning the Principles of Morals.* Edited by J. B. Schneewind. Indianapolis: Hackett, 1983.

Huygens, Christiaan. *Cosmotheoros: Or, Conjectures Concerning the Planetary Worlds, and Their Inhabitants.* Electronic reproduction. Farmington Hills, Mich.: Thomson Gale, 2003.

Irigaray, Luce. *Marine Lover of Friedrich Nietzsche.* Translated by Gillian C. Gill. New York: Columbia University Press, 1991.

———. "Plato's *Hystera.*" In *Speculum of the Other Woman,* 243–365. Translated by Gillian C. Gill. Ithaca, N.Y.: Cornell University Press, 1985.

———. "Sexual Difference." In *An Ethics of Sexual Difference,* 5–19. Translated by Carolyn Burke and Gillian C. Gill. Ithaca, N.Y.: Cornell University Press, 1993.

———. *This Sex Which Is Not One.* Translated by Catherine Porter and Carolyn Burke. Ithaca, N.Y.: Cornell University Press, 1985.

James, William. "Is Life Worth Living?" *International Journal of Ethics* 6, no. 1 (1895): 1–24.

———. *A Pluralistic Universe.* Lincoln: University of Nebraska Press, 1996.

———. *The Varieties of Religious Experience.* New York: Penguin, 1982.

———. "The Will to Believe." In *Pragmatism and Other Writings,* 198–218. Edited by Giles Gunn. New York: Penguin, 2000.

Jantzen, Grace. *Becoming Divine: Towards a Feminist Philosophy of Religion.* Bloomington: Indiana University Press, 1999.

Jarosik, N., and C. L. Bennett. "Seven-Year Wilkinson Microwave Anistrophy Probe (WMAP) Observations: Sky Maps, Systematic Errors, and Basic Results." *Astrophysical Journal Supplement Series* 192 (2011): 1–47. arXiv/1001.4744.

Jastrow, Robert. *God and the Astronomers.* New York: Norton, 1978.

Johansen, T. K. *Plato's Natural Philosophy: A Study of the "Timaeus–Critias."* Cambridge: Cambridge University Press, 2004.

Jones, Howard. *The Epicurean Tradition.* New York: Routledge, 1989.

Kaiser, David I. *How the Hippies Saved Physics: Science, Counterculture, and the Quantum Revival.* New York: Norton, 2012.

Kaku, Michio. *Parallel Worlds: A Journey Through Creation, Higher Dimensions, and the Future of the Cosmos.* New York: Doubleday, 2004.

Kallosh, Renata, Jin U. Kang, Andrei Linde, and Viatscheslav Mukhanov. "The New Ekpyrotic Ghost." *Journal of Cosmology and Astroparticle Physics* 2008 (2008): 1–23. arXiv:0712.2040.

Kant, Immanuel. *Allgemeine Naturgeschichte und Theorie des Himmels.* Berlin: Aufbau, 1995.

———. "The Only Possible Argument in Support of a Demonstration of the Existence of God." In *Theoretical Philosophy, 1755–1770,* edited by David Walford, 107–201. Cambridge: Cambridge University Press, 1992.

———. *Universal Natural History and Theory of the Heavens.* Translated by W. Hastie. Ann Arbor: University of Michigan Press, 1969.

———. *Universal Natural History and Theory of the Heavens.* Translated by Stanley L. Jaki. Edinburgh: Scottish Academic Press, 1981.

Kargon, Robert. "Thomas Hariot, the Northumberland Circle, and Early Atomism in England." *Journal of the History of Ideas* 27, no. 1 (1966): 128–36.

Kashlinsky, A., F. Atrio-Barandela, H. Ebeling, A. Edge, and D. Kocevski. "A New Measurement of the Bulk Flow of X-Ray Luminous Clusters of Galaxies." *Astrophysical Journal Letters* 71 (2010): L81–L85. arXiv:astro-ph/0910.4958.

Kather, Regine. "'The Earth Is a Noble Star': Arguments for the Relativity of Motion in the Cosmology of Nicholaus Cusanus and Their Transformation in Einstein's Theory of Relativity." In *Cusanus: The Legacy of Learned Ignorance*, edited by Peter J. Casarella, 226–50. Washington, D.C.: Catholic University of America Press, 2006.

Kearns, Laurel, and Catherine Keller, eds. *Ecospirit: Religions and Philosophies for the Earth*. New York: Fordham University Press, 2007.

Keller, Catherine. "The Cloud of the Impossible." Manuscript.

———. *Face of the Deep: A Theology of Becoming*. New York: Routledge, 2003.

Keller, Evelyn Fox. "Cognitive Repression in Contemporary Physics." In *Reflections on Gender and Science*, 139–49. New Haven, Conn.: Yale University Press, 1985.

Kepler, Johannes. *Kepler's Conversation with Galileo's Sidereal Messenger*. Translated by Edward Rosen, with introduction and notes. New York: Johnson Reprint, 1965.

———. *Kepler's Somnium: The Dream, or Posthumous Works on Lunar Astronomy*. Translated by Edward Rosen, with commentary. Mineola, N.Y.: Dover, 2003.

———. *Mysterium Cosmographicum: The Secret of the Universe*. Translated by A. M. Duncan. New York: Abaris Books, 1981.

Khoury, Justin, Burt A. Ovrut, Paul J. Steinhardt, and Neil Turok. "Ekpyrotic Universe: Colliding Branes and the Origin of the Hot Big Bang." *Physical Review D* 64 (2001): 1–24.

Kierkegaard, Søren [Johannes Climacus, pseud.]. *Philosophical Fragments, or a Fragment of Philosophy*. Translated by Howard V. Hong and Edna H. Hong. Princeton, N.J.: Princeton University Press, 1985.

Kirk, G. S., and J. E. Raven. *The Presocratic Philosophers: A Critical History with a Selection of Texts*. Cambridge: Cambridge University Press, 1957.

Kirshner, Robert P. *The Extravagant Universe: Exploding Stars, Dark Energy, and the Accelerating Cosmos*. Princeton, N.J.: Princeton University Press, 2002.

Knobe, Joshua, Ken D. Olum, and Alexander Vilenkin. "Philosophical Implications of Inflationary Cosmology." *British Journal for the Philosophy of Science* 57, no. 1 (2006): 47–67.

Koyré, Alexander. *From the Closed World to the Infinite Universe*. Baltimore: Johns Hopkins University Press, 1957.

Kragh, Helge. "Ancient Greek–Roman Cosmology: Infinite, Eternal, Finite, Cyclic, and Multiple Universes." *Journal of Cosmology* 9 (2010): 2172–78.

———. "Big Bang Cosmology." In *Cosmology: Historical, Literary, Philosophical, Religious, and Scientific Perspectives*, edited by Norris S. Hetherington, 371–89. New York: Garland, 1993.

———. *Higher Speculations: Grand Theories and Failed Revolutions in Physics and Cosmology*. Oxford: Oxford University Press, 2011.

———. "Steady State Theory." In *Cosmology: Historical, Literary, Philosophical, Religious, and Scientific Perspectives*, edited by Norris S. Hetherington, 391–403. New York: Garland, 1993.

Krauss, Lawrence. *Quintessence: The Mystery of Missing Mass in the Universe*. Rev. ed. New York: Basic Books, 2000.

———. *A Universe from Nothing: Why There Is Something Rather Than Nothing*. New York: Free Press, 2012.

Krauss, Lawrence, and Robert J. Scherrer. "The End of Cosmology?" *Scientific American*, March 2008, 47–53.

Kube-McDowell, Michael P. *Alternities*. N.p.: iBooks, 2005.

Kurzweil, Ray. *The Singularity Is Near: When Humans Transcend Biology*. New York: Penguin, 2006.

Lachieze-ray, M., and Jean-Pierre Luminet. "Cosmic Topology." *Physics Reports* 254 (1995): 135–214. arXiv:gr-qc/ 9605010 (updated version).

Lactantius. "The Divine Institutes." In *The Ante-Nicene Fathers: Translations of the Writings of the Fathers Down to* A.D. *325*, edited by Alexander Roberts and James Donaldson, 7:bk. 3, chap. 24. Edinburgh: Clark, 1911. Internet Sacred Text Archive. http://sacred-texts.com/chr/ecf/007/0070075.htm.

La Cugna, Catherine. *God for Us: The Trinity and Chrisitan Life*. San Francisco: HarperSanFrancisco, 1991.

Lai, Tyrone. "Nicholas of Cusa and the Finite Universe." *Journal of the History of Philosophy* 11, no. 2 (1973): 161–67.

Lambert, Johann. *Cosmological Letters on the Arrangement of the World-Edifice*. Translated by Stanley L. Jaki. New York: Science History, 1976.

———. *Cosmologische Briefe über die Einrichtung des Weltbaues*. Augsburg: E. Kletts Wittib, 1761.

Lapidge, Michael. "Stoic Cosmology." In *The Stoics*, edited by J. M. Rist, 160–85. Berkeley: University of California Press, 1978.

Laplace, Pierre Simon. *Exposition du système du monde*. Paris: Bachelier, 1835.

———. *The System of the World*. Translated by J. Pond. 2 vols. London: Phillips, 1809.

Lasseter, John, dir. *Toy Story*. Walt Disney Pictures, Pixar Studio, 1995.

Latour, Bruno. "Thou Shalt Not Freeze-Frame; or, How Not to Misunderstand the Science and Religion Debate." In *Science, Religion, and the Human Experience*, edited by James D. Proctor, 27–48. Oxford: Oxford University Press, 2005.

Lehners, Jean-Luc, and Paul J. Steinhardt. "Dark Energy and the Return of the Phoenix Universe." *Physical Review D* 79 (2009): 1–5. arXiv:hep-th/0812.3388.

———. "Dynamical Selection of the Primordial Density Fluctuation Amplitude." *Physical Review Letters* 106 (2011): 1–4. arXiv:hep-th/1008.4567v1.

Leibniz, G. W. *Theodicy: Essays on the Goodness of God, the Freedom of Man, and the Origin of Evil*. Translated by E. M. Huggard. Eugene, Ore.: Wipf and Stock, 2001.

Leibniz, G. W., and Samuel Clarke. *Correspondence*. Edited by Roger Ariew. Indianapolis: Hackett, 2000.

Lemaître, Georges. "The Expanding Universe." *Monthly Notices of the Royal Astronomical Society* 91 (1931): 490–501.

———. "Rencontre avec A. Einstein." *Revue des questions scientifiques* 129 (1958): 129–32.

———. "Un univers homogène de masse constante et de rayon croissant rendant ompte de la vitesse radiale des nébuleuses extragalactique." *Annales de la Société scientifique de Bruxelles* 47 (1927): 49–59.

Lemonick, Michael D. "The End." *Time*, June 25, 2001.

Leslie, John. *Universes*. New York: Routledge, 1996.

Levin, Janna. *How the Universe Got Its Spots: Diary of a Finite Time in a Finite Space*. New York: Anchor, 2003.

Lewis, David. *On the Plurality of Worlds*. Oxford: Oxford University Press, 1986.

Lightman, Alan P. "The Accidental Universe: Science's Crisis of Faith." *Harper's Magazine*, December 22, 2011.

"LIGHTS ALL ASKEW IN THE HEAVENS; Men of Science More or Less Agog over Results of Eclipse Observations. EINSTEIN THEORY TRIUMPHS. Stars Not Where They Seemed or Were Calculated to Be, but Nobody Need Worry." *New York Times*, November, 10, 1919.

Linde, Andrei. "Eternally Existing Self-Reproducing Chaotic Inflationary Universe." *Physics Letters B* 175 (1986): 395–400.

———. "Hard Art of the Universe Creation." October 15, 1991. Available only through arXiv:hep-th/9110037v1.

———. "Inflation in Supergravity and String Theory: Brief History of the Multiverse." March 21, 2012. Text at http://www.ctc.cam.ac.uk/stephen70/talks/swh70_linde.pdf; video at http://sms.cam.ac.uk/media/1228717.

———. "The Inflationary Multiverse." In *Universe or Multiverse?* edited by Bernard Carr, 127–49. Cambridge: Cambridge University Press, 2007.

———. "Inflationary Theory Versus Ekpyrotic/Cyclic Scenario: A Talk at Stephen Hawking's 60th Birthday Conference, Cambridge University, Jan. 2002." In *The Future of Theoretical Physics and Cosmology*, edited by G. W. Gibbons, E. P. S. Shellard, and S. J. Rankin, 801–38. Cambridge: Cambridge University Press, 2003. arXiv:hep-th/0205259.

———. "The Self-Reproducing Inflationary Universe." *Scientific American*, November 1994, 48–55.

Linde, Andrei D., Dmitri Linde, and Arthur Mezhlumian. "Do We Live in the Center of the World?" *Physics Letters B* 345 (1995): 203–10.

Lloyd, G. E. R. "Metaphysics Lambda 8." In *Aristotle's "Metaphysics" Lambda: Symposium Aristotelicum*, edited by Michael Frede and David Charles, 245–73. Oxford: Clarendon Press, 2000.

Locke, John. *An Essay Concerning Human Understanding: Complete and Unabridged.* Milwaukee: WLC Books, 2009.

Lockwood, Louise. *Parallel Worlds, Parallel Lives.* PBS, October 21, 2008.

Lovejoy, Arthur O. *The Great Chain of Being: A Study of the History of an Idea.* Cambridge, Mass.: Harvard University Press, 1976.

Lucretius. *De rerum natura.* Translated by W. H. D. Rouse. Revised by Martin Ferguson Smith. Loeb Classical Library 181. Cambridge, Mass.: Harvard University Press, 1975.

———. *The Nature of Things.* Translated by A. E. Stallings. New York: Penguin, 2007.

Luminet, Jean-Pierre, Glenn D. Starkman, and Jeffrey R. Weeks. "Is Space Finite?" *Scientific American*, October 2002, 58–65.

MacFarlane, Seth, creator; Greg Colton, dir.; and Wellesley Wild, writer. "Road to the Multiverse." *Family Guy*, season 8, episode 1, FOX, September 27, 2009.

Manly, Steven. *Visions of the Multiverse.* Pompton Plains, N.J.: Career Press, 2011.

Mansfield, J. "Providence and the Destruction of the Universe in Early Stoic Thought: With Some Remarks on the 'Mysteries of Philosophy.'" In *Studies in Hellenistic Religion*, edited by M. J. Vermaseren, 129–88. Leiden: Brill, 1979.

Manson, Neil A., ed. *God and Design: The Teleological Argument and Modern Science.* New York: Routledge, 2003.

———. "Introduction." In *God and Design: The Teleological Argument and Modern Science*, edited by Neil A. Manson, 1–23. New York: Routledge, 2003.

May, Gerhard. *Creatio ex Nihilo: The Doctrine of "Creation out of Nothing" in Early Christian Thought.* Translated by A. S. Worrall. Edinburgh: Clark, 1994.

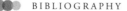

McCutcheon, Russell T. "I Have a Hunch." *Method and Theory in the Study of Religion* 24, no. 1 (2012): 81–92.

McGinn, Bernard. "Maximum Contractum et Absolutum: The Motive for the Incarnation in Nicholas of Cusanus and His Predecessors." In *Nicholas of Cusa and His Age: Intellect and Spirituality: Essays Dedicated to the Memory of F. Edward Cranz, Thomas P. Mctighe, and Charles Trinkaus*, edited by Thomas Izbicki and Christopher M. Bellitto, 149–74. Studies in the History of Christian Traditions 105. Boston: Brill, 2002.

McGrew, Timothy, Lydia McGrew, and Eric Vestrup. "Probabilities and the Fine-Tuning Argument: A Skeptical View." In *God and Design: The Teleological Argument and Modern Science*, edited by Neil A. Manson, 200–208. New York: Routledge, 2003.

McMullin, Ernan. "Indifference Principle and Anthropic Principle in Cosmology." *Studies in History and Philosophy of Science* 24 (1993): 359–89.

Melko, Paul. *The Broken Universe*. New York: Tor Books, 2012.

———. *The Walls of the Universe*. New York: Tor Books, 2009.

Mellor, D. H. "Too Many Universes." In *God and Design: The Teleological Argument and Modern Science*, edited by Neil A. Manson, 221–28. New York: Routledge, 2003.

Melman, Jeff, dir., and Chris McKenna, writer. "Remedial Chaos Theory." *Community*, season 3, episode 4, NBC, October 13, 2011.

Mersini-Houghton, Laura. "Birth of the Universe from the Multiverse." September 22, 2008. Available only through arXiv/0809.3623.

———. "Notes on Time's Enigma." In *The Arrows of Time: A Debate in Cosmology*, edited by Laura Mersini-Houghton and Rüdiger Vaas, 157–68. Fundamental Theories of Physics 172. Berlin: Springer, 2012. arXiv/0909.2330.

———. "Thoughts on Defining the Multiverse." April 27, 2008. Available only through arXiv/0804.4280.

Mersini-Houghton, Laura, and Richard Holman. "'Tilting' the Universe with the Landscape Multiverse: The 'Dark' Flow." *Journal of Cosmology and Astroparticle Physics* 2009 (2009): 1–12. arXiv/0810.5388.

Miao Li, Xiao-Dong Li, and Xin Zhang. "Comparison of Dark Energy Models: A Perspective from the Latest Observational Data." *Science China: Physics, Mechanics, & Astronomy* 53, no. 9 (2010): 1631–45.

Milbank, John. *Theology and Social Theory: Beyond Secular Reason*. Oxford: Blackwell, 1993.

Miller, Clyde Lee. *Reading Cusanus: Metaphor and Dialectic in a Conjectural Universe*. Washington, D.C.: Catholic University of America Press, 2003.

Miller, Dana. "Plutarch's Argument for a Plurality of Worlds in *De defectu oraculorum* 424c10–425e7." *Ancient Philosophy* 17, no. 2 (1997): 375–95.

Misner, Charles, Kip Thorne, and John Archibald Wheeler. "Beyond the End of Time." In *Gravitation*, edited by Charles Misner, Kip Thorne, and John Archibald Wheeler, 1196–217. San Francisco: Freeman, 1970.

Molière. *Théatre complet de Molière*. Edited by Robert Jouanny. Paris: Éditions Garnier Frères, 1956.

Montgomery, R. A. *Space and Beyond*. Choose Your Own Adventure. Waitsfield, Vt.: Chooseco, 2006.

Moravec, Hans. *Robot: Mere Machine to Transcendent Mind*. New York: Oxford University Press, 2000.

More, Henry. *Divine Dialogues Containing Disquisitions Concerning the Attributes and Providence of God*. 3 vols. Glasgow: Robert Foulis, 1743.

Moss, Adam, Douglas Scott, and James P. Zibin. "No Evidence for Anomalously Low Variance Circles on the Sky." *Journal of Cosmology and Astroparticle Physics* 2011 (2011): 1–7. arXiv:astro-ph/1012.1305v3.

"Multiverse." Edited by Terry Gross. *Fresh Air*, NPR, January 24, 2011.

Munitz, Milton. "Introduction." In Immanuel Kant, *Universal Natural History and Theory of the Heavens*, v–xxii. Translated by W. Hastie. Ann Arbor: University of Michigan Press, 1969.

Nancy, Jean-Luc. *Being Singular Plural*. Translated by Robert D. Richardson and Anne E. O'Byrne. Stanford, Calif.: Stanford University Press, 2000.

National Aeronautics and Space Administration (NASA). "Dark Energy, Dark Matter." *NASA Science: Astrophysics*. National Aeronautics and Space Administration. http:// science.nasa.gov/astrophysics/focus-areas/what-is-dark-energy/.

Nave, Rod. "Laminar Flow." HyperPhysics, Georgia State University. http://hyperphysics .phy-astr.gsu.edu/hbase/pfric.html.

Nehemas, Alexander. "The Eternal Recurrence." In *Nietzsche*, edited by John Richardson and Brian Leiter, 118–38. Oxford: Oxford University Press, 2001.

Newton, Isaac. "De gravitatione et aequipondio fluidorum." In *Unpublished Scientific Papers of Isaac Newton*, edited by A. Rupert Hall and Marie Boas Hall, 89–156. Cambridge: Cambridge University Press, 1962.

———. *Four Letters from Sir Isaac Newton to Doctor Bentley, Containing Some Arguments in Proof of a Deity*. London: Dodsley, 1756.

———. *The Principia*. Translated by Andrew Motte. Great Minds Series. Amherst, N.Y.: Prometheus Books, 1995.

Nicholas of Cusa. *De ludo globi/The Game of Spheres*. Edited by Pauline Moffitt Watts. New York: Abaris Books, 1986.

———. *De possest*. Vol. 11 of *Nicolai de Cusa: Opera omnia*, edited by Ernest Hoffman and Raymond Klibansky. Leipzig: Meiner, 1932.

———. *De venatione sapientiae*. Vol. 12 of *Nicolai de Cusa: Opera omnia*, edited by Ernest Hoffman and Raymond Klibansky. Leipzig: Felix Meiner, 1932.

———. *Directio speculantis seu de li non aliud*. Vol. 13 of *Nicolai de Cusa: Opera omnia*, edited by Ernest Hoffman and Raymond Klibansky. Leipzig: Meiner, 1932.

———. *On Learned Ignorance*. In *Nicholas of Cusa: Selected Spiritual Writings*, 85–206. Translated by H. Lawrence Bond. New York: Paulist Press, 1997.

———. *On the Vision of God*. In *Nicholas of Cusa: Selected Spiritual Writings*, 233–90. Translated by H. Lawrence Bond. New York: Paulist Press, 1997.

Nietzsche, Friedrich. *Ecce Homo*. Translated by Walter Kaufman. New York: Vintage, 1967.

———. *The Gay Science*. Translated by Walter Kaufman. New York: Vintage, 1974.

———. *On the Genealogy of Morals*. Translated by Walter Kaufmann. New York: Vintage, 1989.

———. *Thus Spoke Zarathustra*. Translated by Walter Kaufman. New York: Penguin, 1978.

"No Big Bang? Endless Universe Made Possible by New Model." January 30, 2007. Phys. org. http://phys.org/news89399974.html.

Norton, John D. "The Cosmological Woes of Newtonian Gravitation Theory." In *The Expanding Worlds of General Relativity*, edited by Hubert Goenner, Jürgen Renn, Jim Ritter, and Tilman Sauer, 271–322. Boston: Center for Einstein Studies, 1999.

Nozick, Robert. *Philosophical Explanations*. Cambridge, Mass.: Harvard University Press, 1981.

O'Connor, Robert. "The Design Inference: Old Wine in New Wineskins." In *God and Design: The Teleological Argument and Modern Science*, edited by Neil A. Manson, 66–87. New York: Routledge, 2003.

Oord, Thomas, and Richard Livingston, eds. *Creation Options: Rethinking Initial Creation*. New York: Routledge, forthcoming.

Origen. *De principiis*. In *The Ante-Nicene Fathers: Translations of the Writings of the Fathers Down to* A.D. *325*, edited by Alexander Roberts and James Donaldson, 4:239–382. Buffalo, N.Y.: Christian Literature, 1885.

Overbye, Dennis. "Dark, Perhaps Forever." *New York Times*, June 3, 2008.

———. "Gauging a Collider's Odds of Creating a Black Hole." *New York Times*, April 15, 2008.

———. "Rings in Sky Leave Alternate Visions of Universes." *New York Times*, December 13, 2010.

Page, Don N. "Does God So Love the Multiverse?" In *Science and Religion in Dialogue*, edited by Melville Y. Stewart, 1:380–95. Malden, Mass.: Wiley-Blackwell, 2010. arXiv/0801.0246.

———. "Predictions and Tests of Multiverse Theories." In *Universe or Multiverse?* edited by Bernard Carr, 411–29. Cambridge: Cambridge University Press, 2007.

Pagels, Heinz. *The Dreams of Reason*. New York: Bantam, 1989.

Paine, Thomas. *The Age of Reason: Being an Investigation of True and Fabulous Theology*. Electronic reproduction. Farmington Hills, Mich.: Cengage Gale, 2009.

Paley, William. *Natural Theology; or, Evidences of the Existence and Attributes of the Deity. Collected from the Appearances of Nature*. London: Faulder, 1802.

Palmer, Jason. "Cosmos May Show Echoes of Events Before Big Bang." November 27, 2010. BBC News: Science and Environment. http://www.bbc.co.uk/news/science-envi ronment-11837869.

———. "'Multiverse' Theory Suggested by Microwave Background." August 3, 2011. BBC News: Science and Environment. http://www.bbc.co.uk/news/science-environ-ment-14372387.

Panek, Richard. "Out There." *New York Times*, March 11, 2007.

Parry, Richard. "Empedocles." In *Stanford Encyclopedia of Philosophy*. 2005. http://plato. stanford.edu/entries/empedocles/.

Pascal, Blaise. *Pensées*. Paris: Bookking International, 1995.

———. *Pensées*. Translated by A. J. Krailsheimer. New York: Penguin, 1966.

Patton, C. M., and John Archibald Wheeler. "Is Physics Legislated by Cosmogony?" In *Quantum Gravity: An Oxford Symposium*, edited by C. J. Isham, Roger Penrose, and Dennis W. Sciama, 538–605. Oxford: Clarendon Press, 1975.

Paul III. *Sublimus Dei*. 1537. Papal Encyclicals Online, http://www.papalencyclicals.net /Paul03/p3subli.htm.

Peebles, P. J. E., and David T. Wilkinson. "The Primeval Fireball." *Scientific American*, June 1967, 28–37.

Penrose, Roger. *Cycles of Time: An Extraordinary New View of the Universe*. New York: Knopf, 2011.

Penzias, A. A., and R. W. Wilson. "A Measurement of Excess Antenna Temperature at 4080 Mc/S." *Astrophysical Journal* 142 (1965): 419–21.

Perkins, Pheme. "On the Origin of the World: A Gnostic Physics." *Vigiliae Christianae* 34, no. 1 (1980): 36–46.

Peters, Ted. "Cosmos as Creation." In *Cosmos as Creation: Theology and Science in Consonance*, edited by Ted Peters, 45–113. Nashville: Abingdon Press, 1989.

Pius XII. "Modern Science and the Existence of God." *Catholic Mind* 49 (1952): 182–92.

Plato. *Republic*. Translated by G. M. A. Grube. Edited by C. D. C. Reeve. Indianapolis: Hackett, 1992.

———. *Statesman*. Edited by Julia Annas and Robin Waterfield. Translated by Robin Waterfield. Cambridge Texts in the History of Political Thought. Cambridge: Cambridge University Press, 1995.

———. *Timaeus*. In *Timaeus and Critias*, 28–126. Translated by Desmond Lee. New York: Penguin, 1977.

Poe, Edgar Allan. *Eureka*. Edited by Stuart Levine and Susan F. Levine. Champaign: University of Illinois Press, 2004.

Polkinghorne, John. *Beyond Science: The Wider Human Context*. Cambridge: Cambridge University Press, 1996.

———. *The Faith of a Physicist: Reflections of a Bottom-up Thinker*. Minneapolis: Fortress, 1996.

Popper, Karl. *The Logic of Scientific Discovery*. New York: Basic Books, 1959.

Proudfoot, Wayne. "Pragmatism and 'an Unseen Order.' " In *William James and a Science of Religions: Reexperiencing the Varieties of Religious Experience*, edited by Wayne Proudfoot, 31–47. New York: Columbia University Press, 2004.

Pseudo-Dionysius. *The Celestial Hierarchy*. In *The Complete Works*, edited by Colm Luibheid, 143–92. New York: Paulist Press, 1987.

Puech, Henri-Charles. "Gnosis and Time." In *Man and Time: Papers from the Eranos Yearbooks*, edited by Joseph Campbell and R. F. Hull, 38–84. New York: Pantheon, 1958.

Randall, Lisa. *Warped Passages: Unraveling the Mysteries of the Universe's Hidden Dimensions*. New York: HarperCollins, 2005.

Randall, Lisa, and Raman Sundrum. "Large Mass Hierarchy from a Small Extra Dimension." *Physical Review Letters* 83 (1999): 3370–73.

Rees, Martin. "Concluding Perspective." In *New Cosmological Data and the Values of the Fundamental Parameters: Proceedings of the 201st Symposium of the International Astronomical Union Held During the IAU General Assembly XXIV, the Victoria University of Manchester, United Kingdom, 7–11 August, 2000*, edited by Anthony Lasenby and Althea Wilkinson, 415–24. San Francisco: Astronomical Society of the Pacific, 2005. arXiv:astro-ph/0101268v1.

———. "Cosmology and the Multiverse." In *Universe or Multiverse?* edited by Bernard Carr, 57–75. Cambridge: Cambridge University Press, 2007.

———. "Exploring Our Universe and Others." In "The Once and Future Cosmos." Special issue, *Scientific American*, December 1999, 78–83.

———. "In the Matrix." *Edge*, May 19, 2003.

———. *Just Six Numbers: The Deep Forces That Shape the Universe*. New York: Basic Books, 2000.

"Revolution in Science: New Theory of the Universe, Newtonian Ideas Overthrown." *Times* (London), November 17, 1919.

Ricoeur, Paul. *The Symbolism of Evil*. New York: Harper & Row, 1967.

Roark, Rhys W. "Nicholas Cusanus, Linear Perspective, and the Finite Cosmos." *Viator* 41, no. 1 (2010): 315–66.

Ross, Hugh. *Why the Universe Is the Way It Is*. Grand Rapids, Mich.: Baker, 2008.

Ross, J. M. "Introduction." In *Cicero: The Nature of the Gods*, 7–63. New York: Penguin, 1972.

Rowland, Ingrid D. *Giordano Bruno: Philosopher/Heretic*. New York: Farrar, Straus and Giroux, 2008.

Rubenstein, Mary-Jane. "Cosmic Singularities: On the Nothing and the Sovereign." *Journal of the American Association of Religion* 80, no. 2 (2012): 485–517.

———. "Dionysius, Derrida, and the Problem of Ontotheology." *Modern Theology* 24, no. 2 (2008): 725–42.

———. "End Without End: Cosmology and Infinity in Nicholas of Cusa." In *The Trials of Desire: A Festschrift for Denys A. Turner*, edited by Eric Bugyis and David Newheiser. Notre Dame, Ind.: University of Notre Dame Press, forthcoming.

———. "The Fire Each Time: Dark Energy and the Breath of Creation." In *Cosmology, Ecology, and the Energy of God*, edited by Donna Bowman and Clayton Crockett, 26–41. New York: Fordham University Press, 2011.

———. "Myth and Modern Physics: On the Power of Nothing." In *Creation Options: Rethinking Initial Creation*, edited by Thomas Oord and Richard Livingston. New York: Routledge, forthcoming.

———. *Strange Wonder: The Closure of Metaphysics and the Opening of Awe*. New York: Columbia University Press, 2009.

———. "Undone by Each Other: Interrupted Sovereignty in Augustine's *Confessions*." In *Polydoxy*, edited by Catherine Keller and Laurel Schneider, 105–25. New York: Routledge, 2010.

"Runaway Universe." *NOVA*, PBS, November 11, 2000.

Sadakata, Akira. *Buddhist Cosmology: Philosophy and Origins*. Tokyo: Kosei, 2004.

Sagan, Carl. "The Edge of Forever." In *Cosmos: A Personal Voyage*, episode 10. Cosmos Studios, 1980.

Salles, Ricardo. "Chrysippus on Conflagration and the Indestructibility of the Cosmos." In *God and Cosmos in Stoicism*, edited by Ricardo Salles, 118–34. Oxford: Oxford University Press, 2009.

Sallis, John. *Chorology: On Beginning in Plato's "Timaeus."* Studies in Continental Thought. Bloomington: Indiana University Press, 1999.

Schneider, Laurel. *Beyond Monotheism: A Theology of Multiplicity*. New York: Routledge, 2008.

Schneider, Nathan. "The Multiverse Problem." *Seed Magazine*, April 14, 2009.

Schönborn, Christof. "Finding Design in Nature." *New York Times*, July 7, 2005.

Schönfeld, Martin. "The Phoenix of Nature: Kant and the Big Bounce." In "The Copernican Imperative." Special issue, *Collapse* 5 (2009): 361–76.

Schwarzschild, Karl. "On the Gravitational Field of a Point-Mass, According to Einstein's Theory." Translated by Larissa Borissova and Dmitri Rabounski. *Abraham Zelmanov Journal* 1 (2008): 10–19.

"Scientists Believe That They Have Discovered Another Universe." December 15, 2009. Current.com.

Sedley, David. *Creationism and Its Critics in Antiquity*. Berkeley: University of California Press, 2007.

Seneca. *To Marcia on Consolation*. In *Seneca: Moral Essays*, 2–97. Translated by John W. Basore. Loeb Classical Library 254. New York: Putnam, 1932.

Serres, Michel. *The Birth of Physics*. Translated by Jack Hawkes. Manchester, Eng.: Clinamen Press, 2000.

———. *Genesis*. Translated by Genevieve James and James Nielson. Ann Arbor: University of Michigan Press, 1995.

Shelton, Jessie, Washington Taylor, and Brian Wecht. "Generalized Flux Vacua." *Journal of High Energy Physics*, no. 2 (2007): 1–27. arXiv:hep-th/0607015v2.

Shostak, Seth. "The Lugubrious Universe." *Huffington Post*, November 26, 2010. http://www.huffingtonpost.com/seth-shostak/the-lugubrious-universe_b_788478.html.

Simek, Rudolph. *Heaven and Earth in the Middle Ages: The Physical World Before Columbus*. Woodbridge, Eng.: Boydell & Brewer, 1996.

Smith, Walter Chalmers. "Immortal, Invisible God Only Wise." In *The English Hymnal (Second Edition), with Tunes*, edited by J. H. Arnold. London: Oxford University Press, 1933. http://www.oremus.org/hymnal/i/i225.html.

Smolin, Lee. *The Life of the Cosmos*. New York: Oxford University Press, 1997.

———. "Scientific Alternatives to the Anthropic Principle." In *Universe or Multiverse?* edited by Bernard Carr, 323–66. Cambridge: Cambridge University Press, 2007.

———. *Three Roads to Quantum Gravity*. New York: Basic Books, 2001.

———. *The Trouble with Physics: The Rise of String Theory, the Fall of a Science, and What Comes Next*. New York: First Mariner, 2007.

Solmsen, Friedrich. *Aristotle's System of the Physical World: A Comparison with His Predecessors*. Ithaca, N.Y.: Cornell University Press, 1960.

———. "Epicurus and Cosmological Heresies." *American Journal of Philology* 72, no. 1 (1951): 1–23.

———. "Epicurus on the Growth and Decline of the Cosmos." *American Journal of Philology* 74, no. 1 (1953): 34–51.

———. "Love and Strife in Empedocles' Cosmology." *Phronesis* 10, no. 2 (1965): 109–48.

Steinhardt, Paul J. "The Inflation Debate." *Scientific American*, April 2011, 36–43.

———. "Natural Inflation." In *The Very Early Universe*, edited by Gary W. Gibbons, Stephen W. Hawking, and S. T. C. Siklos, 251–66. Cambridge: Cambridge University Press, 1983.

Steinhardt, Paul J., and Neil Turok. "Cosmic Evolution in a Cyclic Universe." *Physical Review D* 65 (2002): 1–20.

———. "The Cyclic Model Simplified." *New Astronomy Reviews* 49, no. 206 (2005): 43–57.

———. *Endless Universe: Beyond the Big Bang—Rewriting Cosmic History*. New York: Broadway Books, 2008.

Stoeger, William R., S.J. "Are Anthropic Arguments, Involving Multiverses and Beyond, Legitimate?" In *Universe or Multiverse?* edited by Bernard Carr, 445–57. Cambridge: Cambridge University Press, 2007.

Stoeger, William R., G. F. R. Ellis, and U. Kirchner. "Multiverses and Cosmology: Philosophical Issues." January 19, 2006. Available only through arXiv/0407329v2.

Super Mario Galaxy. Nintendo, 2007.

Super Mario Galaxy 2. Nintendo, 2010.

Surin, Kenneth. *Theology and the Problem of Evil*. Eugene, Ore.: Wipf and Stock, 1986.

Suskind, Ron. "Faith, Certainty, and the Presidency of George W. Bush." *New York Times Magazine*, October 17, 2004. http://www.nytimes.com/2004/10/17/magazine/17BUSH.html?_r=0.

Susskind, Leonard. "The Anthropic Landscape of String Theory." In *Universe or Multiverse?* edited by Bernard Carr, 247–66. Cambridge: Cambridge University Press, 2007. arXiv:hep-th/0302219v1.

———. *The Black Hole War: My Battle with Stephen Hawking to Make the World Safe for Quantum Mechanics.* New York: Little, Brown, 2008.

———. *The Cosmic Landscape: String Theory and the Illusion of Intelligent Design.* Boston: Back Bay Books, 2006.

Swinburne, Richard. "Argument from the Fine-tuning of the Universe." In *Physical Cosmology and Philosophy,* edited by John Leslie and Paul Edwards, 154–73. New York: Macmillan, 1990.

———. *The Existence of God.* New York: Oxford University Press, 1979.

Taye, Jamgon Kongtrul Lodro. *Myriad Worlds: Buddhist Cosmology in Abhidharma, Kalachakra, & Dzog-Chen.* Boston: Snow Lion, 1995.

Tegmark, Max. "The Multiverse Hierarchy." In *Universe or Multiverse?* edited by Bernard Carr, 99–125. Cambridge: Cambridge University Press, 2007.

———. "Parallel Universes." *Scientific American,* May 2003, 41–51.

Tegmark, Max, Michael A. Strauss, Michael R. Blanton, Kevork Abazajian, Scott Dodelson, Havard Sandvik, Xiaomin Wang, et al. "Cosmological Parameters from SDSS and WMAP." *Physical Review D* 69 (2004): 1–26.

Thomas Aquinas. *Exposition of Aristotle's Treatise "On the Heavens" (Unpublished).* Translated by R. F. Larcher and Pierre H. Conway. Columbus, Ohio: College of St. Mary of the Springs, 1963.

———. *Summa theologiae.* Translated by Fathers of the English Dominican Province. 5 vols. Allen, Tex.: Christian Classics, 1981.

Till, Eric. *The Christmas Toy.* Henson Associates and Sony Pictures, 1986.

Tolman, Richard C. *Relativity, Thermodynamics, and Cosmology.* Oxford: Clarendon Press, 1934.

Trevor-Roper, Hugh. "Nicholas Hill, the English Atomist." In *Catholics, Anglicans, and Puritans: Seventeenth Century Essays,* 1–39. Chicago: University of Chicago Press, 1987.

Tryon, E. P. "Is the Universe a Vacuum Fluctuation?" *Nature,* December 14, 1973, 396–97.

Turner, Denys. *Faith, Reason, and the Existence of God.* Cambridge: Cambridge University Press, 2004.

Turok, Neil. "The Cyclic Universe: A Talk with Neil Turok." *Edge* 210, May 17, 2007. http://www.edge.org/documents/archive/edge210.html#turok2.

Vaas, Rüdiger. "Dark Energy and Life's Ultimate Future." In *The Future of Life and the Future of Our Civilization,* edited by Vladimir Burdyuzha, 231–47. Dordrecht: Springer, 2006. Also at http://philsci-archive.pitt.edu/archive/00003271/.

Van der Meer, Matthieu Herman. "World Without End: Nicholas of Cusa's View of Time and Eternity." In *Christian Humanism,* edited by Alasdair A. Macdonald, R. W. M. von Martels, and Jan R. Veenstra Zweder, 317–37. Leiden: Brill, 2009.

Van Steenberghen, Fernand. *Aristotle in the West: The Origins of Latin Aristotelianism.* Louvain, Belgium: Nauwelaerts, 1970.

Veneziano, Gabriele. "The Myth of the Beginning of Time." *Scientific American,* April 2004, 54–59, 62–65.

Vilenkin, Alexander. "The Birth of Inflationary Universes." *Physical Review D* 27 (1983): 2848–55.

———. "Creation of Universes from Nothing." *Physics Letters B* 117 (1982): 25–28.

———. *Many Worlds in One: The Search for Other Universes.* New York: Hill and Wang, 2006.

Wakeman, Mary K. *God's Battle with the Monster.* Leiden: Brill, 1973.

Warren, James. "Ancient Atomists and the Plurality of Worlds." *Classical Quarterly* 54, no. 2 (2004): 354–65.

Wecht, Brian. "Controversies in Modern Theoretical Particle Physics." Paper presented at the Northeast Conference on Science and Skepticism, New York City, April 22, 2012.

———. "Lectures on Nongeometric Flux Compactifications." August 29, 2007. Available only through arXiv:hep-th/0708.3984v1.

Wegter-McNelly, Kirk. *The Entangled God: Divine Relationality and Quantum Physics.* New York: Routledge, 2011.

Wehus, I. K., and H. K. Eriksen. "A Search for Concentric Circles in the 7-Year WMAP Temperature Sky Maps." *Astrophysical Journal Letters* 733 (2011): 1–6. arXiv:astro-ph/1012.1268v1.

Weinberg, Steven. "Anthropic Bound on the Cosmological Constant." *Physical Review Letters* 59 (1987): 2607–10.

———. "The Cosmological Constant Problems." Lecture given at the Dark Matter 2000 Conference, Marina del Ray, Calif., February 22–24, 2000. arXiv:astro-ph/0005265v1.

———. *The First Three Minutes.* New York: Basic Books, 1977.

———. *Gravitation and Cosmology.* New York: Wiley, 1972.

———. "Living in the Multiverse." In *Universe or Multiverse*, edited by Bernard Carr, 29–42. Cambridge: Cambridge University Press, 2007.

Westfall, Richard S. "Newtonian Cosmology." In *Cosmology: Historical, Literary, Philosophical, Religious, and Scientific Perspectives*, edited by Norris S. Hetherington, 263–74. New York: Garland, 1993.

Wheeler, John. "Genesis and Observership." In *Foundational Porblems in the Special Sciences*, edited by Robert E. Butts and Jaakko Hintikka, 3–33. Dordrecht: Reidel, 1977.

Whitrow, G. J. "Kant and the Extragalactic Nebulae." *Quarterly Journal of the Royal Astronomical Society* 8 (1967): 48–55.

Worthing, Mark. "Christian Theism and the Idea of an Oscillating Universe." In *God, Life, and the Cosmos: Christian and Islamic Perspectives*, 281–301. Burlington, Vt.: Ashgate, 2002.

Wright, Thomas. *An Original Theory or New Hypothesis of the Universe (1750): A Fascimile Reprint Together with the First Publication of "A Theory of the Universe" (1734).* Introduction and transcription by Michael A. Hoskin. New York: American Elsevier, 1971.

Zanstra, Herman. "On the Pulsating or Expanding Universe and Its Thermodynamical Aspect." *Proceedings of the Royal Dutch Academy of Sciences, Series B* 60 (1957): 286–307.

Zizioulas, John. *Being as Communion.* Crestwood, N.Y.: SVS Press, 1985.

Zweernick, Jeffrey A. *Who's Afraid of the Multiverse?* Pasadena, Calif.: Reasons to Believe, 2008.

Zyga, Lisa. "Scientists Find First Evidence That Many Universes Exist." December 17, 2010. Phys.org. http://phys.org/news/2010-12-scientists-evidence-universes.html.

FIGURE 1.1 The Ptolemaic universe. (From Peter Apian, *Cosmographia* [Antwerp, 1524])

INDEX

Numbers in italics refer to pages on which illustrations appear